Aqueous Polymeric Coatings for Pharmaceutical Dosage Forms

Fourth Edition

DRUGS AND THE PHARMACEUTICAL SCIENCES

A Series of Textbooks and Monographs

Series Executive Editor

James Swarbrick

PharmaceuTech, Inc.
Pinehurst, North Carolina

Recent Titles in Series

Aqueous Polymeric Coatings for Pharmaceutical Dosage Forms, Fourth Edition,
Linda A. Felton

Good Design Practices for GMP Pharmaceutical Facilities, Second Edition,
Terry Jacobs and Andrew A. Signore

Handbook of Bioequivalence Testing, Second Edition, Sarfaraz K. Niazi

Generic Drug Product Development: Solid Oral Dosage Forms, Second Edition,
edited by Leon Shargel and Isadore Kanfer

Drug Stereochemistry: Analytical Methods and Pharmacology, Third Edition,
edited by Krzysztof Jozwiak, W. J. Lough, and Irving W. Wainer

Pharmaceutical Powder Compaction Technology, Second Edition,
edited by Metin Çelik

Pharmaceutical Stress Testing: Predicting Drug Degradation, Second Edition,
edited by Steven W. Baertschi, Karen M. Alsante, and Robert A. Reed

Pharmaceutical Process Scale-Up, Third Edition, edited by Michael Levin

Sterile Drug Products: Formulation, Packaging, Manufacturing and Quality,
Michael J. Akers

Freeze-Drying/Lyophilization of Pharmaceutical and Biological Products,
Third Edition, edited by Louis Rey and Joan C. May

Oral Drug Absorption: Prediction and Assessment, edited by Jennifer B. Dressman
and Christos Reppas

Generic Drug Product Development: Specialty Dosage Forms, edited by
Leon Shargel and Isadore Kanfer

Generic Drug Product Development: International Regulatory Requirements
for Bioequivalence, edited by Isadore Kanfer and Leon Shargel

Active Pharmaceutical Ingredients: Development, Manufacturing, and
Regulation, Second Edition, edited by Stanley Nusim

Pharmaceutical Statistics: Practical and Clinical Applications, Fifth Edition,
edited by Sanford Bolton and Charles Bon

Biodrug Delivery Systems: Fundamentals, Applications and Clinical
Development, edited by Mariko Morishita and Kinam Park

Biodrug Delivery Systems: Fundamentals, Applications and Clinical
Development, edited by Mariko Morishita and Kinam Park

A complete listing of all volumes in this series
*can be found at **www.crcpress.com***

Aqueous Polymeric Coatings for Pharmaceutical Dosage Forms

Fourth Edition

Edited by

Linda A. Felton
University of New Mexico
Albuquerque, NM, USA

CRC Press
Taylor & Francis Group
Boca Raton London New York

CRC Press is an imprint of the
Taylor & Francis Group, an **informa** business

CRC Press
Taylor & Francis Group
6000 Broken Sound Parkway NW, Suite 300
Boca Raton, FL 33487-2742

First issued in paperback 2020

© 2017 by Taylor & Francis Group, LLC
CRC Press is an imprint of Taylor & Francis Group, an Informa business

No claim to original U.S. Government works

ISBN-13: 978-1-4987-3208-6 (hbk)
ISBN-13: 978-0-367-73687-3 (pbk)

Visit the Taylor & Francis Web site at
http://www.taylorandfrancis.com

and the CRC Press Web site at
http://www.crcpress.com

To James W. McGinity, who got me interested in film-coating technology and has been a great mentor throughout my career.

Contents

Preface

While the elimination of organic solvents from film-coating systems circumvents problems associated with residual solvents and solvent recovery, the use of aqueous-based coatings presents its own challenges to the pharmaceutical scientist. Aqueous film-coating technology has become a routine process in the pharmaceutical industry, yet its intrinsic complexity has limited our full understanding of all the variables that impact product performance. This fourth edition of *Aqueous Polymeric Coatings for Pharmaceutical Dosage Forms* aims to provide more insight into the factors and parameters that should be considered and controlled for the successful development and subsequent commercialization of an optimized coated product.

Since the third edition of this text was published, considerable advances in aqueous-based film-coating technologies have been made. Publications in the scientific literature have focused on many issues, including the interaction of drugs with functional polymers, the influence of processing parameters on coating quality, and the introduction of new polymeric coating materials. The fourth edition has been revised and expanded to reflect the most recent scientific advancements from the literature. Some of the world's leading experts in aqueous film-coating technology have contributed to this edition. Chapters from the third edition have been updated, and new chapters address subjects such as substrate considerations and proactive troubleshooting. A new introductory chapter provides a broad overview of film coating and is designed to orient the novice to the subject. The contributing authors have attempted to explain in detail, using illustrated examples, appropriate steps to solve and ideally avoid formulation, processing, and stability problems and to achieve an optimized dosage form.

As with the prior editions, the prime objective of this fourth edition is to further expand the number of new researchers to this field of pharmaceutical technology and to stimulate new ideas, concepts, and product opportunities. Trade names and chemical names of commercially marketed coatings are used throughout the text to help familiarize the reader with the various polymers available for pharmaceutical applications. This book will be a valuable resource for anyone in the pharmaceutical industry working in the area of aqueous-based film coating.

I would like to thank the chapter authors for their contributions and our readers who over the past several years have provided many useful comments and suggestions. As usual, your comments and constructive criticism on this edition will continue to be appreciated.

Linda A. Felton

Editor

Linda A. Felton is a professor of pharmaceutics and chair of the department of pharmaceutical sciences in the College of Pharmacy at the University of New Mexico in Albuquerque. She earned a BS in pharmacy and a PhD in pharmaceutics from the University of Texas at Austin. Dr. Felton's research interests are focused on polymeric film-coating technology and modified release systems. She has presented her work at national and international conferences and has published extensively in high-quality, peer-reviewed journals. Dr. Felton is an editorial board member of *Drug Development and Industrial Pharmacy, AAPS PharmSciTech*, and the *Journal of Drug Delivery Science and Technology*. She served as the manufacturing section editor of the 21st and 22nd editions of the well-known reference text *Remington: The Science and Practice of Pharmacy.* Dr. Felton holds a joint appointment with the Department of Veteran's Affairs Cooperative Studies Program with which she oversees the formulation development of clinical trials materials. She is highly active in the American Association of Pharmaceutical Scientists and is a current member of the Controlled Release Society and the International Society for Pharmaceutical Engineering.

Contributors

Roland Bodmeier
Freie Universität Berlin
Berlin, Germany

L. Diane Bruce
Sovereign Pharmaceuticals, LLC
Fort Worth, Texas

Brian Carlin
FMC BioPolymer
Princeton, New Jersey

Tanvi M. Deshpande
School of Pharmacy
University of Maryland
Baltimore, Maryland

Thomas P. Farrell
Colorcon
West Point, Pennsylvania

Linda A. Felton
College of Pharmacy
University of New Mexico
Albuquerque, New Mexico

Philip J. Hadfield
PJH Consultancy
Maidstone Kent, United Kingdom

Stephen W. Hoag
School of Pharmacy
University of Maryland
Baltimore, Maryland

Klaus Knop
Institut für Pharmazeutische Technologie und
    Biopharmazie Heinrich-Heine-Universität
Düsseldorf, Germany

Hiroyasu Kokubo
Cellulose and Pharmaceutical Excipients
    Department
Shin-Etsu Chemical Co., Ltd.
Tokyo, Japan

Shawn A. Kucera
Pain Therapeutics, Inc.
Austin, Texas

Jian-Xin Li
Ferring Pharmaceuticals Inc.
Parsippany, New Jersey

Elena Macchi
College of Pharmacy
University of New Mexico
Albuquerque, New Mexico

James W. McGinity
College of Pharmacy
The University of Texas at Austin
Austin, Texas

Atul M. Mehta
Mehta Consulting LLC
Mahwah, New Jersey

Nasser N. Nyamweya
Novo Nordisk
Copenhagen, Denmark

Sakae Obara
Cellulose and Pharmaceutical Excipients
    Department
Shin-Etsu Chemical Co., Ltd.
Tokyo, Japan

Patrick B. O'Donnell
Daiichi-Sankyo
San Diego, California

Ornlaksana Paeratakul
Srinakharinwirot University
Bangkok, Thailand

Hans-Ulrich Petereit
Evonik Nutrition & Care GmbH
Darmstadt, Germany

Stuart C. Porter
NA Pharma Technical Services and Global
    Film Coating Tech.
Pharmaceutical and Nutrition Specialties,
    Ashland Specialty Ingredients
Wilmington, Delaware

Anisul Quadir
Cellulose and Pharmaceutical Excipients
    Department
Shin-Etsu Chemical Co., Ltd.
Tokyo, Japan

Ali R. Rajabi-Siahboomi
Colorcon
West Point, Pennsylvania

Florence Siepmann
College of Pharmacy
University of Lille
Lille, France

Juergen Siepmann
College of Pharmacy
University of Lille
Lille, France

Brigitte Skalsky
Evonik Nutrition & Care GmbH
Darmstadt, Germany

Shirley Yang
FMC BioPolymer
Princeton, New Jersey

1 Introduction to Aqueous-Based Polymeric Coating

Klaus Knop

CONTENTS

INTRODUCTION

The European Pharmacopoeia defines coated tablets as "tablets covered with one or more layers of mixtures of various substances such as natural or synthetic resins, gums, gelatin, inactive and insoluble fillers, sugars, plasticisers, polyols, waxes, colouring matter … and sometimes flavouring substances and active substances" [1].

The coating of solid dosage forms, such as tablets, capsules, and pellets goes back to the early manufacturing of pharmaceutical preparations. The industrial application of pharmaceutical coating started in the 19th century when special rotating coating pans were constructed [2]. At that time, sugar was used as a coating material and was applied onto cores as aqueous solutions or suspensions and as dry powder in several steps with intermediate drying cycles. The whole procedure took several days and was more art than science. Today, coating with natural or (semi-)synthetic polymers is widely used for solid dosage forms, and the underlying processes are well-controlled.

ADVANTAGES OF COATED SOLID DOSAGE FORMS

Coating is an additional step in the manufacturing of a dosage form. Therefore, it is time-consuming, requires special equipment, and increases production costs. Nevertheless, such a coated formulation can have several advantages. These can be divided into four sections (see Table 1.1).

The chemical stability can be improved by the protection of a sensitive active pharmaceutical ingredient (API) against oxidation. A coating can slow the rate of oxygen penetration into the core [3], and an opaque or colored coat can protect against photolytic degradation, especially from UV radiation [4]. Moisture coming from a humid environment during storage can be reduced with polymer coatings [5] and may be particularly useful for moisture-sensitive APIs, such as esters that undergo hydrolysis in the presence of water. A gastro-resistant coating protects a pH-sensitive API from degradation in the acidic gastric fluid [6–8]. Also interactions between ingredients (two APIs or API and excipient) can be avoided by separating them, one in the core and one in the coat.

TABLE 1.1

Advantages of Coated Solid Dosage Forms

Improvement of	By
Chemical stability	Protection against light, humidity, moisture, pH, and/or oxygen; avoidance of interaction between ingredients
Mechanical stability	Protection against attrition during processing and handling
Drug release	Modified release (controlled release): delayed release (taste masking, gastro-resistant, colon delivery), pulsatile release, prolonged release (extended release, sustained release)
Compliance	Taste masking, odor masking, aesthetic appearance, identification, facilitating swallowing

The mechanical stability of the drug product, which is important for further processing steps, such as packaging and later handling by health care professionals or patients, can be improved by a coating [9]. In this way, attrition as well as breakage tendencies may be reduced.

The control of drug release is a major advantage of coated dosage forms. Taste masking is achievable with a coating that is not soluble in the saliva. Gastric-resistant coating can be the solution for an API that irritates the gastric mucosa [10]. A combination of different polymers in the coat may lead to dosage forms that release the API in the colon [11]. A prolonged release is attainable with insoluble polymers. Even pulsatile release behavior is described for coated dosage forms [12].

The coating is also important for patient compliance. Bitter taste or unpleasant odor can be covered, for example [13]. Coated tablets have a more aesthetic appearance than uncoated ones and a smoother surface, which may facilitate swallowing. Colored coatings make it easier for the patient to identify the right tablet, which is of special interest for multi-morbid patients who take several different medications.

COATING PROCESS

Starting material for coating processes are solid cores of different sizes and shapes. See Chapter 12 for more details on substrate considerations when developing a coated drug product. Usually pellets or tablets are used as core materials, but also capsules, granules, and crystals can be coated for specific applications. A spherical or rounded shape and a smooth surface are advantageous for achieving a uniform coating. A liquid that contains a polymer together with other necessary excipients is sprayed onto the cores and the droplets spread across the surface of the solid, followed by coalescence of the droplets. Simultaneously or intermittently, the liquid evaporates in a warm air stream, and the polymer remains as a more-or-less dry film on the surface of the cores. Figure 1.1

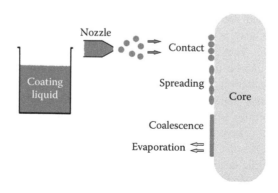

FIGURE 1.1   Schematic illustration of the coating procedure.

shows a schematic illustration of the process. This procedure is performed over a period of time, and a coated dosage form is obtained when sufficient polymer builds up to entirely surround the substrate. The process is stopped when the desired thickness of the coating is achieved. Often a final drying and/or curing step follows to ensure storage stability [14–16].

PROCESS EQUIPMENT

The coating process requires special equipment, and Chapter 3 provides much detail about processing equipment. For the coating of pellets, mini tablets, granules, or crystals, typically fluid bed apparatuses are used. In these, the cores are fluidized by a (warm) air stream, and the coating liquid is sprayed onto them through one or more nozzles. The different types of fluidized bed equipment (e.g., top or bottom spray, Wurster or rotor insert) differ in the position of the nozzle(s) and special inserts. In contrast, the coating of tablets is preferably performed in coating pans or drums, which usually are perforated to allow drying air to go through the wall of the pan and which rotate around a horizontal axis. The pan is filled to a certain level with tablets, and by means of the drum rotation, the tablets are mixed in a radial direction. Baffles are often installed in the pan to achieve axial mixing. Sufficient mixing is necessary to ensure a uniform coating of all tablets. The coating liquid is sprayed onto the moving tablet bed, and simultaneously, (warm) air is passed through the tablet bed to dry the tablets.

AQUEOUS- VERSUS ORGANIC-BASED COATING LIQUIDS

The coating liquid that is used in the process can either be an organic solution or an aqueous solution or dispersion. Aqueous-based coating liquids have several advantages over organic solutions, and therefore, many old organic-based formulations have been reformulated in the more recent past. The costs for the correct usage, recovery, and disposal of organic solvents are increasing more and more due to regulatory, environmental, safety, and health issues. Recovery systems for organic solvents can be installed to fulfill the emission limits, but the explosion risks and the health hazards still remain. When organic solvents are used, a test on residual solvent is necessary, and the requirements of the pharmacopoeia have to be met. The amount of polymer that is dissolved in an organic solution is limited due to the increasing viscosity of the solution. High viscosities limit the solution to be sprayed in conventional nozzles. Increasing solution viscosity with increasing polymer concentration also occurs with water-soluble polymers. In aqueous dispersions, the viscosity only slightly increases with increasing polymer amount. Therefore, dispersions with a higher solid content can be used, which reduces time and energy for the coating process.

On the other hand, aqueous liquids have a few drawbacks. The heat of evaporation is higher than for organic liquids. This leads to longer drying times and/or higher drying temperatures, which both result in higher energy costs. A second drawback is the more complex film formation from aqueous dispersions than from solutions (see below), which makes it often necessary to add a final curing step at the end of the process [14–16].

An alternative to liquid-based coating operations is hot-melt coating [17], a process running without the use of solvents. The coating excipient, usually a lipid or polyethylene glycol, is melted and sprayed onto the cores as a melt. After the lipid droplets hit the core, they spread over the surface and solidify during cooling to form a coating.

A further alternative is the so-called dry powder coating [18]. In this process, the polymer coating is applied as a very fine powder, which adheres to the surface of the substrate. A plasticizer can be sprayed simultaneously onto the cores or can be mixed with the polymer prior to addition. The film formation can be described as coalescence and sintering, and mechanical forces play an additional role in the leveling and consolidation of the film. As is true for conventional aqueous-based coating, a temperature above the glass transition temperature T_g of the polymer or polymer blend is usually needed for film formation.

COATING SOLUTION VERSUS DIFFERENT TYPES OF DISPERSION

Aqueous-based coating liquids can either be solutions or dispersions. Solutions arise when the polymer in the desired concentration is soluble in water. This is the case, for example, for hypromellose, polyvinyl alcohol, and copovidone (see Table 1.2). Film formation from these (colloidal) solutions is a relatively simple process: The water evaporates, and the polymer remains as a homogeneous film.

Aqueous (colloidal) dispersions can be divided into three types: Latices, pseudolatices, and dispersions made from redispersible powders. Latices (e.g., polyvinyl acetate dispersions) can be prepared by emulsion polymerization with which the monomer is emulsified in water by means of a surfactant and then polymerized [19]. For the preparation of pseudolatices, an organic solution

---

TABLE 1.2

Polymers Used for Aqueous-Based Coating as Solution or Dispersion

Type	Example	Abbreviation	Solubility	Trade Names, Examples
Cellulose derivatives	Hypromellose[a,b] (hydroxypropyl methylcellulose)	HPMC	Water soluble	Pharmacoat®, Methocel®, Vivapharm®
	Hypromellose acetate succinate[b]	HPMCAS	Enteric coating	Aqoat®, AquaSolve® HPMCAS
	Hypromellose phthalate[a,b]	HPMCP	Enteric coating	
	Cellulose acetate phthtalate[a]	CAP	Enteric coating	Aquacoat® CPD, Eastman® C-A-P
	Hydroxypropylcellulose[a,b]	HPC	Water soluble	Klucel® HPC, L-HPC
	Carmellose sodium[a] (carboxymethylcellulose sodium[b])	CMC Na	Water soluble	Aqualon®, Blanose®
	Ethylcellulose[a,b]	EC	Insoluble	Aquacoat® ECD, Surelease®
Acrylic resin, poly(meth) acrylates	Ethyl acrylate-methyl methacrylate copolymer[b] (polyacrylate dispersion[a])	PMMA	Insoluble	Eudragit® NE/NM; Kollicoat® EMM
	Ammonio methacrylate copolymer type A/type B[a,b]	PMMA	Insoluble	Eudragit® RS/RL
	Methacrylic acid-methyl methacrylate copolymer 1:1/1:2[a,b]	PMMA	Enteric coating	Eudragit® L/S
	Methacrylic acid-ethyl acrylate copolymer[a,b]	PMMA	Enteric coating	Eudragit® L100-55, Kollicoat® MAE, Acryl-EZE®
	Basic butylated methacrylate copolymer[a] (amino methacrylate copolymer[b])	PMMA	Taste masking	Eudragit® E
	Methyl methacrylate-diethylaminoethyl methacrylate copolymer	PMMA	Taste masking	Kollicoat® Smartseal
Polyvinyl derivatives	Polyvinyl alcohol[a,b]	PVA	Water soluble	Opadry® AMB
	Polyethylene glycol-polyvinyl alcohol graft polymer[a,b]	PVA-PEG	Water soluble	Kollicoat® IR
	Copovidone[a,b]	PVP-VAc	Water soluble	Kollidon® VA 64
	Polyvinyl acetate[a,b]	PVAc	Insoluble	Kollicoat® SR

[a] Listed in Ph.Eur.
[b] Listed in USP/NF.

of the polymer (e.g., ethylcellulose, cellulose acetate) is finely dispersed in an aqueous phase, and afterward, the organic solvent is removed [20]. Redispersible powders consist of micronized polymer particles, which can easily be dispersed in water prior to use. They can be obtained by, for example, spray drying of polymer solutions or dispersions or ultrafine grinding of the solid polymer.

The film formation for all types of dispersions is the same and much more complex than that from solutions. The dispersion is sprayed onto the core and spread on the surface. Water evaporates, and the dispersed particles come in close contact. The polymer particles begin to deform due to capillary forces. The next step is the coalescence of the particles, which only can take place if the temperature exceeds the minimum film formation temperature (MFT) [21]. The MFT is defined as the minimum temperature at which latex vehicles coalesce to a uniform film without visible cracks [22]. Generally, the film formation is not finished at that stage. With time and especially at temperatures above the T_g of the polymer, further coalescence and interdiffusion of the polymer chains can occur and can modify the properties of the film. To obtain a film that is stable during storage, a so-called curing step is often necessary. This is a process in which the dosage form is treated under a defined condition (time, temperature, humidity) to ensure complete film formation.

POLYMERS USED FOR COATING

The polymers used in aqueous-based coating solutions can be classified under various aspects. In Table 1.2, the most common polymers are divided by their chemical structure. Another criterion, which is also listed in Table 1.2, is the solubility of the polymer in aqueous media. The solubility is not only responsible for the type of resulting liquid (solution or dispersion); it is also important for the functionality of the resulting coating.

A dosage form coated with a water-soluble film comprising, for example, hypromellose, polyvinyl alcohol, or copovidone, will act as a protecting film with good handling properties and an attractive appearance, and the release of the API will be largely unaffected. Therefore, such films are sometimes referred to as nonfunctional. The coating can also serve as a moisture barrier [5] to improve storage stability or for aesthetic purposes. If the film is thick enough, taste masking can be achieved because it will take some time for the film to dissolve.

Water-insoluble polymers with acidic carboxylate groups (e.g., methacrylic acid–methyl methacrylate copolymer, Eudragit® L/S) show a pH-dependent solubility due to ionization of the functional groups at higher pH values. Depending on the amount of acidic groups in the polymer chain, they are insoluble at pH values below 6 (Eudragit® L with 50% methacrylic acid units) or 7 (Eudragit® S with about 30% methacrylic acid units) and soluble in slightly alkaline media. These films are called enteric coatings because they are insoluble in the acidic gastric fluid and can protect the API against the acid environment and/or the gastric mucosa from an irritant API. In the more-or-less neutral pH in the small intestine, the polymers dissolve, and the API is released. In addition to the acrylic polymers mentioned above, other polymers for enteric coating are cellulose acetate phthalate, hypromellose phthalate, or hypromellose acetate succinate.

Polymers with basic amino groups (e.g., amino methacrylate copolymer, Eudragit® E) are only soluble in acidic media. With such polymer coatings, taste masking can be achieved [23]. The coating is insoluble in the saliva, which has a pH value around 7 and is soluble in the acidic gastric fluid.

The water-insoluble polymers (e.g., ethylcellulose, polyvinyl acetate, neutral or quaternary poly(meth)acrylates) can be applied for controlled drug delivery, especially for prolonged and sustained release. They are insoluble in the entire gastrointestinal tract and act as a diffusion barrier as long as the coating remains intact. The API is released by a diffusion-controlled mechanism through the polymer film itself (often after swelling) or through pores. The release rate of the drug can mainly be influenced by the type of polymer, the thickness of the film [24], and the addition of excipients, such as pore formers [25,26].

COMPOSITION OF AQUEOUS-BASED COATING LIQUIDS

Apart from the polymer and water, aqueous-based coating liquids usually comprise additional excipients, which are listed in Table 1.3. These excipients are added to the solution or dispersion of the polymer under stirring. Often a homogenizing step follows. When latices or pseudolatices are prepared, high shear forces have to be avoided, however, because they can lead to a coagulation of the dispersed polymer.

The addition of plasticizers, such as triethyl citrate, triacetin, or dibutyl sebacate, lead to more flexible and less brittle films. On the molecular level, they interact with the polymer chains, and in this way, the mobility of the polymer molecules increases and the T_g of the polymer is lowered. They also lower the MFT of the aqueous dispersion and allow film formation at lower temperatures. Hydrophilic plasticizers can easily be mixed with aqueous dispersions whereas lipophilic ones have to be emulsified and stirred for a longer period of time to allow distribution of the plasticizer into the polymer. Some polymers, such as Eudragit® E and Eudragit® NE, show a low T_g and do not require the addition of a plasticizer [21].

Surfactants or emulsifiers, such as sodium lauryl sulfate or polysorbate, are necessary for the preparation of the latex or pseudolatex dispersions (see above). They are also useful for the incorporation of other excipients (e.g., pigments or lipophilic plasticizers) as they improve the wettability of lipophilic particles or droplets by lowering the surface tension of the dispersion. Finally, they ensure the wetting of the cores (pellets, tablets) by the droplets of the sprayed coating liquid and the spreading and uniform distribution over the substrate surface.

In some cases, a stabilizer can be necessary to enhance the stability of the dispersion during storage and application. For example, in Kollicoat® SR 30 D dispersion, povidone acts as a protective colloid for the polymer particles of polyvinyl acetate [19].

Pigments and dyes are used to create dosage forms with a high aesthetic appearance and to protect the API against degradation induced by light. The most appropriate opacifying agent is titanium dioxide. Iron oxides and water-insoluble lakes are widely used colorants [27]. The incorporation of pigments can cause stability problems of the dispersion and can alter the properties of the film to

TABLE 1.3
Composition of Aqueous-Based Coating Solutions or Dispersions

Function	Functionality	Examples
Polymer/polymer blend	Film-forming agent	Cellulose derivatives, acrylic resin, polyvinyl derivatives (see Table 1.2)
Plasticizers	Make film more flexible and less brittle; lower MFT and T_g	Hydrophilic: triethyl citrate, triacetin; lipophilic: dibutyl sebacate
Surfactants/emulsifiers	Improve the wettability of droplets, particles or cores	Polysorbate, sodium lauryl sulfate, sorbitan monooleate
Stabilizers	Stabilize the coating dispersion	Polyvinylpyrrolidone
Pigments/dyes	Make films opaque and colored	Titanium dioxide, iron oxide, aluminum lakes
Glidants/anti-tacking agents	Prevent coated particles from sticking together	Talc, magnesium stearate, glycerol monostearate, precipitated silica
Pore former/release modifier	Will create pores in films to accelerate drug release	Soluble polymers
Anti-foaming agents	Prevent extensive foam formation during handling	Simethicone
Drug (API)	Only in active coatings (see text)	
Solvent/dispersant		Water

a great extent. The critical pigment volume concentration (CPVC) is the maximum concentration of insoluble material that can be incorporated into a film and differs for various polymers and pigments [27].

Polymer films can be more or less sticky. Anti-tacking agents or glidants, such as talcum or glycerol monostearate, minimize or prevent the tablets or pellets from sticking together and adhering to the walls of process equipment during processing and storage [28].

Water-insoluble film coatings often need pore formers to adjust the release rate of the API to the desired release profile, especially when poorly water soluble APIs are used [29,30]. These pore formers dissolve in the gastrointestinal fluids and create water-filled pores in the polymer membrane. Typically, soluble polymers, which lead to solutions with low viscosity (e.g., hypromellose, povidone), are used for this purpose.

The preparation of coating dispersions includes several mixing and/or homogenizing steps with the risk of extensive foam formation. The addition of antifoaming agents, such as simethicone, can be a useful solution to reduce the amount of foam formed during mixing.

A special type of coating is the so-called "active coating" with which an API is incorporated into the polymer film. This is one way to prepare drug-loaded pellets as alternative to a wet granulation process. Small sugar or microcrystalline cellulose spheres (nonpareil beads) are used as starter cores and a solution or suspension containing polymer and API is layered onto them [31]. These coatings can be much thicker than the above described. Active coating is also applicable on tablets. The objective here can be to produce a dosage form that contains an API with controlled release kinetic in the core and another or the same API showing immediate release in the coating. The separation of two drugs that are incompatible or interact with each other can be realized by active coating with one API in the core and the other in the coating. Active coating is one of the most challenging coating processes because the drug content of the individual dosage form depends on the amount of coating on each tablet. A uniform distribution of the coating material over all tablets and a precise end point determination of the coating process are mandatory to fulfill the requirements of the pharmacopoeia on uniformity of dosage units [32].

SUMMARY

Many coated solid dosage forms, including tablets, pellets, and capsules, are on the pharmaceutical market due to the previously mentioned advantages. The preferred way of coating is the use of aqueous-based polymer liquids, which can be solutions or dispersions. The properties of coating material as well as the process of coating and the underlying mechanisms are nowadays well understood and controlled, but due to the introduction of new polymers and new technologies, further investigations and knowledge are necessary in that field. The following chapters provide more detail and describe all aspects of the topic, from polymers and additives to process equipment and conditions as well as properties of the films and finished coated dosage forms.

REFERENCES

1. Ph.Eur. 8.5 (2015). *European Pharmacopoeia 8th Edition*, Supplement 8.5. European Directorate for the Quality of Medicines & Health Care (EDQM).
2. Stock, K. W., and A. Meiners (1973). Süßwaren Dragees. Silesia Confiserie Manual No. 2. Norf, Silesia-Essenzenfabrik: 335–338.
3. Felton, L. A., and G. S. Timmins (2006). A nondestructive technique to determine the rate of oxygen permeation into solid dosage forms. *Pharm. Dev. Technol.* 11(1): 141–147.
4. Béchard, S. R., O. Quraishi, and E. Kwong (1992). Film coating: Effect of titanium dioxide concentration and film thickness on the photostability of nifedipine. *Int. J. Pharm.* 87(1–3): 133–139.
5. Bruce, H. F., P. J. Sheskey, P. Garcia-Todd, and L. A. Felton (2011). Novel low-molecular-weight hypromellose polymeric films for aqueous film coating applications. *Drug Dev. Ind. Pharm.* 37(12): 1439–1045.

6. Macchi, E., L. Zema, A. Maroni, A. Gazzaniga, and L. A. Felton (2015). Enteric-coating of pulsatile-release HPC capsules prepared by injection molding. *Eur. J. Pharm. Sci.* 70: 1–11.

7. Fang, Y., G. Wang, R. Zhang, Z. Liu, X. Wu, and D. Cao (2014). Eudragit L/HPMCAS blend enteric-coated lansoprazole pellets: Enhanced drug stability and oral bioavailability. *AAPS PharmSciTech* 15(3): 513–521.

8. Pyar, H., and K. K. Peh (2014). Enteric coating of granules containing the probiotic Lactobacillus Acidophilus. *Acta Pharm.* 64(2): 247–256.

9. Felton, L. A., N. H. Shah, G. Zhang, M. H. Infeld, A. W. Malick, and J. W. McGinity (1997). Compaction properties of individual nonpareil beads coated with an acrylic resin copolymer. *S.T.P. Pharma Sciences* 7(6): 457–462.

10. Albanez, R., M. Nitz, and O. P. Taranto (2015). Influence of the type of enteric coating suspension, coating layer and process conditions on dissolution profile and stability of coated pellets of diclofenac sodium. *Powder Technol.* 269: 185–192.

11. Maroni, A., M. D. Del Curto, L. Zema, A. Foppoli, and A. Gazzaniga (2013). Film coatings for oral colon delivery. *Int. J. Pharm.* 457(2): 372–394.

12. Maroni, A., L. Zema, G. Loreti, L. Palugan, and A. Gazzaniga (2013). Film coatings for oral pulsatile release. *Int. J. Pharm.* 457(2): 362–371.

13. Chivate, A., V. Sargar, P. Nalawade, and V. Tawde (2013). Formulation and development of oral dry suspension using taste masked ornidazole particles prepared using Kollicoat® Smartseal 30 D. *Drug Dev. Ind. Pharm.* 39(7): 1091–1097.

14. Felton, L. A., and M. L. Baca (2001). Influence of curing on the adhesive and thermomechanical properties of an applied acrylic polymer. *Pharm. Dev. Technol.* 6(1): 53–59.

15. Zheng, W., and J. W. McGinity (2003). Influence of Eudragit® NE 30 D blended with Eudragit® L 30 D-55 on the release of phenylpropanolamine hydrochloride from coated pellets. *Drug Dev. Ind. Pharm.* 29(3): 357–366.

16. Gendre, C., M. Genty, B. Fayard, A. Tfayli, M. Boiret, O. Lecoq, M. Baron, P. Chaminade, and J. M. Peán (2013). Comparative static curing versus dynamic curing on tablet coating structures. *Int. J. Pharm.* 453(2): 448–453.

17. Jannin, V., and Y. Cuppok (2013). Hot-melt coating with lipid excipients. *Int. J. Pharm.* 457(2): 480–487.

18. Sauer, D., M. Cerea, J. DiNunzio, and J. McGinity (2013). Dry powder coating of pharmaceuticals: A review. *Int. J. Pharm.* 457(2): 488–502.

19. Kolter, K., A. Dashevsky, M. Irfan, and R. Bodmeier (2013). Polyvinyl acetate-based film coatings. *Int. J. Pharm.* 457(2): 470–479.

20. Sastry, S. V., W. Wilber, I. K. Reddy, and M. A. Khan (1998). Aqueous-based polymeric dispersion: Preparation and characterization of cellulose acetate pseudolatex. *Int. J. Pharm.* 165(2): 175–189.

21. Nollenberger, K., and J. Albers (2013). Poly(meth)acrylate-based coatings. *Int. J. Pharm.* 457(2): 461–469.

22. ASTM (1998). ASTM D 2354-98 Standard Test Method for Minimum Film Formation Temperature (MFFT) of Emulsion Vehicles. American Society for Testing and Materials.

23. Guhmann, M., M. Preis, F. Gerber, N. Pöllinger, J. Breitkreutz, and W. Weitschies (2015). Design, development and in-vitro evaluation of diclofenac taste-masked orodispersible tablet formulations. *Drug Dev. Ind. Pharm.* 41(4): 540–551.

24. Rekhi, G. S., S. C. Porter, and S. S. Jambhekar (1995). Factors affecting the release of propranolol hydrochloride from beads coated with aqueous polymeric dispersions. *Drug Dev. Ind. Pharm.* 21(6): 709–729.

25. Lippold, B. C., and R. Monells Pagés (2001). Control and stability of drug release from diffusion pellets coated with the aqueous quaternary polymethacrylate dispersion Eudragit® RS 30 D. *Pharmazie* 56(6): 477–483.

26. Frohoff-Huelsmann, M. A., A. Schmitz, and B. C. Lippold (1999). Aqueous ethyl cellulose dispersions containing plasticizers of different water solubility and hydroxypropyl methylcellulose as coating material for diffusion pellets. I. Drug release rates from coated pellets. *Int. J. Pharm.* 177(1): 69–82.

27. Felton, L. A., and S. C. Porter (2013). An update on pharmaceutical film coating for drug delivery. *Expert Opin. Drug Deliv.* 10(4): 421–435.

28. Ammar, H. O., M. M. Ghorab, L. A. Felton, S. Gad, and A. A. Fouly (2015). Effect of antiadherents on the physical and drug release properties of acrylic polymeric films. *AAPS PharmSciTech*: In press.

29. Kranz, H., K. Jürgens, M. Pinier, and J. Siepmann (2009). Drug release from MCC- and carrageenan-based pellets: Experiment and theory. *Eur. J. Pharm. Biopharm.* 73(2): 302–309.

30. Mehta, R. Y., S. Missaghi, S. B. Tiwari, and A. R. Rajabi-Siahboomi (2014). Application of ethylcellulose coating to hydrophilic matrices: A strategy to modulate drug release profile and reduce drug release variability. *AAPS PharmSciTech* 15(5): 1049–1059.

31. Kleinebudde, P., and K. Knop (2007). Direct pelletization of pharmaceutical pellets in fluid-bed processes. In *Granulation (Handbook of Powder Technology, Vol. 11)*, eds. A. Salman, M. Hounslow and J. P. K. Seville, 779–811, Elsevier, Amsterdam.

32. Knop, K., and P. Kleinebudde (2013). PAT-tools for process control in pharmaceutical film coating applications. *Int. J. Pharm.* 457(2): 527–536.

2 Pseudolatex Dispersions for Controlled Drug Delivery

Brian Carlin, Shirley Yang, Jian-Xin Li, and Linda A. Felton

CONTENTS

INTRODUCTION

Reservoir systems, widely used for oral-controlled or sustained drug release, consist of a polymer coating on solid substrates (reservoir), such as powders, beads, granules, capsules, or tablets. Latex or pseudolatex presentations of water-insoluble polymers have largely superseded the use of organic solvents for applying such coatings. Aqueous polymer dispersions are preferred on environmental and safety grounds as solvents are not used during the coating process. These dispersions or aqueous polymer emulsions may be prepared by emulsion polymerization of a monomer (latex) or by emulsification of a polymer (pseudolatex). Dispersions of biopolymer derivatives, such as the cellulosics, can only be prepared as pseudolatices.

A number of emulsification procedures can be used to prepare pseudolatices from pharmaceutically acceptable polymers, avoiding the problem of monomer residues [1]. They are typically prepared by dissolving the polymer in a water-immiscible solvent and emulsifying the organic phase into water. After homogenization, the solvent is removed by vacuum distillation, leaving a 30% solids dispersion in water. Pseudolatices are colloidal dispersions containing spherical solid or semisolid particles in the nanometer to micron range, typically 0.1 to 0.3 µm (Figure 2.1). Because the 30% polymer loading is in suspension rather than in solution, viscosity is low, and the dispersions are free-flowing mobile liquids that can be easily atomized and sprayed. The particle size is also low enough for the particles to be self-suspending.

The Emulsion Polymers Institute, Lehigh University, developed the process for converting water-insoluble polymers into colloidal aqueous dispersions [1], and the Industrial and Physical Pharmacy Department at Purdue University applied the Vanderhoff process to pharmaceutical polymers useful

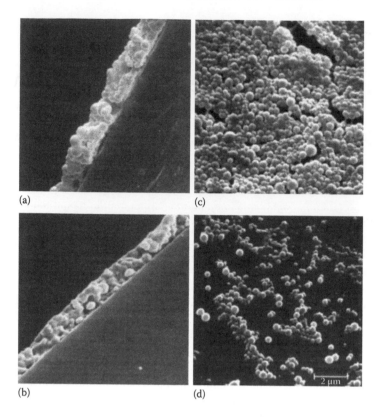

FIGURE 2.1 (a) Cross-section: free Aquacoat® ECD film cast on glass, showing discrete polymer spheres. (b) Cross-section: same Aquacoat film as coalescence proceeds. (c) Top view, free film freshly cast on glass. (d) Liquid latex. (Magnification: a, b, d: 8000×.)

in controlled release technology [2]. Ethylcellulose Aqueous Dispersion NF, JP is commercially available as Aquacoat® ECD. A cellulose acetate phthalate (CAP) dispersion for enteric coating, Aquacoat® CPD, is also available. Both Aquacoat® ECD and CPD are plasticizer-free for maximum stability and to afford flexibility to the formulator in terms of performance and regulatory acceptability.

Aquacoat® ECD is used to illustrate the formulation, manufacture, and utilization of aqueous polymer dispersions for extended release, taste masking, and moisture-barrier applications. Other methods of preparing aqueous dispersions of ethylcellulose have been developed, such as the emulsification of an extrusion melt (ethylcellulose, plasticizer, and oleic acid) into ammoniated water, used for Surelease® (U.S. patents 4,123,403, 4,502,888). Aquacoat® CPD is discussed at the end of the chapter for delayed release (enteric) applications and colonic drug delivery.

ADVANTAGES OF PSEUDOLATEX DISPERSIONS

Aqueous pseudolatex colloidal polymer dispersions offer several advantages over polymers dissolved in organic solvents, including lower spraying viscosities; higher solids loading; higher spray rates; no solvent environmental, toxicity, or flammability issues; and reduced energy requirements relative to aqueous polymer solutions. Wesseling and Bodmeier [3] showed equivalent release profiles of cured plasticized Aquacoat® ECD coatings against the corresponding coatings deposited from an organic solvent.

The viscosity advantage is demonstrated by the concentration–viscosity plot [4] in Figure 2.2. Polymer solution viscosities are dependent on concentration and molecular weight and usually

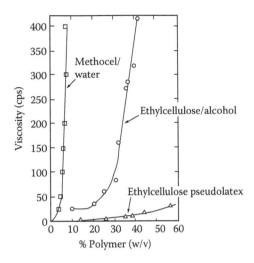

FIGURE 2.2 Concentration–viscosity relationship of ethylcellulose pseudolatex and polymer solutions. (From G. S. Banker, G. E. Peck. *Pharm Technol*, 1981, 5, 4, 55–61.)

limit the maximum loading that can be sprayed. With pseudolatices, viscosity is independent of the molecular weight of the polymer in the dispersed system, and greater (undissolved) polymer concentrations (30%) are possible at extremely low viscosities (<150 mPas).

Water in pseudolatices evaporates more readily compared to aqueous polymer solutions. Figure 2.3 shows a solvent loss–time curve for a pseudolatex versus an idealized curve for a polymer solution. With a pseudolatex, there is a zero-order loss of water independent of the solids concentration. This is due to the film-formation mechanism involving coalescence of discrete sub-micrometer latex spheres. At about 85% water loss, the curve begins to tail off due to particle–particle contact. Then there is a slow exponential water loss during coalescence. In contrast, the rate of solvent loss from a polymer solution, such as ethylcellulose in an organic solvent, is proportional to the vapor pressure of the solvent. As the concentration of the solids in the solution increases, the vapor pressure drops, and there is a concurrent decrease in the rate of solvent loss. Thus, latex dispersions give up water more quickly and completely.

Table 2.1 compares the water vapor transmission rates (WVTRs) of ethylcellulose films from organic solvents against films from a plasticized ethylcellulose pseudolatex as a function of film

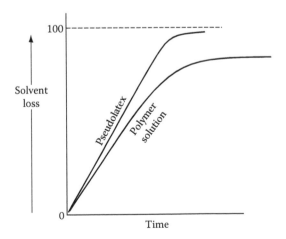

FIGURE 2.3 Solvent loss–time curves for pseudolatex and polymer solution.

TABLE 2.1
Water Vapor Transmission Rates

Plasticized Ethocel® Pseudolatex		Ethocel® 50 cP Organosol		Ethocel-Methocel® E-50 Organosol	
Film Thickness (cm)	WVTR[a] ($\times 10^{-5}$)	Film Thickness (cm)	WVTR[a] ($\times 10^{-5}$)	Film Thickness (cm)	WVTR[a] ($\times 10^{-5}$)
0.0050	3.9480	0.0046	7.5002	0.0070	13.332
–	–	0.0078	6.0013	0.0093	11.844
0.0101	3.6824	0.0094	5.5723	0.0105	11.023
0.0120	3.3350	0.0124	5.1478	0.0116	10.921

[a] Water vapor transmission rate in g/hr cm^2 mmHg.

thickness. The water vapor pressure across the films was 29.0 mmHg at 30°C in each case. The WVTRs of the pseudolatex films were about one half the value of the ethylcellulose polymer film from an organic solvent.

MANUFACTURE OF AQUEOUS POLYMER DISPERSIONS OF ETHYLCELLULOSE

Aquacoat® ECD is used for aqueous film coating of solid dosage forms to extend drug release, taste mask, or protect against moisture. It consists primarily of ethylcellulose with a surfactant and a stabilizer from the emulsion stage (sodium lauryl sulfate [SLS] and cetyl alcohol) as shown in Table 2.2. Traces of dimethylpolysiloxane (<400 ppm) to suppress foaming during distillation may also be present. Ethylcellulose is a cellulose ether made by the reaction of ethyl chloride with alkali cellulose. Each anhydroglucose unit has three replaceable OH groups, some or all of which may react with ethyl chloride. Figure 2.4 shows the molecular formula for ethylcellulose and the method of manufacture is illustrated in Figure 2.5. The ethylcellulose is dissolved in a water-immiscible organic solvent, and cetyl alcohol (cetanol) is added as a dispersion stabilizer. The solution is then emulsified into an aqueous SLS solution. The resulting crude emulsion is passed through a homogenizer to yield a submicron "fine" emulsion, which is then distilled to remove the organic solvent and sufficient water to yield a 30% solids dispersion.

TABLE 2.2
Composition of Aquacoat® ECD

	Solids (%)	Finished Product (%)
Ethylcellulose	87.1	26.1
Cetyl alcohol	8.7	2.6
SLS	4.2	1.3
Water	–	70.0
	100.0	100.0

Note: SLS, sodium lauryl sulfate.

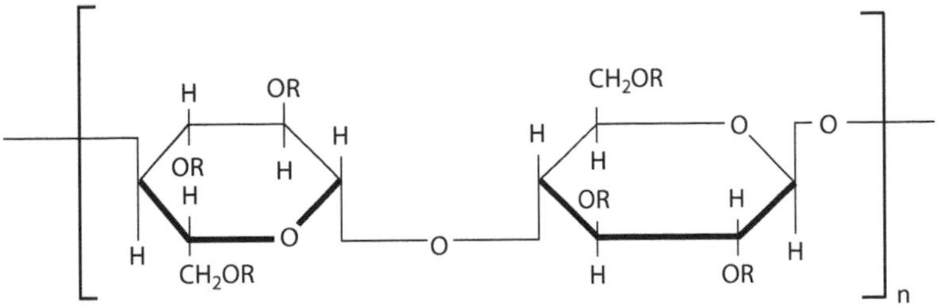

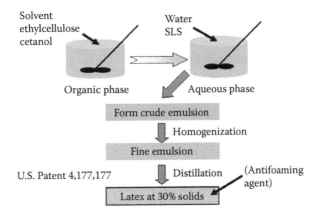

FIGURE 2.4 Ethylcellulose polymer. R = −CH$_2$CH$_3$.

Solvent
ethylcellulose
cetanol

Water
SLS

Organic phase   Aqueous phase

Form crude emulsion

Homogenization

Fine emulsion

U.S. Patent 4,177,177   Distillation   (Antifoaming agent)

Latex at 30% solids

FIGURE 2.5 Manufacturing process of Aquacoat® pseudolatex. SLS, sodium lauryl sulfate.

MECHANISM OF FILM FORMATION

The mechanism of pseudolatex film formation is different from that of polymer deposition from a solvent and must be understood in order to avoid unanticipated effects. Polymer and plasticizer deposited from a solution are intimately mixed on a molecular scale. In contrast, a pseudolatex is initially deposited as discrete polymer spheres, which must coalesce to form a continuous film. A plasticizer is often included in the formulation to promote the coalescence process. Chevalier et al. [5] defined four stages of the film formation process for pseudolatices:

1. Ordering and close packing of the particles due to water evaporation to give a face-centered cubic construction
2. Deformation and filling the voids left by the removal of water to give a foam structure
3. Coalescence or fusion of the particles due to fragmentation of the hydrophilic layers between particle cores, leading to phase inversion in which the remaining water is no longer the continuous phase
4. Polymer interpenetration between cores, forming a continuous polymer matrix and erasing the original particle identity

Figure 2.6 provides an example of film formation from a pseudolatex dispersion [6]. As water evaporates, the spheres come in contact as a close-packed array. The capillary force of the interstitial water then deforms the particles, causing the spheres to fuse, resulting in complete coalescence. The properties of partially coalesced films may be radically different from that of the corresponding fully coalesced films. Partially coalesced films are also inherently unstable, as coalescence typically continues slowly over time, resulting in decreases in the drug-release rates. It is essential to

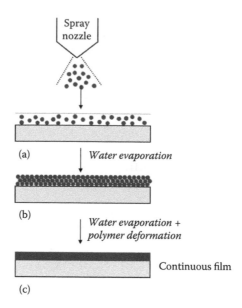

FIGURE 2.6 Film formation from a pseudolatex dispersion. (From L. A. Felton, *Int J Pharm*, 2013, 457, 423–427.)

ensure complete coalescence for long-term stability. Unfortunately, verification of complete coalescence is not always described in the pseudolatex literature, which complicates interpretation of data.

Figure 2.7 illustrates the forces exerted on spherical particles as water evaporation proceeds. Energy required for the coalescence of the spheres results from the surface tension of the polymer generated by the negative curvature of the particle surface as approximated by Frenkel's equation [7,8]:

$$\theta^2 = \frac{3\sigma t}{2\pi\eta r}$$

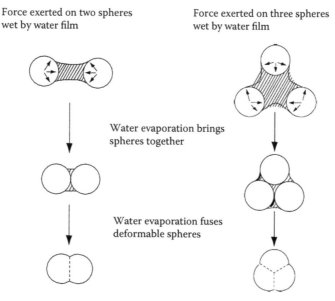

FIGURE 2.7 Particle coalescence during the evaporative phase.

where θ is the half-angle of coalescence (contact angle) at time t, σ is the surface or interfacial tension, r is the radius of a sphere, and η is the viscosity of the spheres. The contact angle is initially zero at the point of first contact and increases as the two particles fuse together.

This equation illustrates the inverse relationship between polymer viscosity of the spheres and the degree of coalescence, which is the rationale for adding a plasticizer to the coating formulation to soften the spheres and promote fusion. The equation also illustrates the utility of smaller (submicron) spheres as less force is required to completely fuse or coalesce the particles. The Frenkel equation uses the air–polymer interfacial tension (dry sintering) as the driving force for coalescence, but Brown [9] proposed that the capillary pressure of interstitial water between the closely packed spheres is the driving force. This is consistent with the presence of surfactants in aqueous polymer dispersions, which would otherwise reduce the driving force implied by the Frenkel equation. According to Brown, when the force due to the capillarity of the interstitial water is large enough to overcome the resistance of the polymer spheres, coalescence to form a continuous film will occur, as shown in Figure 2.7. "Porous, incompletely coalesced films may be formed from many polymers simply by maintaining, during water evaporation, a temperature lower than a certain critical value. It is observed that for certain polymers a higher temperature exists which is insufficient for coalescence of the porous structure previously formed at a lower temperature, but is adequate for complete coalescence if applied during the entire course of water evaporation. In addition to the plasticization of polymers by the water, the water exerts a strong force responsible for coalescence. The role of water in the process is of extreme importance," Brown [9]. Thus, temperature and rate of water evaporation are critical parameters for film formation.

Brown derived the capillary pressure, P_C, for the sphere of radius R, between three contiguous latex particles (Figure 2.8), in terms of the latex particle radii, r:

$$P_C = 2\gamma_w/R = 12.9\gamma_w/r$$

where γ_w = polymer–water interfacial tension.

Whether attributed to dry sintering or capillary pressure, both mechanisms share the same pseudolatex particle size dependency. The smaller the particle size, the greater the driving force for coalescence as shown in Figure 2.9 [9]. Various authors have further refined the Brown equation as reviewed by Steward [10] in his thesis. Steward provides a very comprehensive online review and discussion of the relevant detailed theory on his website [11].

Sperry et al. [12] used pre-dried controls to investigate the role of water in film formation using the transition from an opaque to a clear film (the minimum film-forming temperature [MFFT]). Latices pre-dried at temperatures below their MFFT were compared to wet latices. Using hydrophobic polymers, the dry MFFT was virtually identical to the wet MFFT, indicating that capillary forces contributed little to film formation. However, the author was unable to rule out that capillary forces may have an effect in more hydrophilic systems. Plasticization by water was said to be the cause of hydrophilic polymers yielding wet MFFTs, which were lower than the dry MFFTs by up to 10°C.

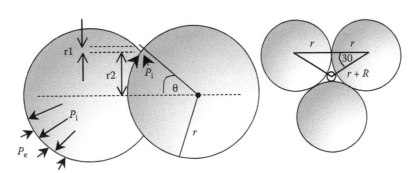

FIGURE 2.8 A cross-section of sintered latex particles and a plane view showing the inter-particle capillary.

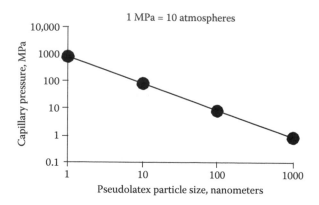

FIGURE 2.9 Capillary pressure is inversely proportional to particle size.

Lissant [13] postulated that closely packed spheres above the maximum packing volume (74%, face-centered cubic configuration) will deform at a constant volume to fill all of the space, forming a rhomboid dodecahedron. Joanicot et al. [14] showed that such polyhedra formed when a latex had lost most of its water, creating a structure similar to that of a foam.

Using small-angle neutron scattering (SANS), Chevalier et al. [5] demonstrated a reversible compression of the latex-in-water dispersion to a latex-in-water foam, followed by irreversible coalescence of the foam with inversion to a water-in-latex topology. Water is involved in both stages. Coalescence depends on the fragmentation of the foam membranes. The phase inversion involves connection of latex domains and fragmentation of water domains, driven by the spontaneous curvature of membranes according to their water content. This work thus differentiated between coalescence, which was defined as the break-up of the hydrophilic layer and subsequent polymer chain interdiffusion.

Nicholson and Wasson [15] divided coalescence mechanisms into two groups: (a) those dependent on sintering or capillarity processes, which dominate when there are polar repulsions present, and (b) those dependent on polymer chain interdiffusion, when there is very little repulsion between particles. According to Voyutskii [16], interdiffusion of polymer chains (autohesion) across what was the interface between discrete polymer spheres is the final step in the formation of integral homogeneous latex films. Voyutskii and Vakula [17] provided a comprehensive review of the effects of self-diffusion and interdiffusion in polymer systems. This is consistent with the strength and cohesiveness of films obtained immediately when deposited from good organic solvents due to complete solvation and maximum extension of polymer chains. Interdiffusion may take longer in latex films, especially if coalescence is not complete. Bradford and Vanderhoff [18] studied the changes in structure occurring in an uncured, continuous, transparent film as a function of film age. Using transmission electron microscopy within hours of casting, vestiges of the original latex particles could be seen, which disappeared over a 14-day period, accompanied by the exudation of material from within the film, assumed to be a hydrophilic stabilizer.

Bradford and Vanderhoff [19] coined the term "further gradual coalescence" and showed that it occurred at the film–substrate and film–air interfaces as well as within the film's interior where a stabilizer was exuded into "pockets." The size of the holes and porosity due to the leaching of surfactant was reduced if the film was aged or heat-treated before testing. Interdiffusion requires temperatures above the glass transition temperature (T_g) as there will be insufficient polymer segment mobility in the glassy state.

Using SANS, Hahn et al. [20,21] demonstrated "massive" interdiffusion of polymer chains from different latex particles during particle coalescence. A 30-fold increase in diffusion coefficients was observed on increasing the "tempering" or curing temperature from 70°C to 90°C. Also using

SANS, Sperling et al. [22] concluded that the rate of coalescence was dependent on where the polymer chain ends lie with respect to the particle surface and that films form faster when the ends lie on the particle surface.

Distler and Kanig [23] postulated that, upon deformation of the particles into a film, hydrophilic surface boundary layers would interdiffuse to form an interconnected hydrophilic "honeycomb," which might inhibit further hydrophobic polymer interdiffusion. The authors pointed to the fact that a normally transparent film may turn opaque or even show Bragg diffraction iridescence, when swollen with water, both of which require latex particulate–sized features to cause the necessary difference in refractive index and crystalline structure, respectively.

The increased water absorbency, reduced surfactant leachability, and reduced tendency to whiten (swell) in water were attributed by Aten and Wassenburg [24] to the redistribution of surfactant molecules from the surfaces of the latex particles to a more even distribution throughout the film, following a period of secondary "drying" above the polymer T_g. Such redistribution was ascribed to the increased polymer chain mobility, which was not apparent in films annealed below the T_g.

PRACTICAL ASPECTS OF PSEUDOLATEX FILM COATING

Pseudolatex film coating is a complex process, and the formulation scientist must carefully consider the coating formulation, the physicochemical properties of both the dosage form and the drug, and the processing parameters used. In addition to the pseudolatex dispersion, a coating formulation often includes plasticizers to enhance the flexibility of the film and facilitate polymer sphere coalescence, anti-adherents to prevent substrate agglomeration during both the coating process and storage, surfactants to promote spreading of the atomized droplets own the substrate surface, and pigments. The addition of other excipients can significantly impact the physical stability of the dispersion, drug release, and film quality. The dosage form should be strong enough to withstand attrition during coating. The drug itself must be stable to the temperatures used during processing. For these aqueous-based systems, the drug should also be stable to the moisture challenge of aqueous film coating or a seal-coat must be used to protect the active. Processing parameters must be carefully controlled to optimize film formation. This section discusses some of the most critical concerns during pseudolatex coating processes.

DRYING AND CURING

Aqueous pseudolatices have the appearance and consistency of milk and are therefore easily sprayable using conventional aqueous coating techniques, such as fluid bed (Wurster) or perforated pan coating. The coated substrate is dried in situ during the coating process and may or may not require subsequent heat treatment (curing) to complete coalescence of the polymer spheres, depending on the coating formulation and conditions employed during coating. This additional curing may sometimes be described as "drying" at elevated temperature.

As discussed previously, the mechanism of film formation from aqueous pseudolatices of water-insoluble polymers, such as Aquacoat® ECD, is very different from simple deposition of a polymer from solvent-based coatings. Water evaporation concentrates the polymer particles in a closely packed arrangement on the substrate surface, and the capillary force of the interstitial water deforms the particles to cause coalescence and produce a dense, continuous film. When the coated substrates are cured at a temperature higher than the MFFT, the interstitial water in the coating layer will ensure an adequate capillary force for the completion of film coalescence. Unfortunately, the warnings against overdrying given by Brown [9] are not always heeded, and overdrying remains a leading cause of partial coalescence and associated problems, particularly decreasing release rates on storage due to further gradual coalescence.

The rate of heat transfer not only affects the rate of evaporation of the solvent but also, in the case of latex and pseudolatex systems, regulates the rate and degree of coalescence of the polymeric material. Rapid drying rates, although generally desirable, may at times have a negative effect. The rapid loss of water will not permit sufficient capillary pressure to develop, and the latex particles cannot coalesce to form coherent films. Excessive drying conditions also do not allow the coating formulation to spread evenly over the substrate and thus inhibit particle deformation and coalescence [25]. The drying rate is determined by several parameters, including the latent heat of vaporization and the relative humidity (RH) of the incoming drying air [26].

Ideally, humidity should be controlled during the coating or curing process itself to avoid such overdrying as illustrated in Figure 2.10. Curing at elevated temperatures using ambient humidity may not be sufficient to complete coalescence of overdried particles. The resulting drug-release profiles may decrease on high-humidity storage. When high-humidity curing gives lower release profiles than the corresponding dry curing, this may be a sign of overdrying during coating, resulting in partial coalescence. High-humidity coating conditions facilitate pseudolatex coalescence and can be created by using low-dispersion solids (e.g., diluting the Aquacoat® ECD to below 15%) and humidifying the inlet drying air. Curing to achieve complete coalescence and provide stable drug-release profiles can thus be minimized or eliminated [27].

To determine if a curing step is necessary, the coated substrates may be challenged with heat and humidity. There should be no decrease in release rate if the film has fully coalesced during coating with the caveat that thermal stressing alone (i.e., without elevated humidity) will not distinguish overdried, partially coalesced from fully coalesced films. Both should give thermostable release profiles, but the overdried profile will be faster and potentially could decrease on long-term storage, especially on humidity challenge. This interplay is demonstrated in Figure 2.11, where the challenge times (24–48 hours) are significantly higher than the curing time used in practice [28]. High humidity was maintained during coating by direct humidification of inlet air and the use of low solids in the coating dispersions (i.e., higher water spray input). The substrate was 70% theophylline pellets coated with Aquacoat® ECD plasticized with triethyl citrate (TEC) (1:4 TEC: Aquacoat® ECD solids). Although a significant degree of release retardation was achieved during coating (40% released at 12 hours), curing with dry heat further reduced the release rate to approximately 10% at 12 hours. This profile shows no further time or temperature dependence as demonstrated by the convergence of the four profiles: 24 hours, 48 hours, 60°C, and 80°C. No further retardation was achieved on high-humidity challenge at the same temperatures, which indicates convergence to a true minimum release rate. In this case, the elevated humidity challenge proved detrimental to coating performance as evidenced by the time- and temperature-dependent increases in release rate. High-humidity curing

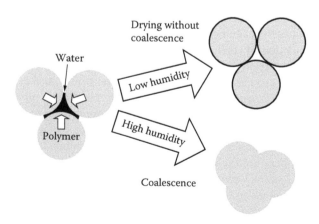

FIGURE 2.10 Film formation from aqueous lattices.

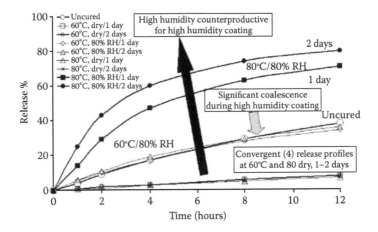

FIGURE 2.11 Effect of curing conditions on theophylline release from coated pellets (15% solids, high-humidity coating, TEC/ECD = 1:4, 4% weight gain). TEC, triethyl citrate. (From J.-X. Li, B. A. Carlin, J. T. Lee et al., AAPS 20th Annual Meeting, San Antonio, Oct 29–Nov 2, 2006.)

has been claimed as being necessary for stable release profiles [29], but if humidity is adequately controlled during coating [30], curing may not be required at all, or simple dry curing may suffice.

Results from a dry counter-example are shown in Figure 2.12. Note the essentially immediate release of the uncured pellets and the sensitivity of the "false" thermostable dry-cured profile to the humidity challenge. Full-strength plasticized Aquacoat® was used without humidification of inlet air.

The ideal release profile should not exhibit time, temperature, or humidity dependence on short-term challenges, as shown in Figure 2.13. This example used low solids loading to maintain high humidity (no humidification of inlet air). Note that during this and the two preceding examples, the coating loading was simultaneously lowered from 4% to 3% to 2% to maximize the amount of drug released. Fully coalesced films of Aquacoat® ECD provide significant release retardation, which may be too much for some poorly permeable drugs, requiring precision coating using very low loadings. To increase permeability in such cases, pore-forming excipients may be added to the coating formulation.

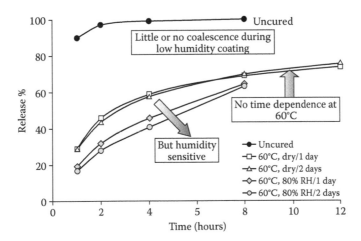

FIGURE 2.12 Effect of low-humidity coating and curing conditions on drug release (35% solids, TEC/ECD = 1:4, 3% weight gain). TEC, triethyl citrate. (From J.-X. Li, B. A. Carlin, J. T. Lee et al., AAPS 20th Annual Meeting, San Antonio, Oct 29–Nov 2, 2006.)

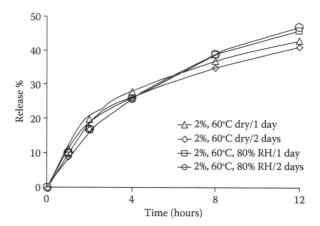

FIGURE 2.13 Drug-release profiles independent of time, temperature, and humidity. 15% solids, low humidity coating, TEC/ECD = 1:4, 2% weight gain. TEC, triethyl citrate. (From J.-X. Li, B. A. Carlin, J. T. Lee et al., AAPS 20th Annual Meeting, San Antonio, Oct 29–Nov 2, 2006.)

Typically, curing for 1 to 2 hours at 60°C will be sufficient as shown in Figure 2.14. Curing can be carried out by oven heating or in situ heating in a fluid-bed coater using increased fluidization to avoid pellet agglomeration. Coating is normally carried out below the T_g of the film to minimize tackiness, especially under the low bulk fluidization conditions in the slowly percolating pellet bed outside the Wurster column. If necessary, a conventional clear or colored (e.g., Opadry®) water-soluble polymer top coating can be applied to enable low fluidization curing above the T_g. If maximum retardation is ensured initially, then the long-term storage stability should be good, including elevated humidity, as shown in Figure 2.15, which is the same batch as in Figure 2.14 retested after storage at 40°C/75% RH for periods of up to 8 months.

It should be noted that Aquacoat® ECD contains SLS, which tends to reside on the surface of the dried polymer spheres. Faster release of nonionic or basic drugs from ECD-coated substrates at high pH is strongly indicative of partial coalescence. Because SLS is insoluble in acid but soluble at neutral pH, pH-dependent SLS channels in partially coalesced ECD films may be observed [3].

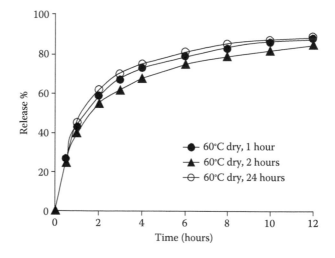

FIGURE 2.14 Drug release as a function of curing time (1.5% weight gain, TEC/ECD = 1:4, 15% solids, high-humidity coating). TEC, triethyl citrate. (From J.-X. Li, B. A. Carlin, J. T. Lee et al., AAPS 20th Annual Meeting, San Antonio, Oct 29–Nov 2, 2006.)

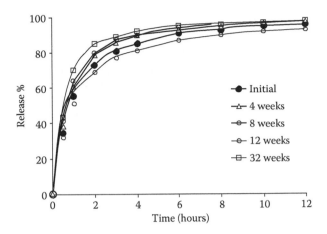

FIGURE 2.15 Drug release as a function of long-term stability at 40C/75% RH (high-humidity coating, 15% solids, 1.5% weight gain, cured at 60°C, dry/1 hour). RH, relative humidity. (From J.-X. Li, B. A. Carlin, J. T. Lee et al., AAPS 20th Annual Meeting, San Antonio, Oct 29–Nov 2, 2006.)

Such pH dependency is not seen in fully coalesced Aquacoat® ECD or ethylcellulose spiked with SLS and deposited from organic solvents.

PLASTICIZERS

Plasticizers are commonly added to film coating formulations to increase the flexibility of the film, decrease the T_g and MFFT, and for pseudolatex dispersions, to facilitate coalescence. The type and level of plasticizer may also affect the drug release by changing the diffusivity of the film. Plasticizer effects are due to a decrease in the cumulative intermolecular forces along the polymer chains (reduction in cohesion), which generally lowers the softening temperature and decreases the T_g [31]. Plasticizers impart flexibility and reduce brittleness as shown in Figure 2.16, in which insufficient plasticizer was used in the batch on the right, and the coating cracked upon drying. Pseudolatex spheres of Aquacoat® ECD have a T_g of ~89°C and must be adequately plasticized to reduce the film-forming temperature to within the processing temperature range.

The basic requirements of any plasticizer in a polymer system, including latex emulsions, are compatibility and permanence. To be compatible, the plasticizer should be miscible with the polymer. To be permanent, plasticizers should be nonvolatile with a high boiling point. The effectiveness of a plasticizer can be evaluated by measuring the T_g of the film. Table 2.3 gives data for six

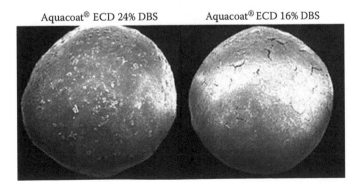

FIGURE 2.16 Effect of sufficient/insufficient plasticizer content on coating morphology. DBS, dibutyl sebacate.

TABLE 2.3
Plasticizer Physical Constant Data

	BP (°C)	Vapor Density (Air = 1)	Vapor Pressure (mmHg)	Water Solubility
DBS	349	10.8	10 at 200°C	Negligible
DEP	298	7.66	100 at 220°C	Insoluble
TEC	294	9.7	1 at 107°C	6.5%
TBC	170 (1 mmHg)	12.4	1 at 170°C	Insoluble
ATBC	173 (1 mmHg)	14.1	0.8 at 170°C	Insoluble
Myvacet 9-45	>500	NA	Nonvolatile	Negligible

Note: ATBC, acetyl tributyl citrate; DBS, dibutyl sebacate; DEP, diethyl phthalate; TBC, tributyl citrate; TEC, triethyl citrate.

plasticizers useful in sustained or prolonged release applications of pseudolatices for oral solid dosage forms. These plasticizers are all high-boiling organic materials, have low vapor pressures, and, with the exception of TEC, are relatively insoluble in water.

Glass Transition Temperature and MFFT

The T_g is the temperature at which a polymer changes from a glassy state to a rubbery state. Below the T_g, the polymer is rigid and glassy with very limited polymer segment movement. Above the T_g, the polymer is in a soft rubbery state, with significant segmental mobility of the polymer chains. If the polymer T_g is higher than the desired operating temperatures for coating, it is necessary to add a plasticizer to the dispersion to obtain good film formation. The formation of a continuous film (i.e., transparent and crack-free) also depends on the MFFT of the polymer film, which, in turn, depends on the elastic modulus (resistance to particle deformation). Above the MFFT, coalescence of latex particles can occur, giving clear films, and friable discontinuous opaque powdery films result when the temperature is below the MFFT. A balance must be struck, however, as too low a T_g or MFFT may cause tackiness and particle adhesion during coating [32].

Figure 2.17 and Table 2.4 show the effect of plasticization on the T_g of an ethylcellulose latex. Aquacoat® ECD films containing various plasticizers were cast, dried at room temperature

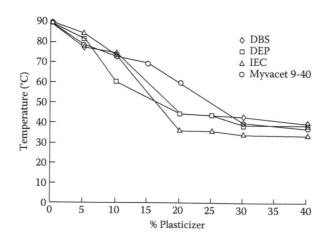

FIGURE 2.17 Glass transition (T_g) of plasticized ethylcellulose latex.

TABLE 2.4
Glass Transition Temperature (T_g) Study for Ethylcellulose Latex

Plasticizer	Temperature (°C)			Myvacet
(%)	DBS	DEP	TEC	9-40
0[a]	89	89	89	89
5	77	81.5	84	78
10	74	60	73	72.5
20	44	44	36	59
25	–	43	35.5	–
30	42.5	38	33.3	39
40	39.5	38	33.3	37

Note: DBS, dibutyl sebacate; DEP, diethyl phthalate; TEC, triethyl citrate.
[a] Ethylcellulose (neat) T_g = 129°C.

overnight, and then oven-dried at 60°C for 8 hours. These films were evaluated after 12-hour equilibration at room temperature. The study was conducted on a Perkin-Elmer TMA7, initially at a 20°C/min heating rate and then at 5°C/min, resulting in more detailed data collection. All measurements were replicated. It can be seen that, as the concentration of the plasticizer was ingcreased, the T_g for the ethylcellulose pseudolatex was lowered, thereby promoting coalescence and film formation. The rank order of effectiveness was TEC > dibutyl sebacate (DBS) > diethyl phthalate (DEP) > acetylated monoglyceride (Myvacet 9-45). For the first three, the optimum level is about 20% to 24%, and for Myvacet 9-45, approximately 30%. These percentages are with respect to Aquacoat® ECD solids (i.e., parts plasticizer to 100 parts Aquacoat® ECD solids) and not the percentages in the final coating formulation. For example, 25% wrt Aquacoat® ECD solids (1:4) is 20% of the total.

Solubility Parameter

The compatibility or miscibility of a plasticizer can be determined by the solubility parameter as investigated by Hildebrand and Scott [33]. These can be calculated for the polymer and plasticizer or found in the literature. In their calculations, Hildebrand and Scott relied on the molar energy of evaporation and the density of cohesive energy to define the solubility parameter of a known plasticizer. For nonpolar systems, the enthalpy of polymer–plasticizer mixing depends on their respective solubility parameters, δ_1 and δ_2. Onions [6] explored in more detail the Hildebrand-Scott and Flory-Huggins approaches to characterizing the extent of polymer–plasticizer affinity. With a known latent heat of evaporation for solvent or plasticizer, Hildebrand proposed that the solubility parameter δ could be calculated as:

$$\delta = \left(\frac{\Delta E_V}{V} \right)^{1/2} - \left(\frac{\Delta H_V RT}{V} \right)^{1/2}$$

where ΔE_V is the molar energy of evaporation of the plasticizer in its pure state, ΔH_V is the latent heat of evaporation of the plasticizer, R is the ideal gas constant, T is the absolute temperature, and V is the molar volume of the plasticizer. The term $\Delta E_V/V$ is usually referred to as the density of cohesive energy and represents the energy required to vaporize 1 cm^3 of liquid.

Once the solubility parameters of the polymer itself (ethylcellulose) and the candidate plasticizer are known, the enthalpy of the mixture (ΔH) can be determined:

$$\Delta H = V_{\mathrm{m}} \left[\left(\frac{\Delta E_1}{V_1} \right)^{1/2} - \left(\frac{\Delta E_2}{V_2} \right)^{1/2} \right]^2 \phi_1 \phi_2 - V_{\mathrm{m}} (\delta_1 - \delta_2) \phi_1 \phi_2$$

where the subscripts 1 and 2 refer to the polymer and plasticizer, respectively. The total molar volume of the mixture is V_{m}, V_1, and V_2 are molar volumes, ΔE_1 and ΔE_2 are molar energies of vaporization, φ_1 and φ_2 are volume fractions, and δ_1 and δ_2 are the respective solubility parameters.

The mixture enthalpy, ΔH, depends on the relative solubility parameters $(\delta_1 - \delta_2)$, and the best theoretical case is a binary mixture miscible in all proportions $(\delta_1 = \delta_2)$. Mixture entropy is positive and the Gibbs free energy $(G = H - T\Delta S)$ is negative. The solubility parameters for a range of plasticizers and ethylcellulose are given in Table 2.5. Plasticizers with solubility parameters close to that of the polymer are generally considered to be more miscible.

Plasticizer Incorporation: Mixing Time

The mechanism of plasticization of pseudolatex dispersions needs to be contrasted with that of solvent systems. In a solvent system, the polymer and plasticizer are dissolved together. On stirring into aqueous pseudolatex dispersions, a water-insoluble plasticizer forms a coarse emulsion due to the presence of the pseudolatex process surfactants, as shown in Figure 2.18. High shear dispersion of a plasticizer is not recommended due to potential destabilization of the pseudolatex.

TABLE 2.5
Solubility Parameters

Polymer/Plasticizer	Solubility Parameter (cal/cm³)¹ᐟ²
Ethylcellulose	8.5–10.1
DEP	8.9–9.9
DBS	7.7–9.2
TEC	8.6–9.5
Glyceryl triacetate (Triacetin)	8.8–9.9
Caster oil	8.53
TBC	9.04

Note: DBS, dibutyl sebacate; DEP, diethyl phthalate; TBC, tributyl citrate; TEC, triethyl citrate.

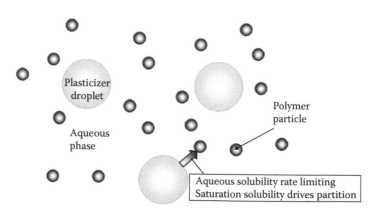

FIGURE 2.18 Plasticizer uptake in aqueous polymer dispersions.

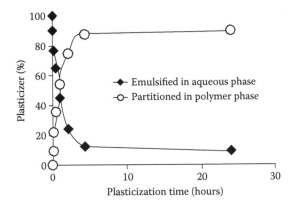

FIGURE 2.19 ATBC uptake in aqueous ethylcellulose dispersion (solids content, 15%; plasticizer/polymer, 1:5). ATBC, acetyl tributyl citrate. (From J. Siepmann, O. Paeratakul, R. Bodmeier, Modeling plasticizer uptake in aqueous polymer dispersions. *Int J Pharm*, 1998, 165, 191–200.)

For the plasticizer to be effective, it must partition into the polymer spheres. Due to their low aqueous solubility, transfer via the aqueous medium is rate limiting. Siepmann et al. [34] quantified the rates of partitioning of various plasticizers into Aquacoat® ECD, as exemplified in Figure 2.19, and the authors considered a minimum uptake of 85% to be reasonable with respect to common curing conditions. The greater the aqueous solubility, the faster the time to reach 85% partitioning (T_{85}). T_{85} values for a range of citrate homologs are shown in Table 2.6. Although Siepmann et al. did not measure a T_{85} for TEC, a reasonable estimate can be made from the other homologs.

Correlation of the aqueous solubility of the plasticizer with the T_{85} is shown in Figure 2.20. The practical significance of extended plasticizer mixing times depends on the degree of coalescence of the plasticized film. If fully coalesced, the degree of partitioning (or plasticizer mixing time) is not of practical significance, and the time allowed for plasticizer mixing with the Aquacoat® ECD does not affect the release rates [35]. Siepmann et al. measured partitioning in an aqueous system but, even if not fully partitioned in the mixing tank, partitioning will still proceed during coating, especially as the water is progressively removed.

Coalescing Agents

Complete coalescence is necessary to achieve long-term stable drug dissolution. Thermal curing after coating, with control of humidity, may be required to achieve a stable dissolution profile [27–30], or humidity can be controlled during coating. It is also possible to add coalescing agents to plasticized Aquacoat® ECD to further facilitate the coalescence. Propylene Glycol Monolaurate NF (PGML, Figure 2.21) has been investigated as a coalescing agent for Aquacoat® ECD-30 coating formulations.

TABLE 2.6
T_{85} Calculated from Homologous Series

Plasticizer	T_{85} (min)	Plasticizer	T_{85} (min)
ATBC	220	ATEC	50
TBC	100	TEC	(25)[a]

Note: ATBC, acetyl tributyl citrate; ATEC, acetyl triethyl citrate; DBS, dibutyl sebacate; DEP, diethyl phthalate; TBC, tributyl citrate; TEC, triethyl citrate.

[a] Predicted

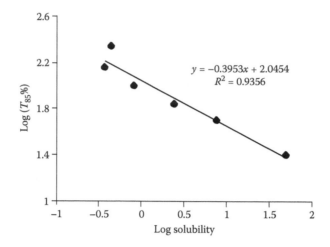

$$y = -0.3953x + 2.0454$$
$$R^2 = 0.9356$$

FIGURE 2.20   Effect of plasticizer solubility on $T_{85}\%$.

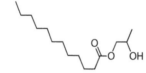

FIGURE 2.21   Structure of propylene glycol monolaurate.

The effect of 4%–16% PGML addition on plasticized Aquacoat® ECD (24% TEC) containing 18% pore former (Kollicoat IR) was investigated using Theophylline pellets (all percentages are with respect to the Aquacoat® ECD solids). As shown in Figure 2.22, a control without PGML exhibited an almost immediate release profile under the coating conditions used and required curing for 2 hours at 60°C/75% RH to achieve the sustained release profile. Addition of PGML eliminated the difference between cured and uncured under the same coating conditions.

PGML is available as type I (>60% monoester) and type II (>90% monoester). Figure 2.23 shows that both types are effective coalescing agents with the Type II giving a slightly faster release. PGML has also been proven to work well with a range of drugs, including Diltiazem and Chlorpheniramine (Figures 2.24 and 2.25).

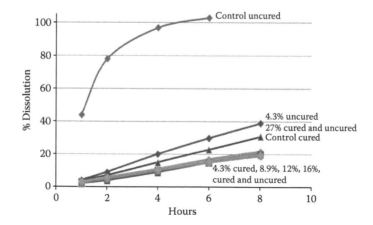

FIGURE 2.22   Effect of PGML on Theophylline release. Curing was 2 hours at 60°C/75% RH.

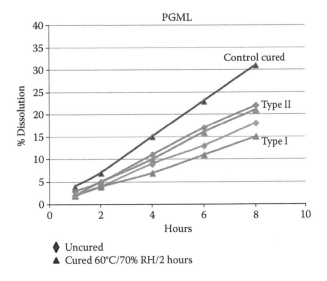

FIGURE 2.23   Effect of PGML type on Theophylline release (8.9% PGML).

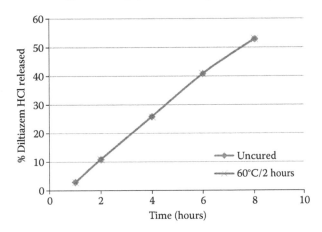

FIGURE 2.24   Release from Diltiazem pellets with 10% coated of Aquacoat® ECD 30 with 12% PGML.

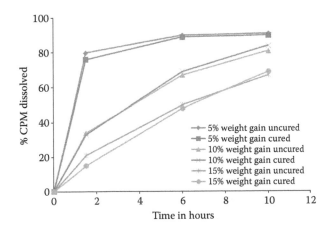

FIGURE 2.25   Release from Chlorpheniramine pellets coated with 5%–15% Aquacoat® ECD 30 with 12% PGML.

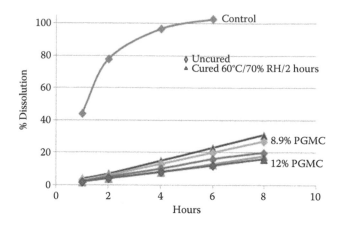

FIGURE 2.26 Effect of PGMC use level on Theophylline dissolution profile. The curing condition was 60°C/75% RH/2 hours.

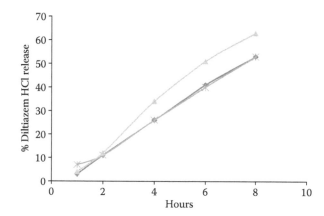

FIGURE 2.27 Dissolution profiles with Diltiazem HCl pellets coated with ECD-30 coating formulated with 12% PGML (♦), SMO (*) or SML (▲).

Propylene Glycol Monocaprylate (PGMC) gave similar results to PGML (Figure 2.26) but a higher level is required. Sorbitan Monooleate (SMO) and Sorbitan monolaurate (SML, Figure 2.27) were also effective.

APPLICATIONS DATA

As shown in the following list, variables that greatly affect the release-rate profiles through a pseudolatex film relate to both the substrate and the drug physicochemical characteristics, most notably solubility. The release patterns for coated beads were analyzed for two model drug systems: phenylpropanolamine (PPA) HCl and anhydrous theophylline. Aquacoat® ECD was applied at various levels and the in vitro drug-release rate was shown to be inversely proportional to film loading (thickness), suggesting that constant drug diffusion through the film is maintained. Such zero-order release is characteristic of a reservoir and rate-limiting barrier as long as a concentration gradient is maintained in the bead.

- Bead size distribution
- Bead diameter/surface area
- Bead surface
- Bead moisture content
- Film continuity
- Drug solubility
- Coated bead sample uniformity
- Film thickness

Flux of drug across the membrane where a water-insoluble membrane encloses a core reservoir (containing the drug) is given by Fick's first law:

$$\frac{\mathrm{d}M}{\mathrm{d}t} = \frac{ADK\Delta C}{l}$$

where A is area, D is the diffusion coefficient, K is the partition coefficient of drug between membrane and core, l is the diffusional path length (film thickness), and ΔC is the concentration difference across the film. The surface area available for drug diffusion is a critical variable with which the mechanism of drug release is diffusion controlled by a thin film membrane, and the kinetics are apparently zero order and Fickian. It is necessary to control particle size and size distribution of the nonpareil beads to be coated, otherwise batch-to-batch differences in release rates might be observed for a given film loading under identical coating conditions. Variability can be minimized by the use of beads of the same sieve fraction, same manufacture, narrowest size distribution, and regular geometry (sphericity). It has been demonstrated by mathematical analysis that as the thickness of coating increases from zero, the release profile will gradually decrease and change from first order to zero order because the release mechanism of the coated sphere changes from a matrix-dominant mechanism to diffusion from a reservoir through a rate-limiting membrane [36,37].

Figure 2.28 shows the effect of various coating (pseudolatex) levels applied to beads of a fairly regular geometry containing anhydrous theophylline. The plasticizer was DBS at a level of 24% (pseudolatex solids: DBS ≈ 4:1). Coating levels of 6% to 8% were necessary on nonpareils of 18 to 20 mesh size (0.84–1.00 mm) in order to sustain apparent zero-order drug release. The level of plasticizer, shown at 24%, is not an arbitrary amount, as seen in Figure 2.29. Here, cumulative release

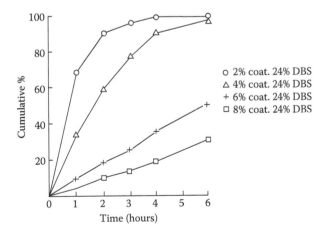

FIGURE 2.28   Effect of film coat level on dissolution

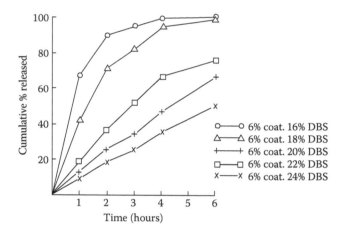

FIGURE 2.29   Effect of plasticizer level on dissolution.

curves for identical beads coated to constant film weight addition (6%) were compared as a function of the level of DBS in the coating formulation. At lower plasticizer levels, there is insufficient plasticizer to soften the ethylcellulose spheres and promote coalescence and film formation at the processing temperatures employed.

PPA HCl represents a more water-soluble drug, which poses an additional dosage design challenge when coating with a water-based polymeric dispersion. Coating conditions employed in the application of an aqueous film to such water-soluble drugs must be modified to minimize partitioning into the coating. An example is given in Table 2.7, in which the ethylcellulose pseudolatex was applied at 30% coating solids to a 10% theoretical coating level. A slow/fast technique was employed, whereby fluid spray rates were held at 2 to 3 ml/min until the beads were sealed and the coating system stabilized; then the rate was increased to up to 10 ml/min. The time, temperature,

TABLE 2.7

Coating Conditions Employed in Application of Aqueous Pseudolatex to Water-Soluble Drug

Process equipment	
Column	Wurster 4 in./6 in.
Nozzle	Spraying systems ¼J series 285070SS
Partition	3/8 in. setting
Pump	Masterflex 16 pump head
Coating conditions	
Bead load (kg)	1.0
Process air temperature (°C)	55–56
Pumping rate (ml/min) (normal)	10
Pumping rate (ml/min) (slow coating)	2–3
Atomizing air (psi)	15
Coating time	
10% film weight (min)	65–73
Slow coating (min)	29–34
Normal coating (min)	32–41
Post-drying (min)	30

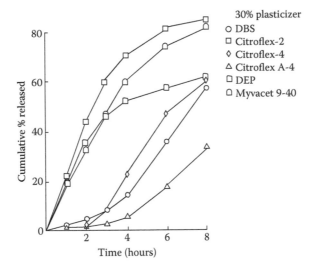

FIGURE 2.30 Effect of various plasticizers on drug release from ethylcellulose latex.

and humidity parameters were not optimized in this comparative study, so the degree of coalescence may have varied.

Six plasticizers were studied at a constant film weight addition and incorporation level (30% based on latex solids) to ascertain effects on in vitro release of PPA HCl as shown in Figure 2.30. The three slowest formulations employed the more hydrophobic butyl ester plasticizers. Faster release was obtained from formulations with ethyl ester or acetylated monoglyceride plasticizers. The stability of drug release after storage at room temperature and at 35°C for 3 and 6 months was determined for three formulations as shown in Figures 2.31 to 2.36.

Figure 2.31 shows the room temperature stability profile for PPA beads coated with an ethylcellulose latex plasticized with tributyl citrate, and elevated temperature (35°C) stability is shown in Figure 2.32. An increase in release was observed when stored for 3 months at 35°C, the profile remaining unchanged between 3 and 6 months of storage. The increase in release rate for these coatings could not be explained by loss of plasticizer (Table 2.8) or changes in film porosity (Table 2.9).

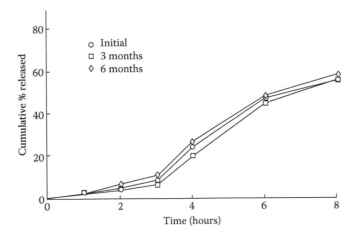

FIGURE 2.31 Stability profile (room temperature) of phenylpropanolamine released from beads coated with TBC-plasticized Aquacoat® ECD. TBC, tributyl citrate.

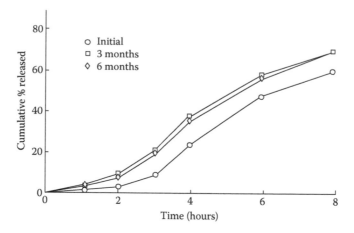

FIGURE 2.32 Elevated temperature (35°C) stability for beads coated with TBC-plasticized Aquacoat® ECD. TBC, tributyl citrate.

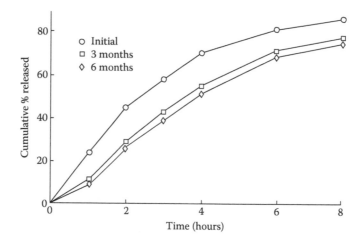

FIGURE 2.33 Stability profile (room temperature) of phenylpropanolamine released from beads coated with TEC-plasticized Aquacoat® ECD. TEC, triethyl citrate.

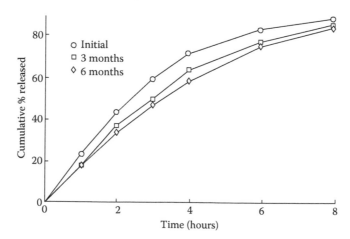

FIGURE 2.34 Elevated temperature (35°C) stability for beads coated with TEC-plasticized Aquacoat® ECD. TEC, triethyl citrate.

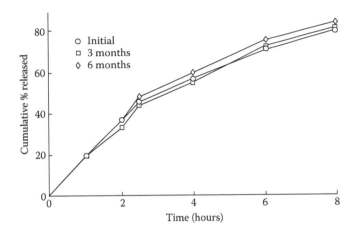

FIGURE 2.35 Stability profile (room temperature) of phenylpropanolamine released from beads coated with acetylated monoglyceride-plasticized Aquacoat® ECD.

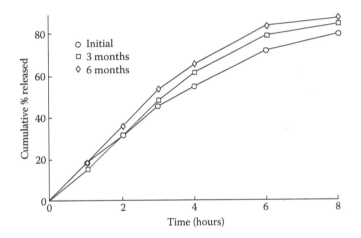

FIGURE 2.36 Elevated temperature (35°C) stability for beads coated with acetylated mono-glyceride-plasticized Aquacoat® ECD.

Figures 2.33 and 2.34 show the corresponding room temperature and 35°C stability profiles for PPA release using TEC as plasticizer. For beads stored at either condition, the drug-release rate slowed at 3 months, with a further slight slowing at 6 months. This decreased release rate on storage is characteristic of an incompletely coalesced film at the time of initial dissolution testing. Pseudolatex coating containing TEC showed the largest loss (Table 2.8) in plasticizer content after 6 months of storage at both room temperature and at 35°C, which did not correlate with the decrease in release rates.

Results using acetylated monoglyceride (Myvacet 9-40) as plasticizer are shown in Figures 2.35 and 2.36. These profiles showed a slight increase in release rates, which was more pronounced at 35°C storage.

To further investigate the differences in drug release on storage, the porosity of the coated beads was measured by mercury porosimetry (Table 2.9). The coated beads had been stored for approximately 6 months at room temperature or 35°C when submitted for analysis. The porosity of the beads was calculated from cumulative pore volume (cm³/g) and particle density (g/cm³) using intrusion porosimetry. The coated beads were of low porosity, varying from 1.6% to 2.0%, with no

TABLE 2.8

Analysis (Gas Chromatography) of Plasticizer Content in Ethylcellulose Pseudolatex Film

Plasticizer (% Remaining)	Initial	3 Months		6 Months	
		RT	35°C	RT	35°C
DBS	100	106	106	99	103
Citroflex-2	100	–	–	92	82

Note: DBS, dibutyl sebacate; RT, room temperature.

TABLE 2.9

Mercury Intrusion Porosimetry Data for Drug Beads Coated with Aqueous Polymeric Dispersion

Plasticizer (30%)		Porosity (E)	Surface Area of Pores (m²/g)
DBS	RT	0.018	2.07
	35°C	0.018	1.97
TEC	RT	0.019	1.94
	35°C	0.016	1.81
TBC	RT	0.020	1.90
	35°C	0.019	1.89
ATBC	RT	0.019	2.04
	35°C	0.018	1.87
DEP	RT	0.019	1.93
	35°C	0.018	1.96
Myvacet 9-40	RT	0.020	2.05
	35°C	0.019	1.81

Note: ATBC, acetyl tributyl citrate; DBS, dibutyl sebacate; DEP, diethyl phthalate; RT, room temperature; TBC, tributyl citrate; TEC, triethyl citrate.

significant difference between samples stored at room temperature or 35°C. The pore surface area (m²/g) generally correlated with the porosity; that is, as the porosity decreases, so does the pore surface area. The pores that were present were very small.

Unanticipated pH-dependent release from aqueous ethylcellulose coatings [38,39] may be attributed to partial coalescence [3]. Dressman et al. [40] demonstrated that heating pellets coated with ethylcellulose above the T_g of the film stabilized the release profile with respect to the pH of the test media. Additional studies were conducted [41] to identify changes in the film that would explain the stabilization of the release profile. Curing converted the film from a surface having a finite contact angle to a surface instantly wetted by the dissolution media. Scanning electron micrography indicated that film morphology changed during curing with latex particles less distinctly after heating, which they called "film relaxation," that is, coalescence. It was concluded that film wetting is an important determinant of the release profile of dosage forms coated with ethylcellulose aqueous dispersions, and these properties are changed when the film is relaxed or coalesced by heating above the T_g. This is consistent with the phase inversion and expulsion of the hydrophilic components (including surfactant) from the coalesced ethylcellulose film [5].

Nesbitt et al. [42] published release rates of pseudoephedrine HCl and diphenhydramine HCl from pellets coated with ethylcellulose pseudolatex. They concluded that drug release through a pseudolatex film occurs through a capillary network whose porosity varies with drying conditions, driven by solubility-dependent osmotic and diffusive forces.

Ozturk et al. [43] cited as possible mechanisms for release solution/diffusion through the continuous polymer phase and/or plasticizer channels, diffusion through aqueous pores, and osmotically driven release through aqueous pores. To distinguish among these mechanisms, the release rate was studied as a function of coating thickness, plasticizer content, and osmotic pressure in the dissolution medium. As the coating thickness was increased from 9 to 50 μm, the rate of release fell from 9.93×10^{-3} to 1.71×10^{-3} g PPA/100 ml/hour (Figure 2.28). Release as a function of plasticizer content was studied over the range of 12% to 24% DBS [29]. At 18% or 24% DBS, the release rates were virtually identical, about 50% in 6 hours. At 12% DBS, over 80% was released in the first hour, and these results were attributed to the presence of cracks in the coating. Release was also studied as a function of the osmotic pressure in the medium (Figure 2.37). A plot of release rate versus osmotic pressure revealed a linear relationship with a nonzero intercept (Figure 2.38). The steep dependency of release rate on osmotic pressure of the medium suggested that osmotically driven release is a major mechanism for release whereas the nonzero intercept indicated some contribution from diffusion mechanisms. For all batches, SEMs indicated that the film exhibited pores approximately

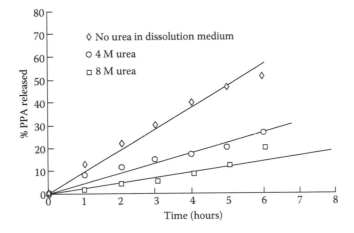

FIGURE 2.37 Effect of osmotic pressure on PPA·HCl release profiles (at a 10% coating loading). PPA, phenylpropanolamine.

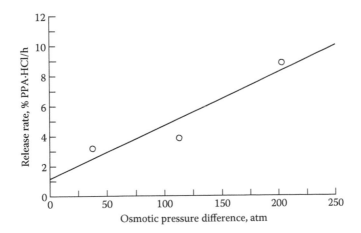

FIGURE 2.38 Effect of osmotic pressure difference on PPA·HCl release rate (at a 10% coating loading). PPA, phenylpropanolamine.

2 μm in diameter, consistent with these mechanisms. Ozturk et al. concluded that the release of PPA from pellets coated with the ethylcellulose-based pseudolatex formulation was mainly driven by osmotic pressure with a minor contribution by diffusion through aqueous pores and perhaps solution/diffusion through the polymer membrane. Osmotic pressure measurements showed that the osmotic pressure generated by both PPA·HCl and the sugar (Nu-pareils) would contribute significantly to the driving force for release. Assuming that these mechanisms operate independently and in parallel, the release of PPA from pellets coated with the ethylcellulose-based film (J) can be mathematically described by an equation that combines these mechanisms:

$$J = \left[\sigma \Delta \Pi + \frac{P_P + P_M}{\delta} \right](C_S - C_B)$$

where σ is the osmotic driving force, $\Delta \Pi$ is the osmotic pressure difference across the coating, P_P and P_M are the permeability coefficients for aqueous pores and membrane, respectively, δ is the film thickness, and C_S and C_B are the core surface and bulk drug concentrations, respectively. The same mechanism is operative over a coating range of 5% to 16%, so film thickness may be used as a means of modifying the release rate without changing the release mechanism (within the range of 10–50 μm). Important factors in determining the release rate from these systems include the volume fraction and size of pores generated during processing, the permeability of the film to water, the rate of core dissolution, and the ability of the core constituents and drug to generate osmotic pressure.

FLUIDIZED BED PROCESSING

Ethylcellulose latices function well not only in Wurster-type coating equipment but also in other types of fluidized bed equipment, for example, conventional air suspension chamber or granulator and the rotary fluid bed coater. Conventional air suspension chambers or fluidized bed granulators are characterized by a random or turbulent movement of particles and by spray nozzles positioned at or near the top of the processing chamber. PPA·HCl beads were coated with an ethylcellulose latex in two types of fluidized bed equipment, for example, Wurster versus top/bottom granulating spray inserts [44]. The coating trials are summarized in Tables 2.10 and 2.11. PPA release from beads coated by the top or bottom spray methods were faster than beads coated by the Wurster method (Figure 2.39). The difference in drug-release profiles between the two coating process

---

TABLE 2.10

Summary of Coating Process Conditions

Constants	Wurster Insert	Granulating Insert
Pump type	Peristaltic	Peristaltic
Atomizing air pressure	1.5 bar	1.5 bar
Inner partition height	⅜ in.	
Port size	1.0 mm	1.2 mm
Nozzle height	Bottom	0.7
Spray angle		0.7
Coating level	10% (2% slow/8% fast)	
Coating suspension	Aquacoat® ECD with DBS 24%[a] applied at 30% solids concentration	

[a] Based on ECD solids.

TABLE 2.11
Batch-Specific Details of Fluid Bed Coating Trials

Equipment (Insert)	Wurster	Granulating	Granulating
Spray mode	Up	Down	Up
Product	1 kg PPA beads	1 kg PPA beads	1.4 kg PPA beads
Inlet set temperature (°C)	55–64	55–64	60–64
Actual temperature (°C)	60–80	52–80	62–81
Outlet temperature (°C)	44–47	44–48	33.5–42
Product temperature (°C)	27–38		
Spray time (min)	46	50.5	51
Dry time (min)	30	30	30
Spray rate (ml/min)—slow	3.4	3.4	5.6
Spray rate (ml/min)—fast	11.7	9.9	12.4
Recovery (%)	99.5	98.6	98.3
RH (%)	10	14	12

Note: PPA, phenylpropanolamine; RH, relative humidity.

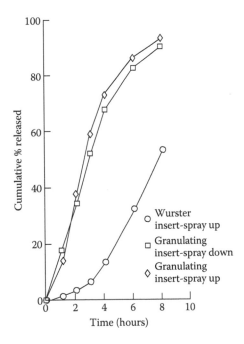

FIGURE 2.39 Release of PPA·HCl from seeds coated by top spray method versus Wurster method. PPA, phenylpropanolamine.

techniques can be explained on the basis of the method of application of coating and on film formation and structure. In the Wurster process, the coating liquid is applied concurrently with the flow of the product. The Wurster system combines a partition (column) and an air distribution plate to organize the flow of particles in close proximity to the spray nozzle. Because the nozzle is immersed in the airflow in order to spray concurrently into the fluidized particles, the dispersion droplets travel only a short distance before impinging on the product. As a result, the film is applied more evenly. On the other hand, spray drying of the coating dispersion is most severe in the counter-current top spray granulating insert. SEM examination showed the top spray samples to be much rougher in surface appearance and more poro us than the Wurster-coated samples, as shown in Figures 2.40 and 2.41.

Ethylcellulose latex dispersions have also been successfully applied to beads by a rotary fluid bed coater. In the rotor (tangential spray) method, the coating dispersion is sprayed tangentially in the same direction as the moving beads in the bed. The beads are rotated in a homogeneous, spiral motion by the combined action of the fluidized air, centrifugal force, and gravity. The differences in action between the two coating process techniques again accounts for the faster release shown in Figures 2.42 and 2.43. Examination of the coated drug beads by SEM showed similar morphological differences as the top spray versus Wurster.

Figure 2.44 shows how release patterns can be modified by the addition of a water-soluble polymer hydroxypropyl methylcellulose (HPMC). However HPMC destabilizes the aqueous ethylcellulose dispersion, which can result in partial coalescence and unpredictable release profiles. Siepmann et al. have identified soluble polymers physically compatible with aqueous ethylcellulose dispersions, such as polyvinyl alcohol (PVA)–polyethylene glycol (PEG) copolymer, propylene glycol alginate (PGA), and carrageenan, which are better suited to giving concentration-dependent release modulations [45–47].

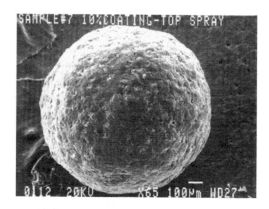

FIGURE 2.40   Surface view of PPA·HCl seed coated by top spray (granulating) method. PPA, phenylpropanolamine.

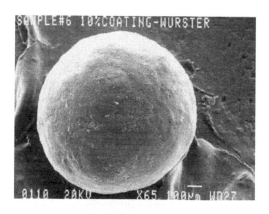

FIGURE 2.41   Surface view of PPA·HCl seed coated by Wurster method. PPA, phenylpropanolamine.

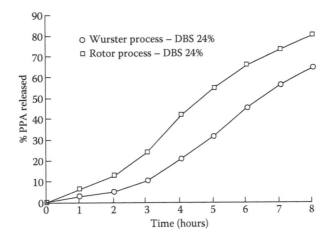

FIGURE 2.42 Release of PPA·HCl from seeds coated by the rotor process versus Wurster process. DBS plasticizer. DBS, dibutyl sebacate; PPA, phenylpropanolamine.

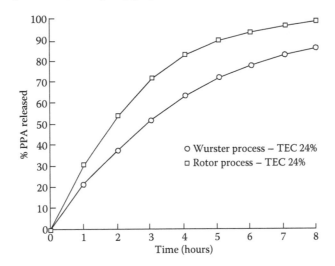

FIGURE 2.43 Release of PPA·HCl from seeds coated by the rotor process versus Wurster process. TEC plasticizer. PPA, phenylpropanolamine; TEC, triethyl citrate.

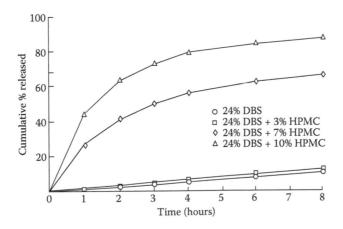

FIGURE 2.44 Effect of water-soluble polymer incorporation on dissolution.

AQUEOUS ENTERIC POLYMER DISPERSIONS

Enteric film-forming polymers, such as CAP, contain ionizable functional groups and exhibit pH-dependent solubility [48–50]. At low pH, the functional groups are unionized, and the film is insoluble. At elevated pH, these functional groups ionize and the polymer becomes soluble. These systems are typically used to protect a drug from the harsh environment of the stomach, to prevent a drug from irritating the stomach mucosa, or to target drug release to the small intestine or colon. Aquacoat® CPD is an aqueous dispersion of CAP and consists primarily of CAP together with a surfactant from the emulsion stage as shown in Table 2.12. Traces of dimethylpolysiloxane to suppress foaming during distillation may also be present. CAP is prepared by reacting a partial ester of cellulose acetate with phthalic anhydride. CAP (Figure 2.45) is a cellulose ester with three hydroxyl groups per glucose unit available for substitution. About half the hydroxyl groups are acetylated, and another quarter esterified with one of the two acid groups of phthalic acid. The dispersion is manufactured by an emulsion process in which the CAP polymer is converted to a pseudolatex in a procedure similar to that used in the production of Aquacoat® ECD.

Table 2.13 defines the processing conditions used to apply plasticized aqueous CAP dispersion to aspirin tablets in a 24-in. Accela-Cota (Thomas Engineering, Hoffman Estates, IL, U.S.A.). Peristaltic pumps are typically used to minimize stress on the latex material and to accurately measure unit fluid rates. Bed temperatures are fairly low (36°C–38°C) for water-based film application.

According to the United States Pharmacopeia, enteric-coated products should resist 0.1 N hydrochloric acid, such that not more than 10% of the active is released in 2 hours. When placed in pH 6.8 phosphate buffer, the film coating should dissolve rapidly to release the active, typically in less than

---

TABLE 2.12

Composition of Aquacoat® CPD

	Solids (%)	Finished Product (%)
CAP	78	23.3
Pluronic F68	22	6.7
Water		70.0

Note: CAP, cellulose acetate phthalate.

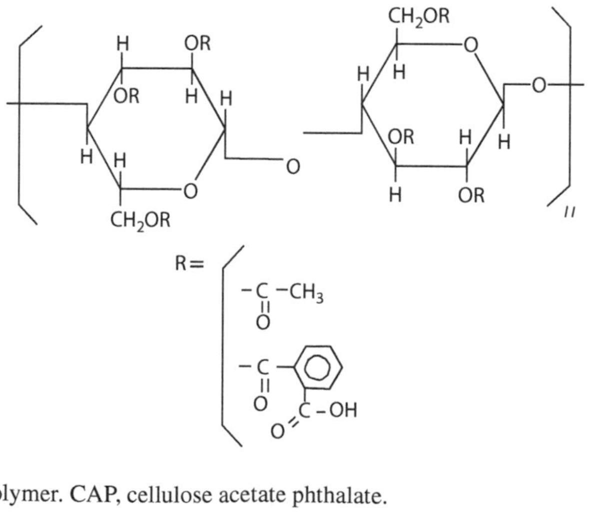

FIGURE 2.45   CAP polymer. CAP, cellulose acetate phthalate.

TABLE 2.13

CAP Pseudolatex Enteric Coating: Equipment and Conditions

Equipment	Accela-Cota 24 in.
Baffles	Four straight bar and four scoop
Pump	Masterflex 7562-10
Pump heads	2 Masterflex 7015
Tubing size	0.1925, 0.3920 in. o.d./in. i.d.
Spray guns	Two spraying systems, 7310-1/4 JAU
Fluid cap	40100 SS
Air cap	134255-45° SS
Conditions	
Tablet charge	9.5 kg
Fluid rate	64 ml/min (total) 32 ml/min per gun
Atomizing air	35 psi per gun
Air temperature	
Inlet	80°C–82°C
Outlet	36°C–38°C
Pan speed	9.5–10.5 rpm
Magnehelic	1.5 in. H_2O
Tablet bed warming	10 min jogging
Coating time	120 min
Postdrying	
Accela-Cota pan	Intermittent jogging
Air temperature, inlet	60°C
Time	60 min
Film weight addition	8.9% w/w

10 minutes. Enteric tablets containing alkaline actives may disintegrate prematurely in acid as the coating solubilizes due to a high pH microenvironment. To prevent the formation of soluble alkali phthalate salts, the substrates can be seal-coated first with HPMC before applying the Aquacoat® CPD.

As with sustained-release coatings, film thickness is of critical importance to the functional performance of enteric coated products. Too thin a coating can result in tablet failure in an acidic environment. Too much enteric coating may lengthen the intestinal disintegration time. In fact, high loadings of Aquacoat® CPD can be utilized for colonic drug delivery systems. A film level of at least 5% w/w was required to ensure the integrity in acid of coatings made from either aqueous CAP dispersion or CAP applied from an organic solvent system (Figure 2.46). At levels higher than 5% film weight, the CAP pseudolatex coatings exhibit slightly faster disintegration times than the corresponding CAP/solvent coatings applied at the same coating level.

In Figure 2.47, the disintegration time for aspirin cores coated with CAP pseudolatex and CAP/ solvent formulations are compared. It was found that at pH 6.4 and higher, no significant differences in disintegration time were noted for aspirin tablets coated with either the aqueous latex or the organic solution of CAP. However, disintegration time increased substantially as the pH dropped below 6.4, and a significant difference in disintegration time was observed between the two film-coating systems. Table 2.14 shows that there was no significant change in disintegration time for the latex product after 12 months of storage at room temperature and 35°C whereas the aspirin product coated from organic solvent exhibited a substantial increase in disintegration time upon aging.

Figure 2.48 shows the disintegration time of tablets coated with Aquacoat® CPD as a function of plasticizer content and plasticizer type. Two plasticizers were investigated: DEP, which is water insoluble, and the water-soluble propylene glycol. Twenty-five percent by weight of either DEP or

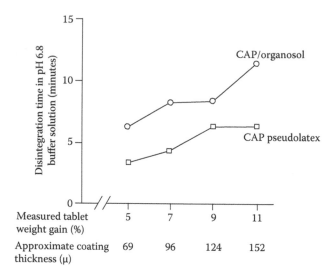

FIGURE 2.46 Disintegration time of aspirin tablets coated with CAP pseudolatex or CAP/organosol in phosphate buffer as a function of film thickness. CAP, cellulose acetate phthalate.

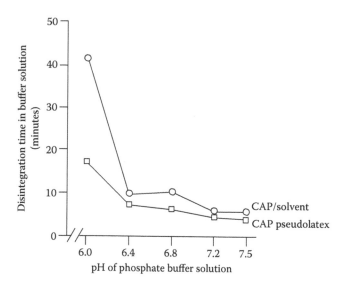

FIGURE 2.47 Disintegration time of aspirin tablets coated with CAP Pseudolatex or CAP/solvent coating as a function of pH. CAP, cellulose acetate phthalate.

propylene glycol was insufficient to achieve adequate film quality, and the coatings failed in 0.1 N HCl. However, when the plasticizer level was increased to at least 30% (based on pseudolatex solids), the coatings were resistant to the low pH test media. At higher (54%) levels of propylene glycol, the enteric film coatings failed in acid medium, which was attributed to plasticizer leaching.

Another study evaluated aspirin release at various pH media using the USP (basket) method I at 100 rpm. Aspirin release was shown to increase with increasing pH. The USP enteric dissolution specification is not more than 10% release of aspirin after 2 hours of testing in a pH 1.5 acid medium. Figure 2.49 shows that aspirin tablets coated with CAP pseudolatex do not show any significant release of aspirin until pH 6, the pH at which the acid functional groups of the CAP polymer begin to ionize [51].

TABLE 2.14
Stability Profiles for Aspirin Tablets Coated with Enteric Latex

		Disintegration Time (min)	
Time	Condition	Aspirin with CAP Pseudolatex Coating	Commercial CAP/ Organosol Coated Aspirin
Initial		6–8	3–4
6 months	RT	5	11.5
	35°	6.5	12
9 months	RT	4–7	6–19
	35°		
12 months	RT	4–6	8–10
	35°	4–6	8–16

Note: CAP, cellulose acetate phthalate.

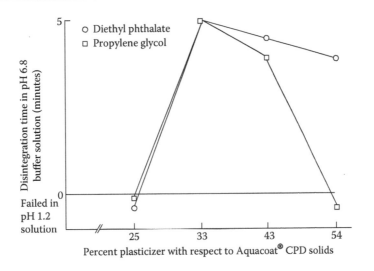

FIGURE 2.48 Disintegration time of aspirin tablets coated with CAP pseudolatex as a function of plasticizer content.

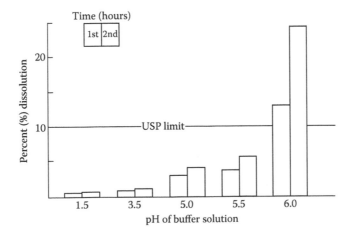

FIGURE 2.49 Dissolution of aspirin tablets coated with CAP pseudolatex as a function of changing pH media.

SUMMARY

Colloidal aqueous dispersions of ethylcellulose and CAP provided effective and versatile rate-controlling membranes in the design of modified-release oral solid dosage forms. Aqueous pseudo-latex coatings avoid the environmental, safety, and toxicological problems associated with organic solvents. The formulation scientist must understand the mechanisms of film formation from such aqueous-based systems in order to achieve stable drug-release rates. Interactions between the coating formulation, substrate, and processing parameters require the formulator to give careful consideration to the entire coating process.

REFERENCES

1. Vanderhoff JW, El-Asser MS. Polymer emulsification process. US Patent 4,177,177, 1979.
2. Ortega AM. Latices of cellulose polymers; manufacture, characterization and applications as pharmaceutical coatings. PhD dissertation, Purdue University, 1977.
3. Wesseling M, Bodmeier R. Drug release from beads coated with an aqueous colloidal ethylcellulose dispersion, Aquacoat® or an organic ethylcellulose solution. *Eur J Pharm Biopharm* 1999; 47(1):33–38.
4. Banker GS, Peck GE. The new water-based colloidal dispersions. *Pharm Technol* 1981; 5(4):55–61.
5. Chevalier Y, Pichot C, Graillat C et al. Film formation with latex particles. *Colloid Polym Sci* 1992; 270(8):806–821.
6. Felton LA. Mechanisms of polymeric film formation. *Int J Pharm* 2013; 457:423–427.
7. Frenkel J. Viscous flow of crystalline bodies under the action of surface tension. *J Phys* 1945; 9(5):385–491.
8. Dillon RE, Matheson LA, Bradford EB. Sintering of synthetic latex particles. *J Colloid Sci* 1951; 6(2):108–117.
9. Brown GL. Formation of films from polymer dispersions. *J Polym Sci* 1956; 22(102):423–434.
10. Steward PA. Modification of the Permeability of Polymer Latex Films. PhD dissertation, Nottingham Trent University, 1995.
11. http://www.initium.demon.co.uk/index.htm.
12. Sperry PR, Snyder BS, O'Dowd ML et al. Role of water in particle deformation and compaction in latex film formation. *Langmuir* 1994; 10(8):2619–2628.
13. Lissant KJ. The geometry of high-internal-phase ratio emulsions. *J Colloid Interface Sci* 1966; 22(5):462–468.
14. Joanicot M, Wong K, Maquet J et al. Ordering of latex particles during film formation. *Prog Colloid Polym Sci* 1990; 81:175–183.
15. Nicholson JW, Wasson EA. Film spreading and film formation by waterborne coatings. Surface Coatings. Vol. 3. Elsevier Applied Science, 1990:91–123.
16. Voyutskii SS. Amendment to the papers by Bradford, Brown, and co-workers: Concerning mechanism of film formation from high polymer dispersions. *J Polym Sci* 1958; 32(125):528–530.
17. Voyutskii SS, Vakula VL. Effect of self-diffusion and inter-diffusion in polymer systems. *Rubber Chem Technol* 1964; 37:1153–1177.
18. Bradford EB, Vanderhoff JW. Morphological changes in latex films. *J Macromol Chem* 1966; 1:335.
19. Bradford EB, Vanderhoff JW. Additional studies of morphological changes in latex films. *J Macromol Sci Phys* 1972; B6:671–694.
20. Hahn K, Ley G, Schuller H et al. On particle coalescence in latex films. *J Colloid Polym Sci* 1986; 264:1092–1096.
21. Hahn K, Ley G, Oberthür R. On particle coalescence in latex films (II). *J Colloid Polym Sci* 1988; 266(7):631–639.
22. Sperling LH, Klein A, Yoo JN et al. The utilization of SANS to solve polymer latex structural problems: Basic science and engineering. *Polym Adv Technol* 1990; 1(3–4):263–273.
23. Distler D, Kanig G. Feinstruktur von Polymeren aus wäßriger Struktur (fine structure of polymers from aqueous structure). *Colloid Polym Sci* 1978; 256(10):1052–1060.
24. Aten WC, Wassenburg TC. Influence of the surfactant distribution on the water resistance of clear latex films. *Plastics Rubber Process Appl* 1983; 3(2):99–104.
25. Ghebre-Sellassie I, Nesbitt RU, Wang J. In: McGinity JW, ed. Aqueous Polymeric Coatings for Pharmaceutical Dosage Forms. 2nd ed. New York: Marcel & Dekker, 1997.

26. Mehta AM. In: McGinity JW, ed. Aqueous Polymeric Coatings for Pharmaceutical Dosage Forms. 2nd ed. New York: Marcel & Dekker, 1997.
27. Li J-X, Carlin BA, Lee JT et al. Effect of high-humidity coating process on the drug release from pellets coated with ethylcellulose aqueous dispersion NF. 33rd Annual Meeting and Exposition of the Controlled Release Society, Vienna, Austria July 22–26, 2006.
28. Li J-X, Carlin BA, Lee JT et al. The effect of high-humidity during aqueous psuedolatex coating. AAPS 20th Annual Meeting, San Antonio, Oct 29–Nov 2, 2006.
29. Oshlack B, Pedi F Jr, Chasin M. Stabilized controlled release substrate having a coating derived from an aqueous dispersion of hydrophobic polymer. US patent 5,273,760, 1993.
30. Li J-X, Carlin BA. A high humidity coating process to produce stabilized controlled release coatings followed by low humidity curing. US Pat App 60518-USA-PROV, 2006.
31. Banker GS. Film coating theory and practice. *J Pharm Sci* 1966; 55:81–89.
32. Talen HW, Hover PF. On film formation by emulsion paints and some properties of these films. *Dtsch Farben Z* 1959; 13:50–55.
33. Hildebrand JF, Scott RL. Solubility of Non-Electrolytes. 3rd ed. New York: Reinhold, 1950.
34. Siepmann J, Paeratakul O, Bodmeier R. Modeling plasticizer uptake in aqueous polymer dispersions. *Int J Pharm* 1998; 165:191–200.
35. Wesseling M, Bodmeier R. Influence of plasticization time, curing conditions, storage time and core properties on the drug release from Aquacoat-coated pellets. *Pharm Dev Technol* 2001; 6(3):325–331.
36. Zhou Y, Li J-X, Carlin BA et al. Analysis of dispersed/dissolved drug release from coated heterogeneous beads in a finite external volume. CRS Annual Meeting and Exposition, Honolulu, Hawaii, June 12–16, 2004.
37. Zhou Y, Li JX, Wu XY. Immediate answer to "what-if" questions in formulation process of coated dosage forms by computer aided design. AAPS 21st Annual Meeting, San Diego, CA, Nov 11–15, 2007.
38. Ozturk AG. Studies on the release of drugs from pellets coated with ethylcellulose-based film. M.Sc. thesis, University of Michigan, 1990.
39. Sutter BK. Aqueous ethylcellulose dispersions for preparation of microcapsules with controlled drug release. PhD thesis, University of Dusseldorf, 1986.
40. Dressman JB, Ismailos G, Jarvis C et al. Influence of plasticizer and drying conditions on the pH-dependency of release from ethylcellulose-coated pellets. Controlled Release Society Symposium on the Controlled Release of Bioactive Materials, Washington, D.C., July 25–30, 1993.
41. Dressman JB, Ismailos G, Naylor LJ et al. Stabilization of ethylcellulose films by oven drying. AAPS Annual Meeting, Orlando, FL, Nov 14–18, 1993.
42. Nesbitt RU, Mahjour M, Mills NL et al. Mechanism of drug release from Aquacoat coated controlled release pellets. Arden House Conference Proceedings, Harriman, NY, January 1986.
43. Ozturk AG, Ozturk SS, Palsson BO et al. Mechanism of release from pellets coated with an ethylcellulose-based film. *J Control Release* 1990; 14:203–213.
44. Wheatley TA, Wieczak JE. FMC Internal Research, 1995.
45. Siepmann F, Hoffmann A, Leclercq B et al. How to adjust desired drug release patterns from ethylcellulose-coated dosage forms. *J Control Release* 2007; 119:182–189.
46. Siepmann F, Wahle C, Leclercq B et al. pH-sensitive film coatings: Towards a better understanding and facilitated optimization. *Eur J Pharm Biopharm* 2007 Jan; 68(1):2–10.
47. Siepmann F, Muschert S, Zach S et al. Carrageenan as an efficient drug release modifier for ethylcellulose-coated pharmaceutical dosage forms. *Biomacromolecules* 2007 Dec; 8(12):3984–3991.
48. Bauer KH, Osterwald H. Film coating—A particulate solid pharmaceutical and emulsions for the process. Ger. Patent 2,926,633, 1981.
49. Bauer KH, Osterwald H. Partial replacement of organic solvents in aqueous organic plastics emulsions. *Pharm Ind* 1979; 41:1203–1207.
50. Osterwald H, Bauer KH. Gegenüberstellung von dünndarmlöslichen Filmüberzügen einiger synthetischer Polymere auf festen Arzneiformen aus wässrigen und aus organischen Umhüllungszubereitungen. *Acta Pharm Technol* 1980; 26(3):201–209.
51. Zatz JL, Knowles BJ. The effect of pH on monolayers of cellulose acetate phthalate. *J Pharm Sci* 1970; 59(12):1750–1751.

Philip J. Hadfield

CONTENTS

INTRODUCTION

The coating of pharmaceutical dosage forms can be traced back at least a thousand years.[1] Sugar coating was introduced relatively recently with pictures depicting the work done using a shallow pan over an open fire. The commonly used sugar-coating pan, which rotates on an inclined axle was introduced in the 1800s, and early images show that it was rotated by a hand crank mechanism.[2]

Fortunately, the electric motor was added, resulting in the more characteristic sugar-coating pan with forced drying air and extraction available to aid the drying process. The sugar-coating process was labor intensive, requiring highly skilled operators to get consistent, high-quality results.[3] There were no baffles in these pans. Mixing of the tablets, when the sugar solution was added, was aided by the operator "stirring" the wet tablet mass by hand until the sugar solution appeared to be uniformly distributed.

---

* This is an updated version of the chapter originally written by Atul M. Mehta.

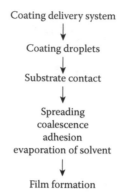

FIGURE 3.1   Dynamics of the coating process.

Film coating, particularly aqueous film coating, is a relatively new process to the pharmaceutical industry.[2] The film-coating process requires a delicately balanced environment (Figure 3.1).

The formation of a high-quality layer of film onto a substrate requires the following:

1. Formation of appropriately sized droplets, delivered at a controlled rate
2. Contact of these droplets with the substrate
3. Adhesion of the droplets to the substrate
4. Spreading and coalescence of the droplets
5. Evaporation of the solvent

An equilibrium must be established such that the coating material adheres and coalesces properly on contact with the surface of the substrate, yet it also must dry rapidly so that core penetration of the solvent and dissolved coating material is minimized and agglomeration of the core material is prevented. To create the necessary environment for such a process to occur, specialized coating equipment and optimal processing conditions are required.

EQUIPMENT

The equipment used in pharmaceutical film coating can be broadly classified into three general categories: solid pans, perforated pans, and fluid bed processes. These systems are used to contain the materials being coated and provide an environment for the coating to dry. They should also provide the means to ensure that an equal amount of coating material is applied evenly to each particle. A delivery system conveys the coating material to the coating equipment in a controlled and reproducible manner. Support equipment contributes to automation and includes the control systems. The available equipment systems vary from simple air handlers and manual process control to automated processes and inlet air dew point control.

Film coating of pharmaceutical tablets was first introduced in the mid 1950s by Abbott Laboratories, utilizing fluid bed technology with a non-aqueous coating system.[4–6] Film coating was not widely adopted until the development of the Accela-Cota (side-vented pan) by Eli Lilly in the mid 1960s, and he licensed the technology to Thomas Engineering (who in turn sublicensed it to Manesty Machines and Glatt). Figure 3.2 shows the principle of airflow through the coating pan.

The development of this technology enabled the drying air to be more intimately directed across and through the tablet bed, resulting in a higher utilization of the available drying capacity. This was important to the development of aqueous film coating due to the relative high latent heat of vaporization of water (539 kcal/kg), which is much greater than that of the then popularly used organic solvents (e.g., 200 kcal/kg for ethanol).

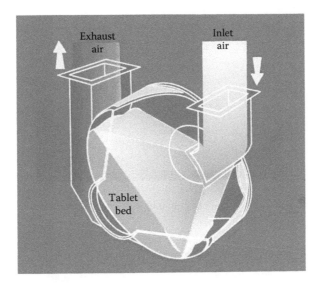

FIGURE 3.2 Schematic of the airflow through a perforated pan. (Courtesy of Bosch Packaging Technology Ltd.)

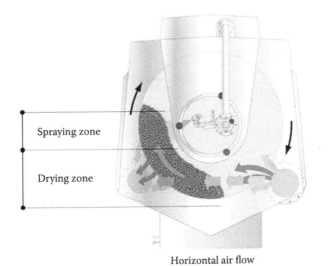

Horizontal air flow

FIGURE 3.3 Schematic of the airflow through a Glatt coating pan. (Courtesy of Glatt GmbH.)

Each of these manufacturers developed variations to the delivery of air to the coating pan. Figure 3.3 shows the variation chosen by Glatt.

Other machine manufacturers chose a different route to enable control of the drying environment in the coating pan, such as Pellegrini with a solid plan which utilized immersed extraction swords with air being blown in from the rear of the pan (Figure 3.4).

Another alternative air flow utilized a semi-perforated pan to circumvent the Eli Lilly patent (Figure 3.5). Both pan types were widely utilized in the pharmaceutical industry.

The introduction of the perforated pan was a significant milestone in the decision making with respect to the equipment used to perform specific film-coating tasks. The perforated pan, by its nature, has holes of approximately 3 mm in diameter through which the air passes. Coating of small particles (nonpareils, drug particles, pellets, etc.) in these pans, unless modified with a mesh insert,

FIGURE 3.4   Film-coating process in the Pellegrini pan with immersion swords. (Courtesy of IMS S.p.A.)

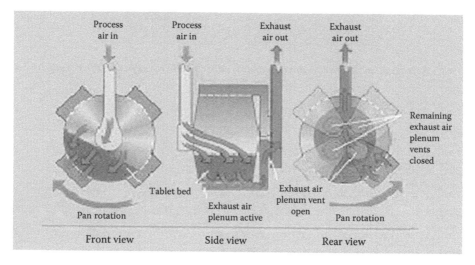

FIGURE 3.5   Schematic of the airflow through a semi-perforated coated pan. (Courtesy of the Vector Corporation.)

was impossible. The solid pan was similarly limited as the exhaust plenum needed to be modified to ensure that the particles were not taken into the exhaust system. The mixing action in a coating pan is relatively gentle, and thus getting good distribution of either drug or modified release coating onto these small particles would take significantly longer than the more aggressive mixing seen in a fluid bed coating system. The fine mesh over the exhaust manifold also has a significant impact on the volume of air that can pass through the pan, which similarly limits the spray rate that can be used. Other factors that limit the suitability of the coating pan for coating small particles is the relatively small surface area offered the coating and potential limitations on the amount atomizing energy available to yield small enough droplets to coat the small particles without incurring sticking and subsequent growth of the pellets. For these reasons, fluid bed coating is the first choice methodology for the coating of small particles of up to 1 mm in diameter. Although special mesh pans can be used in place of the normal perforated pan, they are rarely used in practice to coat small particles as it is significantly easier to do this using a fluid bed coating system fitted with a Würster column. Similarly, it is unusual to see tablets being film-coated at the production level in a fluid bed coating system due to the high energy imparted on the tablets during the coating process. This high energy usually results in the tablets suffering severe physical degradation in the early stages of coating.

The side-vented pan (fully perforated, semi-perforated, and solid wall) thus became the coating pan of choice for film coating tablets. In the 1960s and early 1970s, two distinct methods were used for delivering the coating suspension/solution; airless and airborne spray systems were employed. The airless spray system required the delivery of the coating suspension/solution at very high pressure (typically 10+ bar) through a small orifice that resulted in the atomization of the spray. This system was more suited to non-aqueous systems than aqueous systems. Control of the spray rate was poor, resulting in intermittent spraying being commonly utilized. Aqueous coating systems were considered difficult to work with due to the relatively low solids composition of the coating and thus the high level of water to be evaporated, until the development of lower molecular weight hydroxypropylmethyl cellulose polymers of pharmaceutical quality, for immediate release and the water dispersible acrylic polymers for enteric and sustained release applications. These polymers allowed the development of higher solids coating compositions, facilitating easier aqueous coating processes. Even so, it took some time for aqueous coating to be accepted.

Legislation changes in the 1980s and early 1990s banning the emission of solvents to atmosphere had a significant impact in the move to aqueous coating.[7]

Machine manufacturers also worked on improving the control of the processing air, both in terms of temperature control as well as volume of air through the pan. A further factor was the introduction of inlet air moisture control, which was considered necessary for very moisture-sensitive drugs.

The advent of computer control[8] enabled not only close control of the inlet temperature, but also of the processing air and the balance of air passing through the pan. It also meant that all ancillary factors could also be potentially controlled, such as pan speed, spray rate, atomizing air, fan width air, and inlet air humidity. Computer control also led to the possibility of having a predefined batch process that could be tailored to an individual product. It also meant that these parameters could be logged for later review.

The various manufacturers provide pans in a range of sizes, giving the ability to coat very small batch sizes (<100 g) for development to production batch sizes in excess of 1000 kg (different manufacturers will offer pans of a varying range and size). In essence, the film-coating process can be described as the random impingement of droplets onto the surface of the substrate to be coated. In order to get a uniform coating, the particles need to be moved within the coating chamber. In the case of a coating pan, this is normally achieved using mixing baffles that actively displace the tablets/particles. This brings about the potential for attrition to take place, the more aggressive the mixing, the more likely for damage to occur. The larger the batch size, the larger the mass to be moved, which again impacts the potential for damage to occur. Machine manufacturers strive to improve mixing efficiency[9,10] without incurring damage so that a potentially faster batch coating time can be achieved. A good example of this is the three baffle types that were developed by Bosch Packaging Technology Ltd., the original "rabbit ear" baffle, which mixed the tablets relatively gently; the tubular baffle, which provided a more aggressive mixing; and the ploughshare baffle, which provides less aggressive mixing than the tubular baffle but a higher mixing efficiency than the rabbit ear baffle.

Since the expiry of the original Accela-Cota patent, a number of companies are now selling film-coating machines of a similar design. The choice of machine is complex and is an individual decision, depending on the types of coatings performed, the degree to which the system is to be customized, the materials to be coated, and many more factors to be considered.

MULTI-PAN MACHINES

These were first introduced as development machines, in which a base unit, which included the air-handling plant, could be used to house different size pans (Figure 3.6). The objective was to provide a more flexible option when space was limited and also to reduce the cost of investment in two or more machines. The initial cost, compared to a single pan machine, is more expensive, but extra pans could be added at a relatively small cost. One area that the development scientist needed to be aware of was to adjust the airflow through the pan to suit the pan size selected. However, this extended ability to adjust airflow through a given pan size allowed the development scientist the

15 inch: 2–3 kg

24 inch: 12–15 kg

19 inch: 8–9 kg

30 inch: 45–50 kg

FIGURE 3.6   LabCoat II with four of the available drums. (From the author's library.)

ability to study the flexibility of a film coating process with respect to its impact on coating performance and efficacy. Production-scale multi-pan machines are now available from some suppliers (e.g., Thomas Engineering, O'Hara Technologies, Glatt GmbH).

CONTINUOUS TABLET COATING

Continuous-coating tablet-coating equipment has been available for at least 20 years from equipment manufacturers such as Thomas Engineering and, more recently, O'Hara Technologies. The biggest perceived problem, initially, with such equipment was the cost of start-up and shutdown, with the potential loss of hundreds of kilos of product. Work has been ongoing to resolve this issue. This has resulted in the development of three major different approaches that resolve this issue.

O'Hara Technologies, for example, have developed a process whereby the initial start-up and final close down of the machine are, in essence, a typical batch-coating process with a transition into the continuous process in the middle (Figure 3.7).

Tablets are held in a feed hopper that can have warm air gently blown through the bulk to preheat the tablets prior to being fed into the machine. The flow of tablets into the coater is controlled via a weight feeder system (Figure 3.8). The tablets move through the coater by natural migration; no

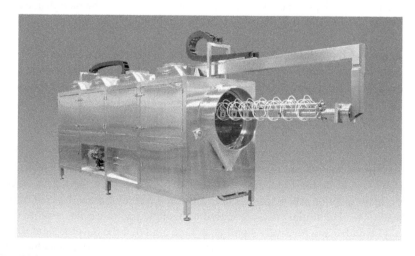

FIGURE 3.7   O'Hara FC-1200 continuous coater. (Courtesy of O'Hara Technologies Inc.)

FIGURE 3.8 Tablet feed into the O'Hara FC C1200 coating pan. (From the author's archive.)

baffles or vanes are used. The quicker the tablets are fed in to the pan, the faster they move through the machine. By controlling the position of the discharge weir, the level of the tablets or residence volume of the coater can be controlled (Figure 3.9). Outputs of up to 1200 kg/hour of coated tablets are achievable with this machine. This machine can also be run as a batch coater, offering a high level of flexibility to the production environment.

Driam Anlagenbau GmbH manufactures the DriaConti-T Pharma continuous coating pan. This pan consists of seven independent chambers and can thus be regarded as a semi-continuous coater (Figure 3.10).

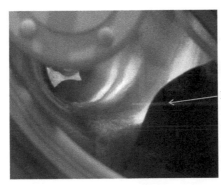

Discharge control weir

FIGURE 3.9 Continuous flow in the FC C1200 O'Hara coating pan. (From the author's archive.)

FIGURE 3.10 Schematic of the DriaConti-T Pharma continuous coater. (Courtesy of Driam Anlagenbau GmbH.)

The bulk tablets are stored in a hopper, which can have warm air gently blown through the mass to preheat the tablets. The tablets are dispensed into the first chamber of the machine via a vibratory feed controlled by load cells to ensure that the correct weight of tablets is dispensed (Figure 3.11). Each of the chambers has a single spray gun with an individually controlled pump. This offers the facility to have a different spray rate in each chamber, a useful feature giving greater flexibility in the coating operation, for example, in the case of aqueous enteric coating with which it is advisable to have a lower initial spray rate to prevent reactivation of the sub-coat. At the end of a predetermined dwell time, the coating stops, the pan slows down, and the transfer gate is actuated, transferring the tablets into the next chamber in one pass (Figure 3.12). The pan speeds up, and coating resumes. As the tablets pass down the machine, a sensor detects which chambers contains tablets and actuates the appropriate spray gun and pump to continue the coating operation. This process is repeated as the tablets pass down the machine until they are discharged. Scale-up is less problematic as a single load can be passed through the pan to determine optimal conditions with the pan effectively operating at a batch size of approximately 18 kg of tablets. Table 3.1 shows the general coating conditions used.

GEA Pharma has approached continuous coating in a different way by including it as part of a continuous processing line. The ConsiGma™ continuous-coating line provides a production line from granulation, compression to film coating (Figure 3.13).

The advantage of this system is that it can be used to produce small batches of product to very large batches, simply by running the process for longer periods. Scale-up and process validation of the whole production process can therefore be achieved at the same time. Another advantage is that the film-coating step, in real terms, does not go through a significant scale-up procedure.

FIGURE 3.11   Tablet feed into the first chamber.

FIGURE 3.12   Tablet inter-chamber transfer.

TABLE 3.1
General Equipment/Process Conditions

Parameter	Details
Coating pan	Driam DriaConti-T, 1000/7-P
Spray guns	7 (one per segment), Schlick 970 (modified body) with ABC air caps
Nozzle opening (mm)	1.0
Gun-to-bed distance (cm)	12
Number of segments	7
Segment width (cm)	18
Quantity of tablets per segment (kg)	18 kg (preheated to set temperature on initial introduction into pan)
Pan diameter (mm)	1000
Process air volume (m³ h⁻¹)*	4000
Segment change	At selected change times, spray switches off, and pan speed drops to 2 rpm
Dwell time at change	In current pan, 2–3 minutes (with new pneumatic dampers, <30 seconds)

*This is the nominal airflow. Process air is delivered to the entire pan. There is a separate exhaust plenum for each segment of the pan, allowing exhaust air to be switched on only when tablets are being coated in that segment. During start-up and shutdown (when all segments will not contain tablets), the inlet air volume is reduced in the ratio of the number of operating segments; for example, at start-up, when tablets have just been introduced into the first segment only, the inlet airflow will be $4000/7 = ~570$ m³ h⁻¹, and when two segments are operating, the air flow will be $(4000/7) \times 2 = ~1140$ m³ h⁻¹, and so on.

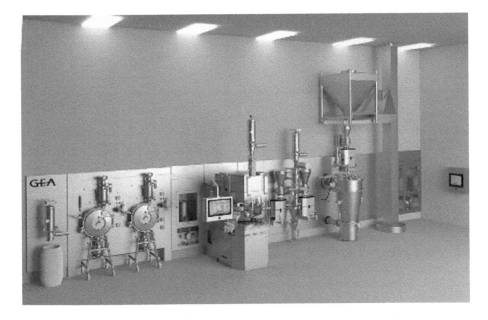

FIGURE 3.13 Continuous granulation, compression, and film-coating line. (Courtesy of GEA Pharma.)

The coating pans (Figure 3.14) are available in three different sizes, 1.5, 3.0, or 6.0 kg approximate capacity (depending on tablet bulk density).

The principle of operation of the film-coating stage is different from a normal drum coating pan (Figure 3.15) in that the pan is charged with tablets and then spun at a relatively high speed causing the tablets to "adhere" to the pan wall by centrifugal force while preheating takes place. Air knives push the tablets away from the side wall to spin through the spray zone, enabling rapid coating. On completion of the coating cycle, the pan slows down, allowing the tablet ring to collapse, and the tablets are then discharged.

FIGURE 3.14   Coating pans. (Courtesy of GEA Pharma.)

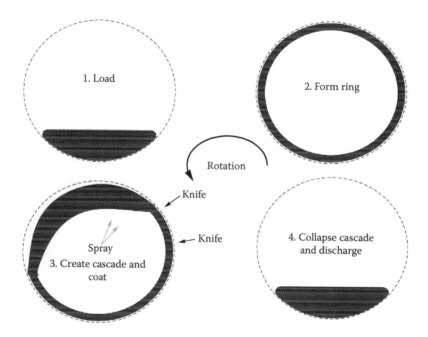

FIGURE 3.15   ConsiGma™ Tablet Coater working principle. (Courtesy of GEA Pharma.)

FLUID BED

The fluid-bed equipment is well known for its drying efficiency, having been used for drying and granulating for many years. A typical fluid-bed system is depicted in Figure 3.16. The use of fluid-bed equipment in applying aqueous coating systems has increased greatly primarily due to (a) improved drying efficiency, (b) improved design considerations, and (c) increased experience.

Fluidized bed equipment is the preferred choice of equipment for aqueous coating systems applied for reasons other than esthetics, such as taste masking, enteric coatings, sustained- or controlled-release coatings, and coatings applied for protection. Aqueous film coating can be applied to the fluidized material by a variety of techniques, including spraying from the

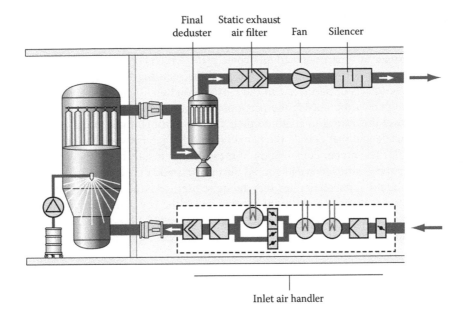

FIGURE 3.16 Typical fluid-bed installation. (Courtesy of Glatt.)

top (granulation or conventional mode), from the bottom (Würster), or tangentially (rotary granulator).

TOP SPRAY (GRANULATOR MODE)

Although it is not applicable for tablets, the top-spray granulator can be used to coat small particles successfully. The films formed in this process are not as uniform, but for releases that are not dependent on membrane thickness or perfection (such as taste masking), it is a viable and simple approach. The substrate is fluidized up to the nozzle, which sprays counter-currently into the material (Figure 3.17). The high particle velocity and efficient heat transfer allow aqueous coating of small particles with little or no agglomeration. Batch sizes range from 0.5 kg to approximately 1000 kg.

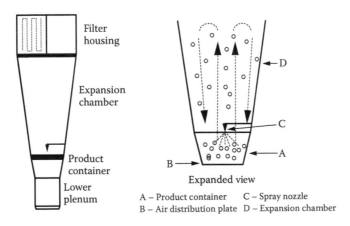

FIGURE 3.17 Top-spray coater. (Courtesy of Glatt Air Techniques, Inc., Ramsey, New Jersey.)

BOTTOM SPRAY (WÜRSTER)

The Würster coating system, invented in the 1950s, has had some success in tablet coating. The flow pattern is formed by a partition and an orifice plate, which control the air flow (Figure 3.18). The majority of the air is diverted through the partition, causing fluidization and upward travel of the cores. As the tablets or particles exit the partition and enter an expansion zone, air velocity decreases, and the cores drop outside the partition. The air in this down-bed acts to cushion the tablets as they travel downward to continue their cycling through the coating zone. The balance between the air inside and that outside the partition and the gap between the orifice plate and the partition are critical. The proper combination of these factors results in a very dense concentration of core material in the coating zone and a rapid down-bed, indicating a short bed cycle time (under these conditions, liquid application rates may be quite high). Additionally, the up-bed height (the distance the tablets rise above the partition) is small and is the key to minimizing the attrition that is usually associated with air suspension tablet coating. Figures 3.19 through 3.22 illustrate tablets coated in a Würster system (Glatt GPCG-5) with three different aqueous polymeric materials.

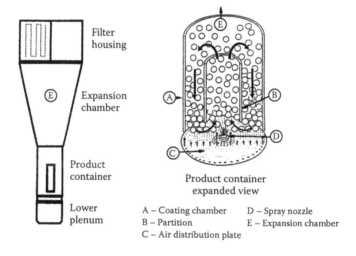

FIGURE 3.18   Bottom-spray (Würster) coater. (Courtesy of Glatt Air Techniques, Inc., Ramsey, New Jersey.)

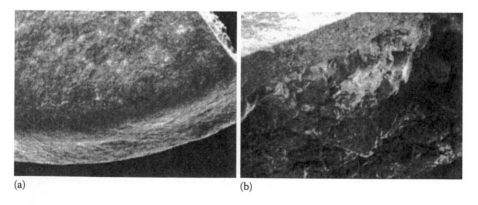

(a)   (b)

FIGURE 3.19   Cross-sections of tablets coated with Eudragit® L 30 D in Würster coaler (Glatt GPCG-5): (a) magnification ×25 and (b) magnification ×100.

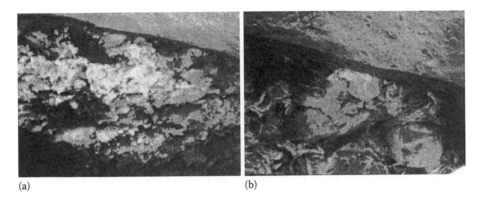

(a)   (b)

FIGURE 3.20 Cross-sectional views of tablets coated with Opadry in Würster coater (Glatt GPCG-5): (a) magnification ×25 and (b) magnification ×100.

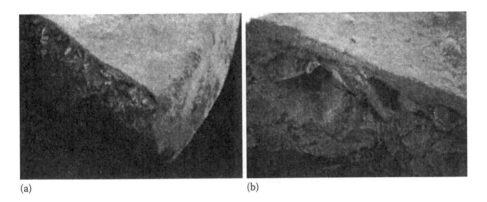

(a)   (b)

FIGURE 3.21 Cross-sections of tablets coated with Aquacoat® in Würster coater (Glatt GPCG-5): (a) magnification ×25 and (b) magnification ×100.

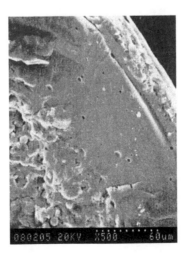

FIGURE 3.22 Cross-section of a drug-containing particle coated with polymers in Würster column. (Courtesy of Elite Laboratories, Inc., Maywood, New Jersey, U.S.A.)

The Würster system is growing in popularity in the coating of smaller particles. It is able to apply droplets to the substrate before much evaporation occurs and rapidly evaporates surface solvent (or water) prior to core penetration. This is evident in Figure 3.22 in which different layers of applied coatings are visible. This is critical for stability as well as end-product performance of the product. Discretely dividing the particles by air suspension allows the application of films to pellets, granules, and materials as fine as 50 μm with little or no agglomeration (depending on the coating substance). The organization of the particles in close proximity to the liquid nozzle and rapid bed cycle times yield uniform distribution of the film. Depending on the vendor, there are variations in the geometry of the total system. However, it is recommended that longer expansion chambers be used to coat small particles (Figure 3.23). The system allows batch sizes from 0.5 kg to approximately 500 kg.

The recently introduced Würster HS™ technology (U.S. Patent 5,236,503) involves the use of a proprietary device to influence the behavior of the coating zone (Figure 3.24). It is designed to keep particles away from the nozzle until the spray pattern is fully developed. As a result, more of

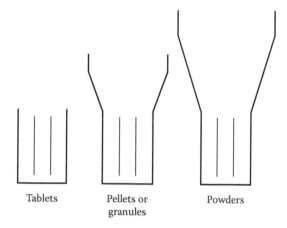

| Tablets | Pellets or granules | Powders |

FIGURE 3.23 Expansion chambers for Würster columns. (Courtesy of Glatt Air Techniques, Inc., Ramsey, New Jersey.)

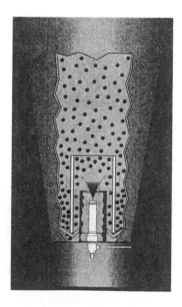

FIGURE 3.24 Würster HS™ system. (Courtesy of Glatt Air Techniques, Inc., Ramsey, New Jersey.)

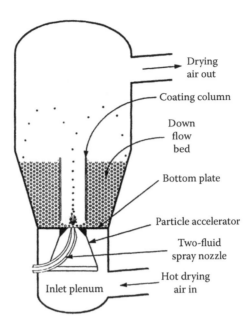

FIGURE 3.25   Precision coater. (Courtesy of Niro Inc., Columbia, Maryland.)

the excess drying capacity can be used, and the application rate can be substantially increased. The high-atomizing air velocities can provide droplet sizes small enough for coating of particles smaller than 100 μm without causing attrition because the velocity decreases prior to contacting the substrate. This is because the product is kept away from the nozzle tip. However, success depends on the type of product to be coated; liquid characteristics, which must be amenable to atomizing to well below 10 μm droplets; and the processing conditions employed. It is to be noted that the high surface area of fine particles requires high amounts of coating to ensure adequate coverage of particles.

Another recently developed approach to improve on the coating application in a Würster column makes use of Swirl Accelerator (particle accelerator; Figure 3.25), a guiding system in which the air is accelerated, stabilized, and given a precise amount of swirl. The objective is to provide a highly controlled airflow pattern in the coating zone. Particles are entrained into the swirling airflow, which leads to greater probability of impact with droplets of atomized coating liquid. This can lead to a reduction of the amount of coating material needed, and the process times are reduced as a result. Again, the product and coating liquid characteristics have to be considered and will dictate the ultimate success of the process.

Precision coaters are configured such as to allow removal and inspection of nozzles during processing—a major advantage over conventional Würster nozzle configuration whereby the process has to be interrupted and the column emptied before access to the nozzles is possible.

TANGENTIAL SPRAY (ROTARY GRANULATOR)

A recently developed fluid-bed system (Figure 3.26) uses a rotating disk to add centrifugal force to the forces of fluidization and gravity and offers very rapid mixing. Applicable for coating of pellets, granules, and particles as small as 200 μm, this device is also capable of producing pellets from seed material or powders. A gap between the rotating disk and the wall of the product container allows for fluidization of air and controls the liquid application rate. This design achieves greater drying efficiency and hence increased spray rates. The rotary type of air suspension system is available for batch sizes ranging from 1 kg to approximately 500 kg. The particle cycling time in tangential spray fluidized bed equipment is very rapid so that the films are uniform in thickness as are those applied using the processes discussed previously.

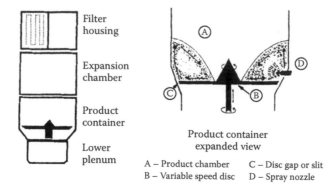

Filter housing

Expansion chamber

Product container

Lower plenum

Product container expanded view

A – Product chamber　　C – Disc gap or slit
B – Variable speed disc　D – Spray nozzle

FIGURE 3.26 Rotor tangential spray coater. (Courtesy of Glatt Air Techniques, Inc., Ramsey, New Jersey.)

The evaluation and advantages and disadvantages of each of these fluidized bed techniques have been reported in the literature[11,12] and are summarized in Table 3.2.

Figures 3.27 through 3.29 illustrate similar morphological characteristics for caffeine pellets coated with the aqueous system[12] in all three fluid-bed techniques, corresponding to similar release profiles (Figure 3.30).

TABLE 3.2
Comparison of Three Fluid Bed Coating Systems

Processing Method	Advantages	Disadvantages	Applications
Top spray coating (granulator mode)	Large batch sizes Simple setup Easy access to nozzle	Limited applications	For aesthetics and enteric coatings; not recommended for sustained-release products or tablet coating
Bottom spray (Würster)	Moderate batch sizes Uniform and reproducible film characteristics Widest application range	Tedious setup Impossible to access nozzles during process Tallest fluid bed machine for fine particle coating	Sustained release, enteric release, layering, aesthetics; not recommended for tablet coating
Tangential spray (rotary mode)	Simple setup Nozzle access during process High spray rate Shortest machine	Mechanical stress on product	Very good for layering, sustained-release and enteric-coated products; not recommended for friable products and tablets

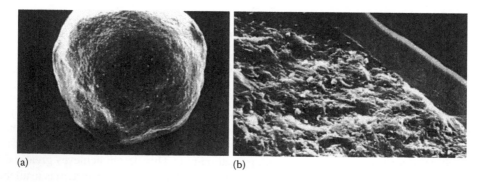

(a)　　　　　　　　　　　　　　　　　(b)

FIGURE 3.27 Caffeine pellets coated to 5% w/w using an aqueous system (Eudragit® L 30 D) and the top-spray method: (a) magnification ×70 and (b) cross-section, magnification ×1000. (From A. M. Mehta, D. M. Jones, *Pharm. Technol.*, 1985, 9, 6, 52–60.)

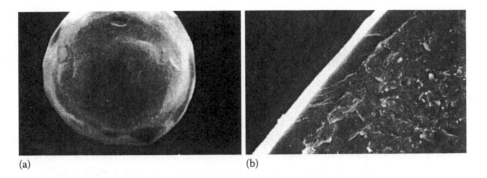

(a) (b)

FIGURE 3.28 Caffeine pellets coated to 5% w/w using an aqueous system (Eudragit® L 30 D) and the bottom-spray method: (a) magnification ×70 and (b) cross section, magnification ×1000. (From A. M. Mehta, D. M. Jones, *Pharm. Technol.*, 1985, 9, 6, 52–60.)

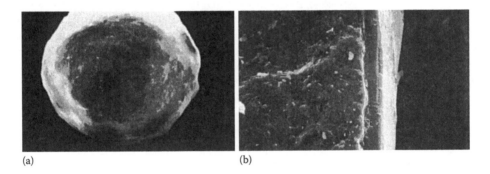

(a) (b)

FIGURE 3.29 Caffeine pellets coated to 5% w/w using an aqueous system (Eudragit® L 30 D) and the tangential spray method: (a) magnification ×70 and (b) cross section, magnification ×1000. (From A. M. Mehta, D. M. Jones, *Pharm. Technol.*, 1985, 9, 6, 52–60.)

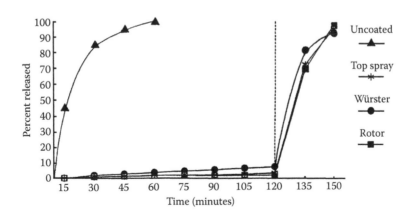

FIGURE 3.30 Dissolution profiles of caffeine pellets coated to a level of 5% w/w using an aqueous system (Eudragit® L 30 D). For 0–120 min, pH = 1.2; for 120–150 min, pH = 6.8. (From A. M. Mehta, D. M. Jones, *Pharm. Technol.*, 1985, 9, 6, 52–60.)

DELIVERY SYSTEMS

The fluid-delivery systems used in film coating consist of the pumping system and the nozzles. Several types of pumps and nozzles are available.

PUMPS

There are several types of pumps used in coating applications, and the choice may depend on the type of coating material to be applied. The peristaltic pump is the simplest and easiest to clean. It uses a multi-lobed, adjustable-speed head to deliver liquid through a flexible hose, which is usually silicon rubber. The disadvantages of this pump include pulsation as the lobes change, low liquid pressure, inability to pump viscous liquids, and fluctuations in the liquid delivery rate when run at low speed. Some of the disadvantages can be overcome, but with many coating formulations, the peristaltic pump is the system of choice. Pulsation can be dampened by selecting a nozzle port that offers some back-pressure to the liquid supply. Selecting the appropriately sized pump tube will insure that the pump is running in its optimal speed range for fluid delivery. An advantage of these pumps is that they can be fitted with multiple drive heads (one for each spray gun, Figure 3.31); thus, if one gun fails, it does not affect the delivery to the other guns, which could cause localized over-wetting to occur. Because the pump does not develop much pressure, it is ideal for latex and pseudolatex coatings, which are low in viscosity and will coagulate when subjected to the high pressure, or shear, that exists in other types of pumps such as gear and piston pumps.

The gear pump consists of a cavity of specific volume in which two gears mesh at very close tolerance. This results in very smooth and precise liquid delivery—the major advantage of this type of pump.

Cleaning is a bit more difficult, but not overly so. The disadvantage of this system is that the close tolerance between the gears can present a problem when using liquids that contain undissolved solids, particularly iron oxide pigments that can be quite abrasive to the gears. In addition, the pressure developed between the gears may makes it unsuitable for use with latex and pseudolatex materials.

The piston pump, used in both air (pneumatic) and airless (hydraulic) systems, uses adjustable stroke length or speed to control flow rate (Figure 3.32). It has the greatest number of parts and is therefore the most difficult pump to clean.

The advantage of the piston pump is its ability to clear minor clogs in nozzles due to its pressure "reserve." A disadvantage is that there is pulsation in the flow as the piston changes direction (severity varies with the check valve type and travel of the piston). A pulse damper may minimize this

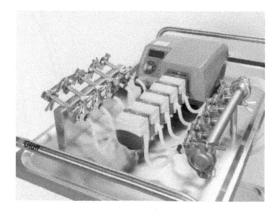

FIGURE 3.31   Multi-headed peristaltic pump. (Courtesy of Glatt GmbH.)

FIGURE 3.32 Picture of a piston pump. (Courtesy of GS.)

effect. Although the line pressure found in the air-spray systems is low, the pressure between the contact points in the check valves may cause coagulation of latex and pseudolatex materials, reducing the effectiveness of the valves and resulting in loss of precision of liquid delivery.

In yet another type of system, the container in which the coating liquid is prepared is a pressure vessel. This system is easy to clean and supplies the liquid in a very smooth manner. The delivery rate is controlled by vessel pressure and a flow controller to the nozzle. There are no high-pressure points in this type of "pump"; therefore, it can be used for latex and pseudolatex coating materials.

Whichever pump system is used with multiple spray guns, it is necessary to ensure that each spray gun delivers the same amount of coating fluid within the anticipated range of spray rate that is expected to be used. A simple method of checking this is to measure the volume of fluid delivered from each gun (without atomization) in a set time, typically 1 minute (Figure 3.33).

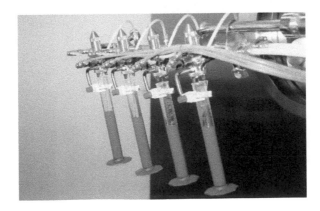

FIGURE 3.33 Checking the spray rate from multiple guns. (Courtesy of O'Hara Technologies Inc.)

SPRAY GUNS

The function of the spray gun[13,14] is to bring about the atomization of the coating solution/suspension in a reproducible and uniform manner throughout the coating operation.

The spray guns that were initially used for the coating of pharmaceutical tablets originated from the spray-painting industry as these were the only devices available. Two types of fluid atomization were explored: airless spray and airborne spray. The airless spray required the delivery of the coating fluid under very high pressure through a small (typically <0.5 mm) nozzle, which resulted in self atomization of the fluid. To some extent, the spray rate was controlled by the applied pressure; this had the effect that when changing the spray rate the degree of atomization of the coating suspension could also change. These spraying systems were originally used for non-aqueous coating systems but rapidly lost favor to the airborne spray gun, particularly for aqueous coating systems with which a high degree of control of the spray rate was readily achievable using the low-pressure pumps available. Various manufacturers of spray guns were used, such as DeVilbiss, Walther, and Binks. These guns were not ideally suited for the pharmaceutical industry, but were used for many years. They were similar in design, usually having one or two compressed air inlet ports and one fluid inlet port. The compressed air used for atomization was split via a needle valve to obtain a degree of control over the fan width; this meant that it was difficult to accurately and reproducibly obtain the same atomizing condition from batch to batch when a single spray gun was used or across a number of spray guns when multiple spray guns were used in a coating pan. Examples of these guns are shown below (Figures 3.34 and 3.35).

The development of spray guns specifically designed for the coating of tablets led to having separate air pressure controls for the main atomizing air, secondary atomizing air (fan width), and the control air (Figure 3.36). This meant that accurate and reproducible control over the atomizing condition for each spray gun could be accurately and reproducibly controlled.

Figure 3.37 shows the internal components of a spray gun. It is important that the O-rings are regularly checked for damage (and replaced, if necessary) and correctly lubricated so that the spray gun will function reproducibly.

Development of the Anti-Bearding Cap by Düsen Schlick GmbH (Figure 3.38) in the early 2000s (patent applied for in 2001) virtually eliminated the fault of "bearding" when material would be

FIGURE 3.34 Binks 460 Spray Gun. (From the author's archive.)

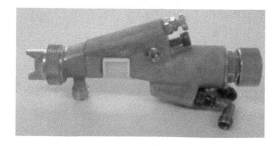

FIGURE 3.35 Walther WAX-V pilot spray gun. (From the author's archive.)

FIGURE 3.36 Schlick 970 spray gun with independent air supply for atomizing and fan air pressure. (From the author's archive.)

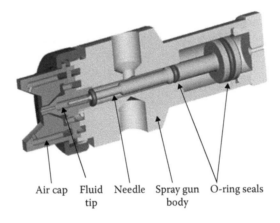

Air cap   Fluid   Needle   Spray gun   O-ring seals
          tip   body

FIGURE 3.37 Internal components of a spray gun. (Courtesy of Bosch Packaging Technology Ltd.)

FIGURE 3.38 Schlick ABC spray gun. (Courtesy of Düsen Schlick GmbH.)

deposited across the spray nozzle, resulting in major changes to droplet formation and in extreme cases of blockage of the fluid nozzle.

Smoke tests show that the airflow over the newly designed anti-bearding cap (ABC) spray gun is significantly different from the classical spray gun that utilizes "horns" to deliver the fan width air. As can be seen from the pictures (Figure 3.39) the air is not pulled into the spray gun in the case of the

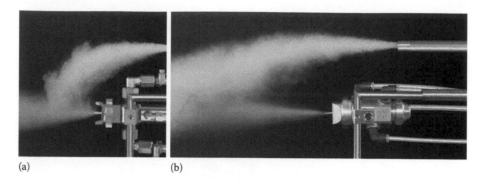

(a)   (b)

FIGURE 3.39   Airflow over the newly designed anti-bearding cap (ABC) spray gun (b) in comparison to the classical spray gun (a). (Courtesy of Düsen Schlick GmbH.)

anti-bearding cap spray gun whereas air and potentially particles or fine droplets can be drawn back onto the face of the conventional spray gun resulting in blockage of the various orifices (bearding).

The number of spray guns used in a coating pan is typically dependent on the surface area available for coating; however, spray guns from the different manufacturers are not identical, and thus different atomizing conditions may be required to achieve the same droplet size for a given coating suspension. It may also be necessary to use more spray guns from one supplier than another to achieve the same coverage in a given coating pan. When multiple spray guns are employed in a coating pan, it is important that the spray rate from each gun is in close tolerance and that the atomization condition from each gun is well controlled so that there is little variance in the amount of fluid arriving across the coating area so that localized over-wetting is avoided.

Spray guns are manufactured to very precise tolerances; are made from 316 stainless steel, which is relatively soft; and should be treated with great care. Any damage to the spray gun can result in defective atomization of the coating suspension or failure of the control needle to function correctly. It is therefore advisable to clean these guns on an individual basis and only use the tools recommended by the manufacturer to assemble and disassemble them.

FILTERS

A typical fluid-bed machine uses filter bags made of a variety of materials and mesh sizes. They are generally mechanically shaken and can be designed so that the batch continues to fluidize during the shaking mode. This is of particular advantage during coating of small particles to avoid agglomeration and allows continuous spraying. However, the disadvantages include (a) tedious setup and clean up; (b) filters that can rupture, resulting in product loss; and (c) coating material that can deposit into the filter, causing occlusion of filters leading to loss in fluidization.

Alternately, fluid-bed machines can be fitted with a cartridge filter system (Figure 3.40). It can filter down to 2 μm, resulting in higher batch yields. They too can provide the ability to perform continuous fluidization. They use a pneumatic pulse design rather than mechanical shaking to reintroduce the product in the process. Their biggest advantage lies in their ease of removal, and they can be designed to provide clean-in-place capability. Their disadvantage may include occlusion, difficulty in cleaning during the process, and the possibility of the product adhering to the outer surface. They too can affect the fluidization pattern during pulse mode.

SUPPORT EQUIPMENT AND OPTIONS

Success in the reproducibility of the film-coating process is dependent on the instrumentation, automation, and control systems of the selected equipment.

FIGURE 3.40 Fluid-bed equipment showing cartridge filters. (Courtesy of Vector Corporation, Marion, Iowa.)

AIR HANDLING

Holding process variables such as spray rate, inlet temperature, and air volume constant without controlling the process air dew point can lead to problems in all types of coating, both aqueous and solvent. The use of higher inlet air temperatures tends to minimize the problem (usually at the expense of a higher proportion of spray drying), but some products and coatings have thermal sensitivity, and film formation may be adversely affected by excessive heat. Therefore a source of clean air with controlled humidity, temperature, and volume is recommended for all coating operations. The incoming and exhaust air also needs to be filtered and, if organic solvents are used, these solvents are removed from the air before it is returned to the atmosphere.

This area of air handling is frequently overlooked, particularly in development laboratories in which the incoming air is drawn from an air-conditioned room and then vented through an exhaust system to the outside world. This gives a relatively constant inlet air condition, which may not be the case when the process undergoes scale-up and the air is drawn in from the ambient external air. Air drawn from the exterior of a building varies greatly in both temperature and humidity, depending on the season and where in the world the equipment is situated. This variance in temperature and humidity can be problematic as it affects the drying capacity of the system unless the air is treated so that it arrives at the machine with consistent humidity and temperature. A typical modern air-handling plant is shown in Figure 3.41, in which the incoming air is filtered, dehumidified (if necessary, re-humidified), and then heated to the desired processing temperature. Inlet air temperature control of $< \pm 1°C$ is now possible for both laboratory and production equipment. Larger equipment will also have a heater bypass system to allow the rapid cooling of the tablet bed to facilitate faster end of process turnaround times.

The exhaust air system (Figure 3.42) is a little simpler for aqueous film coating systems in that the dust particles need to be removed from the outgoing airstream prior to exhausting to atmosphere. The two systems need to work in harmony with one another and be controlled so that the required constant stream of air through the coating pan with the required negative air pressure in the pan is maintained.

This control of the processing air gives the a better opportunity to define the film-coating process for each product and to be able to understand and define the limitations of the process through such techniques as design of experiment. From a production perspective, it means that the operator does not

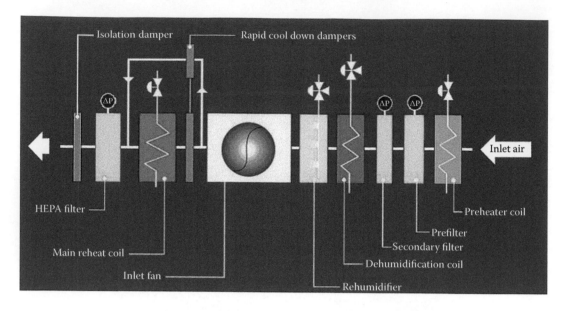

FIGURE 3.41 Typical inlet air handling plant. (Courtesy of Bosch Packaging Technology Ltd.)

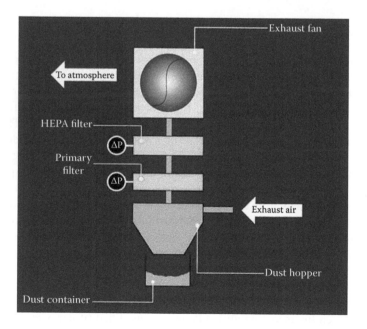

FIGURE 3.42 Typical exhaust handling plant. (Courtesy of Bosch Packaging Technology Ltd.)

constantly need to be aware of the atmospheric condition changes that occur both daily and seasonally as the inlet air conditioning plant will be providing the correct quality of air to the film-coating pan.

AUTOMATION

There are many types of automation packages available from timers to computer control systems accessed via touchscreen control (Figure 3.43). The simplest involve the use of manual controls for the inlet air temperature and air volume. More sophisticated installations will have controls for pan depression, process air volume, spray rate control, atomizing and fan width

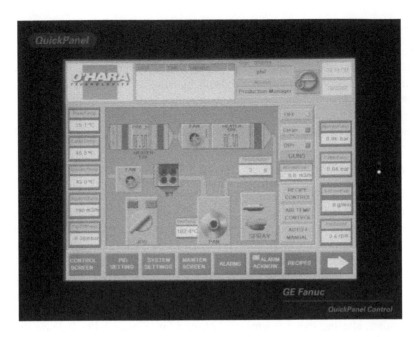

FIGURE 3.43 Picture of the touchscreen control from an O'Hara LabCoat II X. (From the author's library.) Note that the bed temperature probe was not in operation and on this machine defaults to a nominal value of 102.4°C.

air pressure control, pan speed, inlet air humidity, and temperature controls together with the ability to monitor and record these and other parameters for later evaluation. Each controlled parameter will be controlled by specific proportional integral derivative (PID) settings that should bring about changes rapidly without excessive overshoot or hunting (oscillating around the fixed point) so that a stabilized condition is achieved. The systems may be used both manually for process development and process troubleshooting or can be enabled so that a specific process can be run automatically.

The computer system will also monitor the various safety interlocks on the machine to help to prevent damage or accidents occurring.

Changes to any of the control parameters are usually very simple to do and are quickly and effectively put into place. The advent of computerized control has meant that for each defined process it is possible to set tolerance limits, which, if exceeded, can either notify the operator or at a secondary level perform a controlled shutdown of the coating operation. This adds a higher level of security to the coating operation, and it also requires that a full understanding of the process is in place so that the correct tolerance limits can be set.

These computer-controlled systems lend themselves to the coating of highly potent products, with which the machine (and thus coating operation) can be run automatically with the operator working in the normal environment in an adjacent room.

The computer-controlled systems can now be linked to the company computer system so that the data can be securely stored and accessed by the appropriate people.

MATERIAL HANDLING

Several material-handling options are available for both pans and fluidized-bed systems. The fluid bed may be vacuum or gravity fed and discharged by a hoist or a turning bottom. Pans may be rear-loaded or front-loaded by a bin and unloaded by a scoop attached to the pan or by a flap in the drum that opens and discharges into a storage hopper under the drum.

CLEAN-IN-PLACE SYSTEM

Although the fluidized bed does not lend itself to clean-in-place systems because of the exhaust filter, low-porosity bottom screen, and gaskets, the perforated pan does. Spray nozzles are positioned inside the pan housing spray the outside of the drum, and a spray ball cleans the inside (Figure 3.44). Cleaning solutions are injected into the water supply, and drain valves can be kept closed to allow the drum to tumble in a pool of cleaning solution. After draining the cleaning solution, a clean water rinse is applied, and the turbines are engaged to dry the pan and prepare for new product.

FACTORS AFFECTING EQUIPMENT CHOICE

Many factors affect the choice of process or equipment. The physical characteristics of the product, such as surface area, shape, and friability, all affect final dosage form performance. The surface area of the tablets is reproducible because they are compressed to the same size continuously. It is not as easy to achieve uniformity with small particles. Most, if not all, types of equipment that are used to make pellets or granules result in a product that varies from batch to batch. To help reduce surface-area variations, a narrow sieve cut is used in products coated for controlled release, but with many materials, it is still not enough. Variation in surface porosity and friability may result in poor reproducibility of release. Attrited particles scuffed from the surface become entrapped in layers of coating, altering the characteristics of the film (Figure 3.45). Release that is triggered by other mechanisms (enteric, taste mask) is not so severely affected. Fines from this attrition may be embedded early in the coating process; if it is possible to include an overage of coating substance, this can minimize the potential deleterious effect of these embedded fines. With this in mind, a look at the delivery provided by each type of machine is in order. In coating pans and the top-spray fluidized bed, droplets travel through the drying air before impinging on the product, spreading, and drying. In the Würster and rotary systems, the nozzle is immersed in the fluidized particles, which are sprayed concurrently with the substrate flow. A scanning electron microscope analysis[15] reveals that the most uniform films are those that are applied wet to the surface but under conditions whereby the solvent or water is evaporated before core penetration becomes a problem

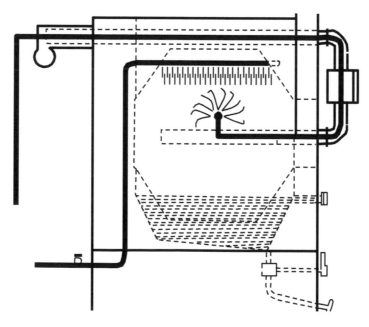

FIGURE 3.44   Clean-in-place system. (Courtesy of Glatt AG, Pratteln, Switzerland.)

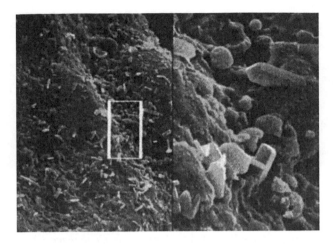

FIGURE 3.45   Scanning electron photomicrograph showing drug particles in the coating; magnification ×250/×1000.

(Figures 3.46 through 3.49). If droplets are applied after too much of the liquid has evaporated, spreading is inhibited, and imperfections in the coating are seen.

Additional coating can eventually result in the desired release profile, but reproducibility may be difficult to achieve. For this reason, controlled-release coating of small particles from aqueous or solvent systems should be limited to the Würster or rotary granulator fluid-bed systems.

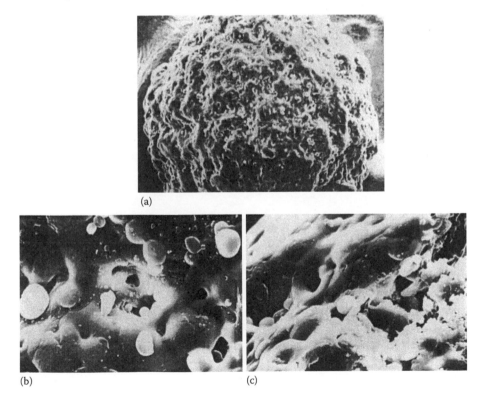

FIGURE 3.46   Pellets coated using an aqueous system in a conventional pan: (a) magnification ×100; (b) magnification ×1000 and (c) cross-section, magnification ×1000. (From A. M. Mehta, Scale-up considerations in the fluid bed process for controlled release products, Presented at *Pharm. Tech. Conference*, Cherry Hill, NJ, Sept. 16–18, 1986.)

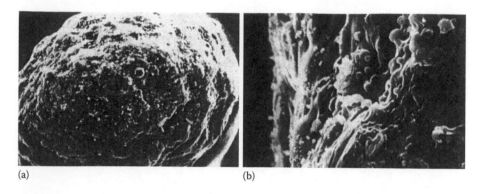

FIGURE 3.47 Pellets coated using an aqueous system in a modified perforated pan: (a) magnification ×100 and (b) cross-section, magnification ×1500. (From A. M. Mehta, Scale-up considerations in the fluid bed process for controlled release products, Presented at *Pharm. Tech. Conference*, Cherry Hill, NJ, Sept. 16–18, 1986.)

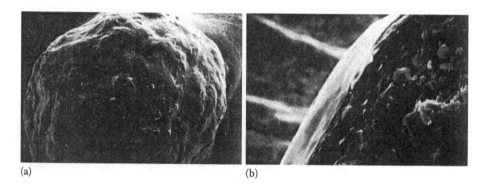

FIGURE 3.48 Pellets coated using an aqueous system in a laboratory-scale fluidized bed using the top-spray method: (a) magnification ×100 and (b) cross-section, magnification ×1000. (From A. M. Mehta, Scale-up considerations in the fluid bed process for controlled release products, Presented at *Pharm. Tech. Conference*, Cherry Hill, NJ, Sept. 16–18, 1986.)

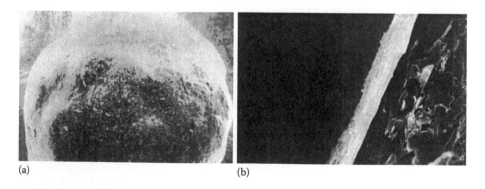

FIGURE 3.49 Pellets coated using an aqueous system in a pilot-scale Würster coater using the bottom-spray method: (a) magnification ×100 and (b) cross-section, magnification ×1000. (From A. M. Mehta, Scale-up considerations in the fluid bed process for controlled release products, Presented at *Pharm. Tech. Conference*, Cherry Hill, NJ, Sept. 16–18, 1986.)

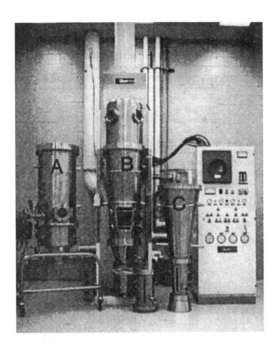

FIGURE 3.50   Fluid-bed processing equipment design for three coating processes: (A) rotor coater; (B) top-spray coater and (C) Würster coater. (Courtesy of Glatt Air Techniques, Inc., Ramsey, New Jersey.)

Water-based enteric and taste mask coatings can be applied in top-spray equipment and possibly with perforated pans adapted for coating small particles. For these reasons, it may be worthwhile to evaluate the effect of different spray modes on product performance during the product development phase in equipment that allows different spray modes such as the one shown in Figure 3.50.

PROCESS VARIABLES AND SCALE-UP CONSIDERATIONS

In addition to the method of spraying, many other variables are involved in the film-coating process. It may be necessary to prioritize these variables in order of significance for a given product and process to avoid the expenditure of an enormous amount of time in the product development phase as well as the scale-up phase. The most significant variables are summarized in Diagram 3.1. The significance of these variables and scale-up factors is highly dependent on the type of equipment and process.[16–18] Often, the scale-up factor selected for a given piece of equipment and process may not be applicable to other equipment and/or processes. As a result, it is very difficult to generalize and discuss these variables in terms of scale-up. However, scale-up considerations for specific processes are reported in the literature.[15–17]

Spray rate, coupled with droplet size, are probably the most significant considerations in the aqueous film-coating process. As stated earlier, film coating requires both uniform deposition of the film and controlled drying of the coating fluid. These two operations occur simultaneously during the coating process. They occur independently but are also interrelated. The drying rate for this process is determined by the rate of heat transfer from the air to the solvent and the rate of mass transfer of the solvent to the coating surface. Because pharmaceutical coatings are typically quite thin, the rate of heat transfer is critical. The rate of heat transfer not only affects the rate of evaporation of the solvent but also, in the case of latex and pseudolatex systems, regulates the rate and degree of coalescence of the polymeric material. The drying rate is determined by several parameters, including the latent heat of vaporization of the solvent, the surface area of the material being dried, the temperature and relative humidity of the incoming drying air, the velocity and direction of the airstream, and the geometry of the drying chamber. For the most part, these are established

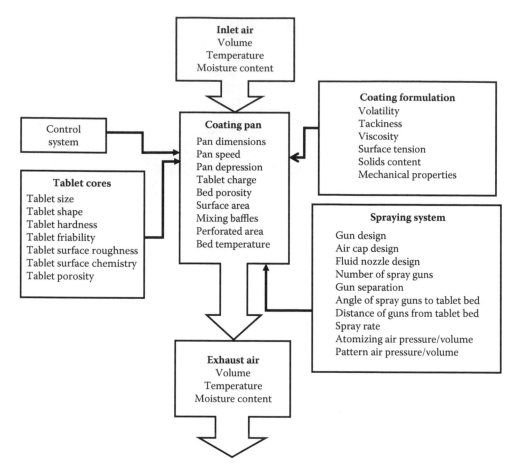

DIAGRAM 3.1   Operational boundaries of a typical pan-coating process.

by the choice of coating equipment and are therefore not easily varied. It is these parameters that dictate the maximum drying ability of the system. Certainly, it would be impossible to dry more solvent than the drying air can accept.

It is important to determine and understand the physical limitations of the coating system that are dictated by these factors and to work within them.[19] It is probably wise to monitor the inlet air temperature (and humidity), the outlet air temperature, the surface bed temperature, and the spray rate (with the atomizing and fan width pressures) as a minimum. The outlet temperature will give an indication as to the overall drying conditions, and the bed temperature will indicate drying conditions at the substrate surface.

One of the problems of scale-up is a lack of understanding of the limitations of the proposed production equipment and how it relates to the development equipment being used. Development equipment has frequently got a greater range for each processing variable than the production equipment, and control of some of these variables is much easier on the development equipment. It is often tempting to devise a film-coating process in development that works "perfectly" that will be difficult (or sometimes impossible) to replicate in production. It is advisable, therefore, that the development scientist, observe and understand the potential limitations of the proposed production equipment compared to the development coating pan.

One of the many problems that can be encountered on the migration of a film-coating process to production is that the preceding steps are also undergoing scale-up. In the case of a film-coated tablet, both granulation and compression are involved. Each of these processes can have an impact on the physical properties of the tablet to be coated. It is worthwhile to take a representative sample of the product produced at the production scale and test it in the laboratory equipment to insure the same process response and quality of product is achieved.

Another of the more easily overlooked problems is that of exceeding the stress limits of the substrate. As the size of the coating equipment is increased, so, obviously, is the weight of the substrate load. This also results in an increase in the stress applied to the individual substrate particles during the coating cycle. As the bed moves in the equipment—for example, in a coating pan—tablets tumble onto each other. Depending on the configuration of the equipment, the stress on the tablets can become quite severe. Tablets that survive well in a 10-kg batch size may not survive in a 500-kg load. The manner in which the equipment manufacturer has increased the capacity of the equipment becomes an important consideration, assuming that the same equipment manufacturer is used in both development and production. This is not always the case, so these considerations take on a higher level of importance. If the diameter of the pan has been increased significantly, particularly if there has not been at least a corresponding increase in the pan depth, the stress on the tablets will be greatly increased. Changes in the geometry of the coating pan affect more than just the stress on the tablets; the geometry is also very important to both the mixing and drying characteristics. Changes in the dimensions of the coating equipment can also affect the airflow patterns. A change in equipment design (particularly that of how the airflow is managed) may present further challenges to scale-up.

As the size of coating equipment is changed, there is also a change in the equipment's air-handling capacity. Changes in airflow volume have dramatic effects on the drying capacity. As the volume of air is increased, so is the drying capacity. The air volume, temperature, and humidity have the greatest impact on the drying capacity and, therefore, the spray rate. It is tempting to use the maximum possible inlet air temperature in order to more efficiently evaporate water, which has a high heat of vaporization, allowing for a greater spray rate. However, it has been demonstrated in the literature[19] that the dissolution rate of the drug can be affected by the spray rate. The use of a very high inlet temperature can also cause problems, such as decreased yield if the product remains too dry, which may subject it to attrition, and with certain thermoplastic polymeric systems, it may cause agglomeration. The most desirable inlet air temperature setting is the one that allows an equilibrium between the application of liquid and subsequent evaporation so that proper film formation occurs. The effect of moisture, also known as the "weather effect," has been discussed in the literature.[20] It is a known fact that the heat content of moist air is higher than that of dry air. However, the variation in heat content could result in different release profiles, depending on the solvents and types of polymeric systems used. Residual water in the coating layers may affect the film-formation process. It is therefore recommended that the effect of ambient air dewpoints be examined as a part of the scale-up program in any coating operation. The effect of process variables must be examined not only individually but also in combination with the formulation variables.

A thermodynamic model for aqueous film coating[21–25] may enable the prediction of the behavior of a tablet-coating process under different environmental conditions. An example of the available software is shown in Figure 3.51, in which it is possible to change certain critical factors, such as inlet temperature or dewpoint and obtain a predicted impact on the process.

The use of design-of-experiment software can also help provide an understanding of the critical process parameters for a given tablet coating process on an individual machine.[26] This knowledge can be used to help understand the scale-up process; however, the same set of trials should ideally be done in the production environment as well. This will determine the critical process parameters and boundaries in the production equipment that may ultimately save time (and money) in the long run.

The data in Figure 3.52 illustrate the predictive value of surface response analysis for optimizing processing parameters. In this case, the influences of inlet air temperature and coating suspension solids (for a study conducted in a laboratory-scale Accela-Cota in which 325-mg aspirin tablets were coated with an 8% w/w enteric coating applied as an aqueous polyvinyl acetate phthalate [PVAP] coating system) on enteric performance are illustrated. These data clearly indicate that, for

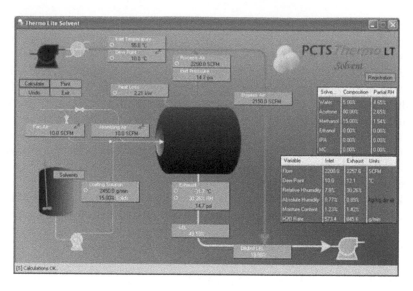

FIGURE 3.51   Thermodynamic program. (Courtesy of PCTS Thermo LT.)

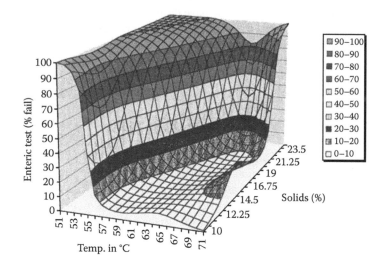

FIGURE 3.52   Enteric test surface response analysis for tablets coated with Sureteric®. (Courtesy of Colorcon Inc., West Point, Pennsylvania.)

this coating pan, the fixed process conditions shown (spray rate, 70 g/min; atomizing air pressure, 35 psi), the best enteric results are achieved when the suspension solids content is kept below 20% w/w and the inlet air temperature is maintained above 57°C. The optimum results are obtained when the coating solids and drying air are set at 10% w/w and 71°C, respectively.

SUMMARY

The development of relatively high solids aqueous film coating systems for both immediate and controlled release applications has allowed the widespread use of these materials. The significance of legislative changes to emission control should not be overlooked in its influence on industry in general and in particular in its effect on the transformation of pharmaceutical film coating in the last

30 years. As a result, aqueous film coating has been widely adopted by the pharmaceutical industry and is now regarded as the system of first choice due to the advancement of our understanding of the film coating process, the development of equipment that is designed for this application, and the application of computer control and automation.

REFERENCES

1. RH Sutaria. The art and science of tablet coating. *Man. Chem. and Aerosol News.* June, 1969; 37–42.
2. SS Behzadi, S Toegel, H Viernstein. Innovations in coating technology. *Recent Pat Drug Deliv Formul.* 2008; 2: 209–230.
3. G Cole, JE Hogan, M Aulton. Pharmaceutical Coating Technology. Taylor & Francis, Boca Raton, FL.
4. G Banker, G Peck, S Jan, P Pirakitikulr, D Taylor. Evaluation of hydroxypropyl cellulose and hydroxypropyl methyl cellulose as aqueous based polymers. *Drug Dev. Ind. Pharm.* 1981; 7(6): 693–716.
5. BJ Munden, HG DeKay, GS Baker. Evaluation of polymeric materials I Screening of selected polymers as film coating agents. *J. Phar. Sci.* April, 1964; 10: 1002.
6. JE Hogan. Aqueous versus organic solvent film coating. *Int. J. Pharm. Tech. Prod. Mfr.* 1982; 3(1): 17–20.
7. BJ Seelig, DA Troy. Change to water based tablet coating prevents 24 tons/yr of air pollution. *Chem. Proc.* August, 1987; 96–98.
8. GC Cole, G May, PJ Neale, MC Olver. The design and performance of an instrumented system for aqueous film coating in an industrial tablet coating machine. *Drug Dev. Ind. Pharm.* 1983; 9(6): 909–944.
9. GW Smith, GS Macleod, JT Fell. Mixing efficiency in side vented coating equipment. *AAPS PharmSciTech.* 2003; 4(3): Article 37.
10. TM Leaver, HD Shannon, RC Rowe. A photometric analysis of tablet movement in a side vented perforated drum (Accela-Cota). *J. Phar. Pharmacol.* 1985; 37: 17–21.
11. AM Mehta, MJ Valazza, SE Abele. Evaluation of fluid bed processes for enteric coating systems. *Pharm. Technol.* 1986; 10(4): 46–56.
12. AM Mehta, DM Jones. Coated pellets under the microscope. *Pharm. Technol.* 1985; 9(6): 52–60.
13. GS Macleod, JT Fell. Different spray guns used in pharmaceutical film coating. *Pharm. Tech. Eur.* 2002; 25–33.
14. R Mueller, P Kleinebudde. Comparison of a laboratory and a production coating spray gun with respect to scale up. *AAPS PharmSciTech.* 2007; 8(1): Article 3.
15. AM Mehta. Scale-up considerations in the fluid bed process for controlled release products. Presented at Pharm. Tech. Conference, Cherry Hill, NJ, Sept. 16–18, 1986.
16. EJ Russo. Typical scale-up problems and experiences. *Pharm. Technol.* 1984; 8(1): 46–56.
17. SC Porter, LF D'Andrea. The effect of choice of process on drug release from non-pareils film-coated with ethylcellulose. Presented at the 12th International Symposium on Controlled Release of Bioactive Materials, Geneva, Switzerland, July 8–12, 1985.
18. DM Jones. Factors to consider in fluid bed processing. *Pharm. Technol.* 1985; 9(4): 50–62.
19. E Van Ness, B Schad, T Riley, B Cheng. Achieving process understanding and control in film coating. *Pharm. Tech.* 2012; s18–s20.
20. TL Reiland, JA Seitz, JL Yeager, RA Brusenback. Aqueous film coating vaporization efficiency. *Drug Dev. Ind. Pharm.* 1983; 9(6): 945–958.
21. GC Ebey. A thermodynamic model for aqueous film coating. *Pharm. Technol.* 1987; 11(4): 40–50.
22. P Panday, R Turton, N Joshi, E Hammerman, J Ergun. Scale up of a pan coating process. *AAPS PharmSciTech.* 2006; 7(4): Article 102.
23. R Turton. The application of modelling techniques to film coating processes. *Drug Dev. Ind. Pharm.* 2010; 36(2): 143–151.
24. R Turton. Challenges in the modelling and prediction of coating of pharmaceutical dosage forms. *Powder Tech.* 2008; 181: 186–194.
25. M Tanya am Ende, A Berchielli. A thermodynamic model for organic and aqueous tablet film coating. *Pharm. Dev. Tech.* 2005; 1: 47–58.
26. SC Porter, RP Verseput, CR Cunningham. Process optimization using design of experiments. *Pharm. Tech.* October 1997.

4 Mechanical Properties of Polymeric Films Prepared from Aqueous Dispersions

Linda A. Felton, Patrick B. O'Donnell, and James W. McGinity

CONTENTS

INTRODUCTION

Pharmaceutically acceptable polymers used in the film coating of solid dosage forms are primarily based on acrylic or cellulosic polymers. Many of these polymers for controlled release applications have been formulated into aqueous colloidal dispersions (e.g., latices or pseudolatices) in order to overcome the high costs, potential toxicities, and environmental concerns associated with the use of organic polymer solutions [1–3]. Film coating has been successfully utilized to control the release of active ingredients, prevent interaction between ingredients, increase the strength of the dosage form to maintain product integrity during shipping, and protect the drug from the environment [3–8].

Coating formulations usually contain additives, in addition to the polymer, that aid in processing, appearance, and product performance. Most formulations contain plasticizers that impart flexibility to the films and reduce the incidence of crack formation [9,10]. Pigments may be added to alter the appearance of the final product [11], and anti-tacking agents may be required to prevent agglomeration of the coated substrates [12]. Numerous polymer blends for controlled drug release have been investigated [5,6,13], and the release characteristics of these coated dosage forms are strongly

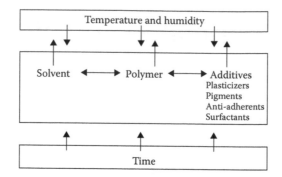

FIGURE 4.1   Factors affecting the mechanical properties of polymeric films.

dependent on certain properties of the film, for example, permeability and mechanical strength [14–17]. The amount and type of plasticizer in the film and the presence of other additives in the coating can significantly impact the film's mechanical properties [18–22]. In addition, factors such as storage conditions and processing temperature will influence coalescence and film formation from latex and pseudolatex dispersions and thus product performance [23–25]. Figure 4.1 illustrates the relationship between these factors.

FILM PREPARATION METHODS

Studies to investigate the mechanical properties of polymers may be conducted using free films or films applied to a substrate. Free films can be obtained by the casting method, in which a polymeric solution or dispersion is cast onto a nonstick substrate and the solvent is evaporated [26–28]. In formulations containing solid particles, however, sedimentation may occur during the drying stages, resulting in non-uniform films. The preparation of multilayered films by the cast method is also difficult because the solvent present when casting secondary layers may dissolve or interact with previous layers [29].

To avoid different film surfaces that may result from casting polymeric dispersions, a spray atomization technique may be employed. This type of spray box apparatus, shown in Figure 4.2, consists of a rotating drum inset into a box with heat introduced to facilitate solvent evaporation. The polymeric material is then sprayed onto the nonstick surface of the rotating drum [30,31]. This technique better simulates coating processes and produces more uniform surfaces [29].

In contrast to the study of free films, the evaluation of applied films has been gaining in popularity [32–34]. Applied films can be used to investigate substrate variables, processing parameters, storage

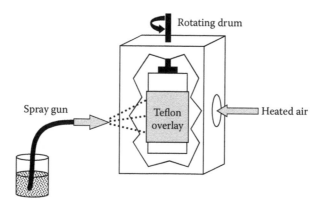

FIGURE 4.2   Schematic of a spray box apparatus used to prepare free films.

conditions, and physical aging in addition to coating formulation factors. Because most solid dosage forms are designed to dissolve in water-based biological fluids and the majority of coating systems used today are aqueous-based, the dissolution of the outermost surfaces of the substrate occurs during the coating process, permitting physical mixing at the film–tablet interface, which could lead to migration of drug or excipient into the film [35,36]. This physical mixing and migration of components into the coating can affect the mechanical, adhesive, and drug-release properties of the polymer film [37,38].

MECHANICAL TESTING TECHNIQUES

STRESS–STRAIN TESTING OF FREE FILMS

Stress–strain testing in the tensile mode has been a popular and widely used mechanical test for polymeric films. The tensile test is practical, and analysis of its data is relatively straightforward. The tensile test gives an indication not only of the elasticity and strength, but also of the toughness of the film. In the development of a film-coating system, evaluation of the mechanical properties of free films can readily characterize the fundamental properties of the coating [39]. However, polymers are viscoelastic, and their mechanical behavior is dependent on many factors, including environmental conditions and experimental testing parameters.

A tensile-testing instrument, such as an Instron (Norwood, Maine) mounted with a load cell, may be used for the measurements. According to the American Society for Testing Materials (ASTM) guidelines, data for tensile properties may be acquired in the form of a load–time (elapsed) profile or, more typically, a load–displacement or stress–strain profile as shown in Figure 4.3. The data collected for a load–time or load–displacement profile can be converted mathematically to a stress–strain curve. Mechanical properties, including tensile strength, elongation, work of failure, and Young's modulus are then computed. The theory behind the computation of these parameters is well documented [22,40]. The equations that define each of these parameters are presented below.

Tensile Strength

Tensile strength is the maximum stress applied to a point at which the film specimen breaks (Figure 4.3d). Tensile strength can be computed as the applied load at rupture divided by the cross-sectional area of the film, as described in Equation 4.1:

$$\text{Tensile strength} = \frac{\text{Load at failure}}{\text{Film thickness} \times \text{Film width}} \tag{4.1}$$

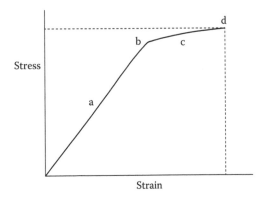

FIGURE 4.3 Example of a stress–strain profile generated from tensile testing of free films. (a) Region of elastic deformation, where stress is proportional to strain; (b) yield point; (c) region of plastic deformation, where polymer chains orient themselves; (d) film breaks.

Tensile strength measurement alone is not useful in predicting the mechanical performance of films; however, higher values of tensile strength are indicative of abrasion resistance [28].

Strain

A film sample will stretch under applied stress, which is referred to as strain. Strain can be calculated as the deformation in the film divided by the original dimension of the sample. Strain is typically reported as percent elongation at fracture and is calculated using Equation 4.2:

$$\text{Percent elongation} = \frac{\text{Increase in length of film}}{\text{Initial length of film between the grips}} \times 100 \qquad (4.2)$$

Elongation of a film will generally increase as the plasticizer level in the coating is increased [21,41].

Work of Failure

Work of failure is a function of the work required to break the film specimen and represents the film toughness. It can be calculated from the area under the curve of the stress–strain diagram, cross-head speed, and the film dimensions as described in Equation 4.3:

$$\text{Work of failure} = \frac{\text{Area under curve} \times \text{Cross-head speed}}{\text{Film thickness} \times \text{Film width}} \qquad (4.3)$$

Young's Modulus

Young's modulus, sometimes referred to as elastic modulus, is the most basic and structurally important of all mechanical properties and is a measure of the stiffness of the film. It is the ratio of applied stress and corresponding strain in the region of approximately linear elastic deformation and can be computed using Equation 4.4:

$$\text{Young's modulus} = \frac{\text{Slope}}{\text{Film thickness} \times \text{Cross-head speed}} \qquad (4.4)$$

Because most amorphous polymers behave as viscoelastic materials, their mechanical properties depend on the temperature and the application rates of stress and strain. The profile in Figure 4.4 shows typical changes in polymer chain arrangement that occur during tensile testing of a free film. Initially, there is a linear portion in the stress–strain profile (Figure 4.3a), in which elongation is directly proportional to applied stress, and polymer chains are randomly oriented (Figure 4.4a). The slope of this straight-line portion of the graph is used to calculate Young's modulus. The greater the slope of the curve, the higher the Young's modulus. As the stiffness and the strength of the film increase, more stress will be required to produce a given amount of deformation.

At the yield point (Figures 4.3b and 4.4b), polymer chains begin to orient themselves to the applied stress and are completely aligned during the plastic deformation stage of stress–strain analysis (Figures 4.3c and 4.4c). Finally, the film specimen fractures (Figures 4.3d and 4.4d).

Not all polymers behave in a typical manner, and depending on the mechanical response of the polymer, a family of stress–strain profiles can be obtained to clearly define elasticity, tensile strength, and film elongation at the break of the plasticized polymer. Some examples are illustrated in Figure 4.5. Hard and brittle films exhibit a high tensile strength and Young's modulus with little elongation. In contrast, a soft and tough film will possess a low tensile strength but much greater elongation and a higher area under the curve (toughness).

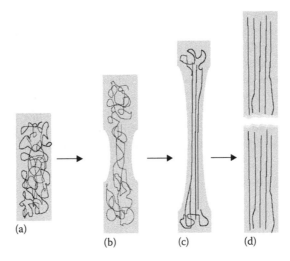

FIGURE 4.4 Schematic of the changes in polymer chain arrangement that occur during tensile testing of a free film. (a) Elastic deformation stress ∝ strain, (b) yield point, (c) plastic deformation, and (d) break.

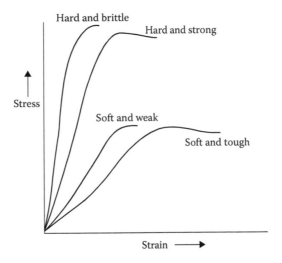

FIGURE 4.5 Examples of characteristic stress–strain profiles obtained from tensile testing of free films.

Tensile Strength to Young's Modulus Ratio

The ratio of tensile strength to Young's modulus has been used to predict the tendency of an applied film to crack with higher values indicative of fewer such defects [19,42].

STRESS–STRAIN TESTING OF APPLIED FILMS

Compression testing of applied films is similar to tensile testing of free films in that uniform displacement rates are applied to a sample, and force and displacement values are recorded. The primary difference between the two techniques is in the direction of the applied stress. The substrate has been shown to significantly influence the mechanical strength of applied films [43,44]. The effects of processing parameters, storage conditions, and physical aging of the applied film can be evaluated using compression testing [23,34]. In addition, compression testing of applied films can provide qualitative information on adhesion of the coating to the substrate [44] with simultaneous fracture of the substrate and film indicating good adhesion.

Knowledge of the compression properties of applied films is critical if the coated substrate is to be tableted. If the compressional force exceeds the coating strength, the film will fracture and faster dissolution will result [33]. The formation of matrix tablets when tableting coated pellets has also been reported, resulting in slower drug release as the polymer coatings fuse during compression [45,46]. To reduce friction during compression and to prevent direct contact of the coating, readily compressible excipients are often blended with the coated pellets prior to tableting [47].

GLASS TRANSITION TEMPERATURE

The glass transition temperature (T_g) is the temperature at which the mechanical behavior of a film changes. Below this temperature, the polymer exists in a glassyg state that is characterized by a substructure in which there is minimal polymer chain movement. Above the T_g, the polymer is in a rubbery state, which is characterized by increased polymer chain movement and polymer elasticity. The T_g is typically measured using a differential scanning calorimeter (DSC), with which a sample and reference pan are heated at a programmed rate, and thermal transitions, with which more energy is absorbed or emitted, are determined. There are numerous examples in the literature of determining T_g values to evaluate polymer properties and interactions with excipients [47–49]. The DSC instrument can also be used to determine melting temperature, detect polymorphism, study polymer miscibility, and investigate oxygen degradation [50–53]. A number of variations in DSC testing have been developed, including a triple-cell system for more precise measurements of enthalpy, temperature-modulated units to separate reversing and nonreversing transitions, and high-sensitivity models [54,55].

DYNAMIC MECHANICAL ANALYSIS

Dynamic mechanical analysis (DMA) is another type of test used to study the mechanical properties of polymeric films. In DMA testing, a free film is placed between two grips, one stationary and the other oscillatory. The free film is then deformed by torsion oscillation as a function of temperature. The storage modulus, loss modulus (dissipated energy), and damping coefficient (ratio of loss modulus to storage modulus) are determined. Several different modes are available, including fixed frequency, creep relaxation, and stress relaxation. DMA can be used to determine the T_g as well as other smaller, sub-T_g transitions that can provide some indication of polymer structure [56]. Modifications to the instrument have permitted the mechanical properties of polymeric films applied to individual pellets to be determined [43].

EFFECTS OF PLASTICIZERS IN THE COATING FORMULATION ON MECHANICAL PROPERTIES

Many polymers used in film coating of pharmaceutical dosage forms display brittle properties at ambient temperature and humidity conditions, and the addition of a plasticizer is essential to achieve effective coatings without cracks or splitting defects [10]. Plasticizers are added to polymeric solutions or dispersions to increase the workability or flexibility of the polymer and reduce brittleness, improve flow, and increase toughness and tear resistance of the films. These effects are the result of the plasticizer's ability to weaken intermolecular attractions between the polymer chains and allow for more movement. Several theories have been proposed to explain the mechanism by which the plasticizing agents impart flexibility to polymeric films [57]. According to the lubricity theory, the plasticizer functions as an internal lubricant and facilitates movement of the polymer chains. The gel theory proposes that the unplasticized polymer exists as a three-dimensional gel and that the plasticizer functions by cleaving the intermolecular bonds within the gel. Finally, the free volume theory states that plasticizers increase the free space around the polymer chains, providing a greater area for movement of the polymer molecules.

In addition to enhancing the flexibility of the film, plasticizers influence permeability and drug release [14]. Many compounds can plasticize polymeric films, including water, drugs, and excipients [37,58]. The selection of a plasticizer, therefore, is an important decision in the development of any coated product, especially a modified release system as plasticizer type and concentration influence film permeability. Ideally, the plasticizer level in the film should be optimized to reduce the brittle character of the film without adding excess plasticizer. Higher levels of plasticizer can cause sticking or agglomeration of the coated product during storage, which may compromise the release properties of the drug from such dosage forms [59].

The effectiveness of a plasticizing agent is dependent to a large extent on the amount of the plasticizer added to the coating formulation and the extent of polymer–plasticizer interactions. The interaction of a plasticizer with the polymer decreases the T_g of the film and is considered a common measure of plasticizer effectiveness with the more effective plasticizers producing greater decreases in T_g. The influence of water-soluble and water-insoluble plasticizers on the T_g of Eudragit® L 100-55 is shown in Figure 4.6 [60]. The presence of 10% plasticizer in the polymer film caused a dramatic decrease in the T_g for all the plasticizers studied. This decrease continued with the water-soluble plasticizers at levels greater than 10%. For the water-insoluble plasticizers, however, a plateau in the T_g of the polymer was observed due to the immiscibility of the plasticizer with the polymer at the higher levels.

Plasticizers will also influence the elastic modulus of the polymer and, as shown in Figure 4.7, there was a decrease in the elastic modulus of Eudragit L 100-55 as the level of plasticizer increased for the water-soluble plasticizers [60]. As mentioned earlier, Young's modulus is a measure of the stiffness of the film or the ability of the film to withstand high stress while undergoing little elastic deformation. The softening effect did not decrease with the insoluble plasticizers at levels greater than 10%, presumably due to the immiscibility of the higher levels of plasticizer with the polymer.

To be effective, a plasticizer must diffuse into the polymer phase and disrupt the intermolecular interactions of the polymer while having minimal or no tendency for migration or volatilization.

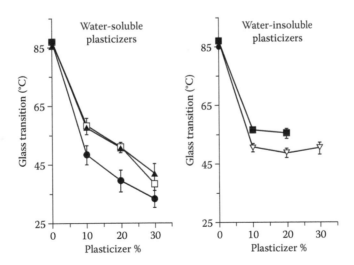

FIGURE 4.6 Effect of different levels of plasticizers on the glass transition temperature of Eudragit® L 100-55 films stored for 60 days at 23°C, 50% RH followed by 30 days at 23°C, 0% RH ($n = 5$). Water-soluble plasticizers: (●) TRI, triacetin (□) TEC, (▲) ATEC; water-insoluble plasticizers: (▽) TBC, (■) ATBC. ATBC, acetyl tributyl citrate, ATEC, acetyl triethyl citrate, RH, relative humidity, TBC, tributyl citrate, TRI, TEC, triethyl citrate. (From J. C. Gutierrez-Rocca, J. W. McGinity, *Int J Pharm.*, 1994, 103, 293–301.)

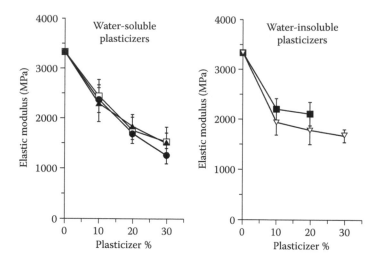

FIGURE 4.7   Effect of different levels of plasticizers on the glass transition temperature of Eudragit® L 100-55 films stored for 60 days at 23°C, 50% RH followed by 30 days at 23°C, 0% RH ($n = 6$). Water-soluble plasticizers: (●) TRI, triacetin (□) TEC, (▲) ATEC; water-insoluble plasticizers: (▽) TBC, (■) ATBC. ATBC, acetyl tributyl citrate; ATEC, acetyl triethyl citrate; RH, relative humidity; TBC, tributyl citrate; TRI; TEC, triethyl citrate. (From J. C. Gutierrez-Rocca, J. W. McGinity, *Drug Dev Ind Pharm.*, 1993, 19, 3, 315–332.)

Dramatic changes in the mechanical and dissolution properties may result when a plasticizer evaporates or leaches from within a polymeric film [27]. Plasticizers that are soluble in the solvent phase can be added directly to the mixture or may be dissolved first in the solvent prior to addition of the polymer. Plasticizers that are not water soluble should first be emulsified in water using latex-compatible emulsifiers and then appropriately agitated with the entire mixture until an equilibrium plasticizer distribution occurs between the water and polymer phases [61].

The incorporation of a plasticizer into an aqueous polymeric dispersion is crucial, and sufficient time must be allowed for the plasticizer to partition into the polymer phase prior to initiation of the coating process [62,63]. The rate and extent of plasticizer partitioning for an aqueous dispersion is dependent on the solubility of the plasticizer in water and its affinity toward the polymer phase. Equilibration of plasticizer distribution in an aqueous polymeric dispersion for water-soluble plasticizers has been shown to occur rapidly whereas the time required to achieve equilibrium distribution for water-insoluble agents requires substantially longer mixing times [27,61]. If insufficient time is allowed for the plasticizer to partition into the polymer phase, the unincorporated plasticizer droplets as well as the plasticized polymer particles will be sprayed onto the substrate during the coating process. Uneven plasticizer distribution within the film could result and potentially cause changes in the mechanical properties of the film over time during storage.

EFFECTS OF OTHER ADDITIVES IN THE COATING FORMULATION ON MECHANICAL PROPERTIES

PIGMENTS

The addition of pigments into a coating formulation may improve the aesthetic appearance of the final product [64]. Opacifiers, such as titanium dioxide, may be used in coatings to protect photosensitive drugs from exposure to light, thus improving product stability [65]. The addition of pigments in the coating may significantly influence the mechanical, adhesive, and drug-release properties of the resulting film [66,67]. As the concentration of an insoluble pigment is increased, the amount of polymer necessary to completely surround the particles increases. At a specific concentration,

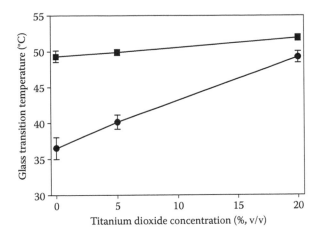

FIGURE 4.8   Influence of titanium dioxide concentration in the coating formulation on the glass transition temperature of applied Eudragit® L 30 D-55 films. (●) 0% wax in the tablet core; (■) 30% wax in the tablet core. (From L. A. Felton, J. W. McGinity, *Drug Dev Ind Pharm.*, 1999, 25, 5, 599–606.)

known as the critical pigment volume concentration (CPVC), the polymer present is insufficient to surround all of the insoluble particles, and marked changes in the mechanical properties of the film will occur [68]. The CPVC is a characteristic of specific polymer–filler combinations, and theoretic determinations of this value are practically impossible [69].

Aulton et al. [22] showed that hydroxypropyl methylcellulose (HPMC) films became more brittle as the concentration of titanium dioxide increased as evidenced by a decrease in elongation and an increase in Young's modulus. Hsu et al. [70] showed that the addition of titanium dioxide to polyvinyl alcohol also resulted in a decrease in tensile strength.

Felton and McGinity [67] investigated the influence of titanium dioxide concentration on the T_g of Eudragit L 30 D-55 films plasticized with 20% triethyl citrate. Increased concentration of the pigment in the film resulted in a significant increase in the T_g when the polymeric dispersion was applied to hydrophilic tablet compacts, as shown in Figure 4.8. Interestingly, only small, incremental increases in the T_g of the polymeric film with increased titanium dioxide concentration were noted when applied to hydrophobic tablets. These findings demonstrate not only that concentration of a pigment in the coating formulation can influence the mechanical properties of the film, but also that the substrate can affect polymer properties.

ANTI-ADHERENTS

Stickiness or tackiness of polymeric films is a concern during both the coating process and storage. The extent of product agglomeration may be influenced by processing temperature, curing temperature, plasticizer content, and polymer type [59]. To minimize product agglomeration, anti-adherents (anti-tacking agents) may be incorporated into the coating formulation. Talc and glyceryl monostearate (GMS) are the most commonly employed anti-adherents in film-coating formulations [71]. These excipients, however, are not water-soluble, and they have been shown to influence the mechanical and drug-release properties [12,21,72,73].

SURFACTANTS

As mentioned earlier, incorporation of water-insoluble plasticizers into aqueous polymeric dispersions requires that the plasticizer first be emulsified in water with an appropriate surfactant. In addition, surfactants have been added to film coating formulations to improve the spreadability of the coating material across tablet surfaces [74] and to modulate drug release [75]. The addition of these compounds to film coating formulations has been shown to influence the mechanical properties of the films. Felton et al. [20] showed that increased concentrations of both sorbitan monooleate and polysorbate 80 significantly lowered the T_g of Eudragit L 30 D-55 films plasticized with the hydrophobic tributyl citrate, and no significant changes in T_g were noted when the polymeric dispersion was plasticized with the water-soluble triethyl citrate.

MECHANICAL PROPERTIES OF WET AND DRY FILMS

Upon ingestion, the polymer coating will become hydrated, and the mechanical properties of the film may not be the same as in the dry state [47,76]. Water may plasticize the film, and plasticizers may leach from the coating upon exposure to biological fluids. To assess the mechanical behavior of films in their hydrated state, a puncture test can be employed [16,77]. As shown in Figure 4.9, the apparatus consists of a platform assembly containing a free film that is submerged in a dissolution bath. A puncture probe attached to a load cell is then driven into the film. Data determined from this experiment include the puncture strength (force at puncture divided by the cross-sectional area of the dry film) and the percentage of elongation at puncture.

Bodmeier and Paeratakul [78] demonstrated that the mechanical properties of dry and wet films were dependent on the polymeric material used to form the film (Table 4.1). The ethylcellulose pseudolatexes, Aquacoat® and Surelease®, were found to be brittle in the dry state and weak in the wet state. In contrast, films of Eudragit L 30 D were shown to be brittle in the dry state yet very flexible in the wet state, presumably due to the plasticization effect of water, while Eudragit NE 30 D films were found to be very flexible in both wet and dry states.

The plasticizer used in the polymeric dispersion can also significantly influence the mechanical strength of polymeric films in both dry and wet states [76,78]. As shown in Table 4.2, wet Eudragit RS 30 D polymer films containing water-insoluble plasticizers were significantly more flexible than the corresponding wet films plasticized with water-soluble plasticizers. These results were attributed to the leaching of the water-soluble plasticizers from the films during exposure to the aqueous medium whereas the water-insoluble plasticizers were almost completely retained within the wet films. Leaching of the plasticizer created pores in the films, with higher concentrations of the water-soluble plasticizers increasing the porosity of the films [76].

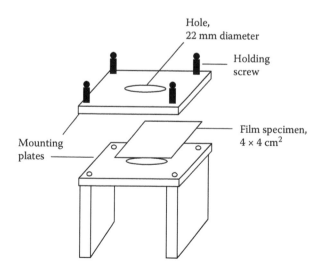

FIGURE 4.9 Schematic of the film holder used in the puncture test device. The holder is submerged in a dissolution bath to hydrate the film. (From R. Bodmeier, O. Paeratakul, *Int J Pharm.*, 1993, 96, 129–138.)

TABLE 4.1

Mechanical Properties of Dry and Wet Films and the Water Content of Wet Films Prepared from Different Polymer Dispersions Plasticized with Triethyl Citrate (20% w/w)

Polymer Dispersion (Film Thickness, μm)	Puncture Strength (MPa)		Elongation (%)		Water (g)/ Polymer (g)
	Dry	Wet	Dry	Wet	
Aquacoat® (309)	0.34 (0.11)	0.10 (0.02)	1.34 .(0.18)	0.13 (0.02)	0.506 (0.032)
Surelease® (394)	0.23 (0.04)[a]	0.74 (0.10)	0.62 (0.12)	4.89 (0.90)	0.100 (0.006)
Eudragit® NE 30 D (314)	2.16 (0.19)	1.58 (0.10)	>365.00	>365.00	0.268 (0.014)
Eudragit® RD 30 D (309)	1.99 (0.23)	0.93 (0.04)	142.83 (4.32)	38.41 (4.65)	0.331 (0.008)
Eudragit® RL 30 D (316)	1.81 (0.11)	1.60 (0.14)	126.31 (8.04)	13.02 (2.45)	0.807 (0.008)
Eudragit® L 30 D (264)	0.83 (0.05)	1.78 (0.09)	0.46 (0.25)	>365.00	0.722 (0.023)

Source: R. Bodmeier, O. Paeratakul, *Pharm Res.*, 1994, 11, 6, 882–888.

Note: SD in parentheses; *n* = 3.

[a] Films did not rupture.

TABLE 4.2

Mechanical Properties of Dry and Wet Eudragit® RS 30 D Films Plasticized with Different Plasticizers (20% w/w)

Plasticizer + (Film Thickness, μm)	Puncture Strength (Mpa)		Elongation (%)		Plasticizer Remaining (% of Original)
	Dry	Wet	Dry	Wet	
TEC (309)	1.99 (0.22)	0.93 (0.05)	142.8 (4.3)	38.4 (4.6)	56.29 (1.79)
Triacetin (302)	1.82 (0.38)	0.61 (0.07)	120.9 (6.0)	6.8 (0.6)	35.92 (1.06)
ATBC (314)	4.30 (0.09)	1.11 (0.13)	77.8 (7.6)	85.2 (3.6)	101.84 (1.67)
ATEC (323)	4.01 (0.18)	1.01 (0.02)	86.9 (5.5)	64.3 (8.5)	90.38 (0.05)
DBP (327)	3.18 (0.47)	0.88 (0.19)	93.2 (12.6)	106.9 (9.2)	99.95 (1.88)
DBS (324)	2.37 (0.09)	0.79 (0.04)	91.8 (2.0)	59.7 (3.6)	88.34 (0.66)
DEP (324)	2.47 (0.40)	0.91 (0.03)	91.1 (3.2)	51.0 (3.8)	95.27 (1.53)
TBC (319)	2.37 (0.40)	0.86 (0.03)	113.5 (1.8)	86.6 (3.4)	97.79 (2.06)

Source: R. Bodmeier, O. Paeratakul, *Pharm Res.*, 1994, 11, 6, 882–888.

Note: SD in parentheses, *n* = 3; ATBC, acetyl tributyl citrate; ATEC, acetyl triethyl citrate; DBP, dibutyl phthalate; DBS, dibutyl sebacate; DEP, diethyl phthalate; TBC, tributyl citrate; TEC, triethyl citrate.

EFFECT OF ENVIRONMENTAL STORAGE CONDITIONS ON MECHANICAL PROPERTIES

The mechanical behavior of polymeric films is dependent on a number of variables, including the temperature and humidity of the environment. The following section highlights some of the effects that temperature and humidity can exert on polymer properties during processing, curing, and storage.

During film formation from an aqueous polymeric dispersion, individual polymer particles must coalesce and fuse together to form a continuous film. The degree of coalescence is dependent on the intensity of the primary driving forces, surface tension, and capillary forces that are generated upon water evaporation and the time exposed to such forces. Complete coalescence of latex particles occurs when the polymer chains located at the interface between adjacent particles interpenetrate due to viscous flow. Incomplete film coalescence can result in significant changes in polymer properties over time and thus has been extensively studied [1,13,63,79–81].

Temperature may significantly influence the completeness of coalescence [82]. Temperatures used during the coating process must be above the minimum film-forming temperature. Processing parameters used during coating, however, must be carefully controlled to ensure an appropriate balance between the rate of water removal, critical for the development of capillary forces, and the bed temperature of the coating apparatus. Low spray rates of aqueous polymeric dispersions, especially when combined with higher bed temperatures, can result in spray drying, where the solvent evaporates before the polymer chains coalesce, and brittle films are produced [83]. In contrast, high spray rates can over-wet the substrate [26] and cause surface dissolution of the product, with a potential for drug/excipient migration into the resulting film coat [35,36].

The ratio of tensile strength to elastic modulus of free films has been correlated to their in situ performance with lower values of this ratio correlating with increased coating defects [42]. Because coating with minimal defects is critical to provide and maintain consistent and reproducible drug-release rates from the coated controlled-release dosage forms, a higher value of this ratio is desirable. Figure 4.10 shows the ratio of tensile strength to elastic modulus plotted as a function of coalescence temperature for films cast from two commercially available aqueous ethylcellulose dispersions (Surelease plasticized with dibutyl sebacate or glycerol tricaprylate/caprate). An increase in the coalescence temperature up to 50°C for both polymeric formulations led to an increase in this ratio with a slight decrease in the ratio at higher temperatures. Because the highest tensile strength to elastic modulus ratio for the films cast from both Surelease formulations was observed at a drying temperature of 50°C, it may be presumed that films coalesced at or around this temperature are less susceptible to physical defects.

Following the completion of the coating process, the coated dosage forms are often stored at temperatures above the T_g of the polymer to promote further coalescence of the film. Storage at elevated temperatures can also ensure a homogeneous distribution of plasticizer within the film [84]. During this storage or curing stage, the microstructure of the polymer is altered [85], and the mechanical properties of the film as well as permeability and drug release are correspondingly affected [79]. Formulations containing high plasticizer concentrations generally require lower processing temperatures and less time for film coalescence and curing [63,86]. Drying air volume and humidity used during film coating processes have also been shown to influence film formation and hence polymer properties [25,87,88].

Chemical, mechanical, and dissolution profiles of drug products are generally determined within a short time after manufacturing. Products may be placed under stress conditions of high

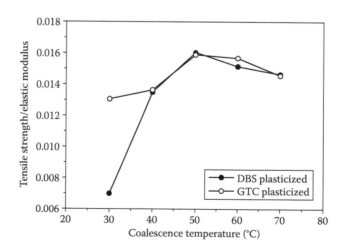

FIGURE 4.10 Tensile strength to elastic modulus ratio of free films of Surelease® plasticized with dibutyl sebacate or glycerol tricaprylate/caprate as a function of coalescence temperature. (From N. H. Parikh, S. C. Porter, B. D. Rohera, *Pharm Res.*, 1993, 10, 6, 810–815.)

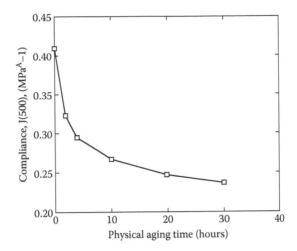

FIGURE 4.11   Effect of physical aging on the 500-s creep compliance of Aquacoat®-free films plasticized with 15% diethyl phthalate. (From J.-H. Guo, R. E. Robertson, G. L. Amidon, *Pharm Res.*, 1993, 10, 3, 405–410.)

temperature and/or relative humidity, and the data extrapolated to predict shelf life. Pharmaceutical products, however, may be exposed to a number of different environmental conditions during normal shelf life, and these changes in storage conditions can affect the mechanical properties of polymeric films, ultimately affecting drug release. Wu and McGinity [23] showed that the mechanical properties of Eudragit RS 30 D/RL 30 D polymer blends containing methyl paraben as a nontraditional plasticizer were dependent on the humidity of the storage environment. A decrease in tensile strength and Young's modulus was noted when coated beads were stored at 84% relative humidity, which was attributed to the absorbed water further plasticizing the film. In contrast, coated beads stored at 0% relative humidity exhibited brittle fracture failure during compression testing.

There are two major issues involved in changes that occur in polymer properties over time. The first and obvious one, based on the previous discussion, is that incomplete coalescence of the film will exert a significant affect on film properties during storage. The other major concern is physical aging. Most polymers used in pharmaceutical products are amorphous and are not at thermodynamic equilibrium at temperatures below their T_g. Over time, amorphous polymers undergo a slow transformation toward a thermodynamic equilibrium. As temperatures are cooled to below the T_g, the free volume of the polymer will slowly relax toward a lower free energy state, a process referred to as physical aging. Although this equilibration process is slow at ambient conditions, physical aging may produce significant changes in polymer properties. Guo et al. [89] used DMA to demonstrate that creep compliance (ratio of the relative creep extension to the applied stress) of Aquacoat films decreased over time, as shown in Figure 4.11. For these experiments, Aquacoat films were equilibrated at 5°C above the T s for 15 minutes, quenched to 25°C, then annealed at this temperature for up to 30 hours. The observed changes in creep compliance were attributed to a decrease in free volume and the further gradual coalescence of latex particles in the films. Physical aging and approaches to reduce or eliminate these problems [13,73,90] are discussed in more detail in another chapter.

CONCLUSIONS

The mechanical properties of free films and applied films prepared from aqueous polymeric dispersions provide valuable information to help the pharmaceutical scientist predict the stability and drug-release properties of film-coated solid dosage forms. The plasticizers in the film coating enhance flexibility, lower the T_g, enhance the coalescence of the colloidal polymeric particles, and minimize the formation of cracks or defects. Thus, plasticizers are essential additives for most polymers of pharmaceutical interest. Sufficient mixing time is required for the plasticizer to partition into the polymer phase with longer equilibration times needed for water-insoluble plasticizers. The permanence of plasticizers in both the dry and the wet state is an important consideration, and leaching of a plasticizer from a film-coated dosage form leads to a porous membrane, which will impact drug release. The addition of other excipients in film coating formulations, such as pigments to enhance product appearance and talc to reduce tackiness of the coating, influence the mechanical properties of polymeric films. Temperature, humidity, and processing parameters as well as physical aging can also have a significant affect on the mechanical properties of polymer films, ultimately affecting drug release from coated dosage forms.

REFERENCES

1. Lorck CA, Grunenberg PC, Junger H, Laicher A. Influence of process parameters on sustained-release theophylline pellets coated with aqueous polymer dispersions and organic solvent-based polymer solutions. *Eur J Pharm Biopharm*. 1997;43:149–57.
2. Bajdik J, Regdon JG, Marek T, Eros I, Suvegh K, Pintye-Hodi K. The effect of the solvent on the film-forming parameters of hydroxypropyl cellulose. *Int J Pharm*. 2005;301:192–8.
3. Bando H, McGinity JW. Relationship between drug dissolution and leaching of plasticizer for pellets coated with an aqueous Eudragit S100:L100 dispersion. *Int J Pharm*. 2006;323:11–7.
4. Cuppok Y, Muschert S, Marucci M, Hjaertstam J, Siepmann J et al. Drug release mechanisms from Kollicoat SR:Eudragit NE coated pellets. *Int J Pharm*. 2011;409:30–7.
5. Siepmann F, Siepmann J, Walther M, MacRae RJ, Bodmeier R. Polymer blends for controlled release coatings. *J Control Rel*. 2008;125:1–5.
6. Gil EC, Colarte AI, Sampedro JL, Bataille B. Subcoating with Kollidon VA 64 as water barrier in a new combined native dextran/HPMC-cetyl alcohol controlled release tablet. *Eur J Pharm Biopharm*. 2008;69:303–11.
7. Bley O, Siepmann J, Bodmeier R. Protection of moisture-sensitive drugs with aqueous polymer coatings: Importance of coating and curing conditions. *Int J Pharm*. 2009;378:59–65.
8. Felton LA, Timmins GS. A nondestructive technique to determine the rate of oxygen permeation into solid dosage forms. *Pharm Dev Technol*. 2006;11:1–7.
9. Rowe RC. The cracking of film coatings on film-coated tablets—A theoretical approach with practical implications. *J Pharm Pharmacol*. 1981;33:423–6.
10. Felton LA. Film coating of oral solid dosage forms. In: Swarbrick J, editor. Encyclopedia of Pharmaceutical Technology. New York: Informa Healthcare; 2006. p. 1–21.
11. Felton LA, Wiley CJ. Blinding controlled-release tablets for clinical trials. *Drug Dev Ind Pharm*. 2003;29(1):9–18.
12. Erdmann H, Gebert S, Kolter K, Schepky G. Studies on modifying the tackiness and drug release of Kollicoat EMM 30 D coatings. *Drug Dev Ind Pharm*. 2003;29(4):429–40.
13. Kucera S, Shah NH, Malick AW, Infeld MH, McGinity JW. Influence of an acrylic polymer blend on the physical stability of film-coated theophylline pellets. *AAPS PharmSciTech*. 2009;10(3):864–71.
14. Rohera BD, Parikh NH. Influence of plasticizer type and coat level on Surelease film properties. *Pharm Dev Technol*. 2002;7(4):407–20.
15. Piao ZZ, Lee KH, Kim DJ, Lee HG, Lee J, Oh KT et al. Comparison of release-controlling efficiency of polymeric coating materials using matrix-type casted films and diffusion-controlled coated tablet. *AAPS PharmSciTech*. 2010;11(2):630–6.
16. Bussemer T, Peppas NA, Bodmeier R. Time-dependent mechanical properties of polymeric coatings used in rupturable pulsatile release dosage forms. *Drug Dev Ind Pharm*. 2003;29(6):623–30.
17. Lecomte F, Siepmann J, Walther M, MacRae RJ, Bodmeier R. Polymer blends used for the coating of multiparticulates: Comparison of aqueous and organic coating techniques. *Pharm Res*. 2004; 21(5):882–90.
18. Tang X, Tai LY, Yang XG, Chen F, Xu HM, Pan WS. In vitro and in vivo evaluation of gliclazide push-pull osmotic pump coated with aqueous colloidal polymer dispersions. *Drug Dev Ind Pharm*. 2013;39:67–76.
19. Bruce HF, Sheskey PJ, Garcia-Todd P, Felton LA. Novel low-molecular-weight hypromellose polymeric films for aqueous film coating applications. *Drug Dev Ind Pharm*. 2011;37:1439–45.

20. Felton LA, Austin-Forbes T, Moore TA. Influence of surfactants in aqueous-based polymeric dispersions on the thermo-mechanical and adhesive properties of acrylic films. *Drug Dev Ind Pharm.* 2000;26(2):205–10.

21. Okhamafe AO, York P. Mechanical properties of some pigmented and unpigmented aqueous-based film coating formulations applied to aspirin tablets. *J Pharm Pharmacol.* 1986;38:414–9.

22. Aulton ME, Abdul-Razzak MH. The mechanical properties of hydroxypropylmethycellulose films derived from aqueous systems Part 2: The influence of solid inclusions. *Drug Dev Ind Pharm.* 1984; 10(4):541–61.

23. Wu C, McGinity JW. Influence of relative humidity on the mechanical and drug release properties of theophylline pellets coated with an acrylic polymer containing methylparaben as a non-traditional plasticizer. *Eur J Pharm Biopharm.* 2000;50:277–84.

24. Howland H, Hoag SW. Analysis of curing of a sustained release coating formulation by application of NIR spectroscopy to monitor changes physical-mechanical properties. *Int J Pharm.* 2013;452:82–91.

25. Williams III RO, Liu J. Influence of processing and curing conditions on beads coated with an aqueous dispersion of cellulose acetate phthalate. *Eur J Pharm Biopharm.* 2000;49:243–52.

26. Krogars K, Heinamaki J, Antikainen O, Karjalainen M, Yliruusi J. A novel amylose corn-starch dispersion as an aqueous film coating for tablets. *Pharm Dev Technol.* 2003;8(3):211–7.

27. Gutierrez-Rocca JC, McGinity JW. Influence of aging on the physical-mechanical properties of acrylic resin films cast from aqueous dispersions and organic solutions. *Drug Dev Ind Pharm.* 1993;19(3):315–32.

28. Parikh NH, Porter SC, Rohera BD. Tensile properties of free films cast from aqueous ethyl cellulose dispersions. *Pharm Res.* 1993;10(6):810–5.

29. Obara S, McGinity JW. Properties of free films prepared from aqueous polymers by a spraying technique. *Pharm Res.* 1994;11(11):1562–7.

30. Obara S, McGinity JW. Influence of processing variables on the properties of free films prepared from aqueous polymeric dispersions by a spray technique. *Int J Pharm.* 1995;126:1–10.

31. Frohoff-Hulsmann MA, Lippold BC, McGinity JW. Aqueous ethyl cellulose dispersions containing plasticizers of different water solubility and hydroxypropyl methylcellulose as coating material for diffusion pellets. II. properties of sprayed films. *Int J Pharm.* 1999;48:67–75.

32. Porter SC, Felton LA. Techniques to assess film coatings and evaluate film-coated products. *Drug Dev Ind Pharm.* 2010;36(2):128–42.

33. Dashevsky A, Kolter K, Bodmeier R. Compression of pellets coated with various aqueous polymer dispersions. *Int J Pharm.* 2004;279:19–26.

34. Felton LA, Shah NH, Zhang G, Infeld MH, Malick AW, McGinity JW. Compaction properties of individual non-pareil beads coated with an acrylic resin copolymer. *STP Pharm Sci.* 1997;7(6):457–62.

35. Felton LA, Perry WL. A novel technique to quantify film-tablet interfacial thickness. *Pharm Dev Technol.* 2002;7(1):1–5.

36. Barbash D, Fulghum JE, Yang J, Felton LA. A novel imaging technique to investigate the influence of atomization air pressure on film-tablet interfacial thickness. *Drug Dev Ind Pharm.* 2009;35(4):480–6.

37. Wu C, McGinity JW. Non-traditional plasticization of polymeric films. *Int J Pharm.* 1999;177:15–27.

38. Missaghi S, Fassihi R. A novel approach in the assessment of polymeric film formation and film adhesion on different pharmaceutical solid substrates. *AAPS PharmSci Tech.* 2004;5(2):Article 29.

39. Deshpande AA, Shah NH, Rhodes CT, Malick W. Evaluation of films used in development of a novel controlled-release system for gastric retention. *Int J Pharm.* 1997;159:255–8.

40. Dowling NE. Mechanical behavior of materials. 1st ed. New Jersey: Prentice Hall; 1993.

41. Fulzele SV, Satturwar PM, Dorle AK. Polymerized rosin: Novel film forming polymer for drug delivery. *Int J Pharm.* 2002;249:175–84.

42. Rowe RC. Correlations between the in-situ performance of tablet film coating formulations based on hydroxypropyl methylcellulose and data obtained from tensile testing of free films. *Acta Pharm Technol.* 1983;29(3):205–7.

43. Bashaiwoldu AB, Podczeck F, Newton JM. Application of dynamic mechanical analysis (DMA) to the determination of the mechanical properties of coated pellets. *Int J Pharm.* 2004;274:53–63.

44. Felton LA, Shah NH, Zhang G, Infeld MH, Malick AW, McGinity JW. Physical-mechanical properties of film-coated soft gelatin capsules. *Int J Pharm.* 1996;127:203–11.

45. Lopez-Rodriguez FJ, Torrado JJ, Torrado S, Escamilla C, Cadorniga R, Augsburger LL. Compression behavior of acetylsalicyclic acid pellets. *Drug Dev Ind Pharm.* 1993;19(12):1369–77.

46. Abdul S, Chandewar AV, Jaiswal SB. A flexible technology for modified release drugs. Multiple-unit pellet system (MUPS). *J Control Rel.* 2010;147(1):2–16.

47. Gruetzmann R, Wagner K. Quantification of the leaching of triethyl citrate/polysorbate 80 mixtures from Eudragit RS films by differential scanning calorimetry. *Eur J Pharm Biopharm.* 2005;60:159–62.

48. Gupta VK, Beckert T, Deusch N, Hariharan M, Price JC. Investigation of potential ionic interactions between anionic and cationic polymethacrylates of multiple coatings of novel colonic delivery system. *Drug Dev Ind Pharm.* 2002;28(2):207–15.

49. Nyamweya N, Hoag SW. Assessment of polymer-polymer interactions in blends of HPMC and film forming polymers by modulated temperature differential scanning calorimetry. *Pharm Res.* 2000;17(5):625–31.

50. Felton LA, Shah K, Yang J, Omidian H, Rocca JG. A rapid technique to evaluate the oxidative stability of a model drug. *Drug Dev Ind Pharm.* 2007;33(6):683–9.

51. Yoshihashi Y, Yonemochi E, Terada K. Estimation of initial dissolution rate of drug substance by thermal analysis: Application for carbamazepine hydrate. *Pharm Dev Technol.* 2002;7(1):89–95.

52. Chidavaenzi OC, Buckton G, Koosha F, Pathak R. The use of thermal techniques to assess the impact of free concentration on the amorphous content and polymorphic forms present in spray dried lactose. *Int J Pharm.* 1997;159:67–74.

53. Zheng W, McGinity JW. Influence of Eudragit NE 30 D blended with Eudragit L 30 D-55 on the release of phenylpropanolamine hydrochloride from coated pellets. *Drug Dev Ind Pharm.* 2003;29(3):357–66.

54. Hill VL, Craig DQM, Feely L. The effects of experimental parameters and calibration on MTDSC data. *Int J Pharm.* 1999;192:21–32.

55. McDaid FM, Barker SA, Fitzpatrick S, Petts CR, Craig DQM. Further investigations into the use of high sensitivity differential scanning calorimetry as a means of predicting drug-excipient interactions. *Int J Pharm.* 2003;252:235–40.

56. Lafferty SV, Newton JM, Podczeck F. Dynamic mechanical thermal analysis studies of polymer films prepared from aqueous dispersion. *Int J Pharm.* 2002;235:107–11.

57. Sears JK, Touchette NW. In: Krostwitch JI, editor. Concise Encyclopedia of Polymer Science and Engineering. New York: John Wiley & Sons, Inc.; 1990. p. 734–44.

58. Bruce LD, Petereit HU, Beckert T, McGinity JW. Properties of enteric coated sodium valproate pellets. *Int J Pharm.* 2003;264:85–96.

59. Wesseling M, Kuppler F, Bodmeier R. Tackiness of acrylic and cellulosic polymer films used in the coating of solid dosage forms. *Eur J Pharm Biopharm.* 1999;47:73–8.

60. Gutierrez-Rocca JC, McGinity JW. Influence of water soluble and insoluble plasticizers on the physical and mechanical properties of acrylic resin copolymers. *Int J Pharm.* 1994;103:293–301.

61. Bodmeier R, Paeratakul O. The distribution of plasticizers between aqueous and polymer phases in aqueous colloidal polymer dispersions. *Int J Pharm.* 1994;103:47–54.

62. Iyer U, Hong W-H, Das N, Ghebre-Sellassie I. Comparative evaluation of three organic solvent and dispersion-based ethylcellulose coating formulations. *Pharm Technol.* 1990;14:68–86.

63. Wesseling M, Bodmeier R. Influence of plasticization time, curing conditions, storage time, and core properties on the drug release from aquacoat-coated pellets. *Pharm Dev Technol.* 2001;6(3):325–31.

64. Rowe RC, Forse SF. The refractive indices of polymer film formers, pigments and additives used in tablet film coating: Their significance and practical application. *J Pharm Pharmacol.* 1983;35:205–7.

65. Bechard SR, Quraishi O, Kwong E. Film coating: Effect of titanium dioxide concentration and film thickness on the photostability of nifedipine. *Int J Pharm.* 1992;87:133–9.

66. Maul KA, Schmidt PC. Influence of different-shaped pigments on bisacodyl release from Eudragit L 30 D. *Int J Pharm.* 1995;118:103–12.

67. Felton LA, McGinity JW. Influence of pigment concentration and particle size on adhesion of an acrylic resin copolymer to tablet compacts. *Drug Dev Ind Pharm.* 1999;25(5):599–606.

68. Felton LA, McGinity JW. Influence of insoluble excipients on film coating systems. *Drug Dev Ind Pharm.* 2002;28(3):225–43.

69. Gibson SHM, Rowe RC, White EFT. Determination of the critical pigment volume concentrations of pigmented film coating formulations using gloss measurement. *Int J Pharm.* 1988;45:245–8.

70. Hsu ER, Gebert MS, Becker NT, Gaertner AL. The effects of plasticizers and titanium dioxide on the properties of poly(vinyl alcohol) coatings. *Pharm Dev Technol.* 2001;6(2):277–84.

71. Nimkulrat S, Suchiva K, Phinyocheep P, Puttipipathachorn S. Influence of selected surfactants on the tackiness of acrylic polymer films. *Int J Pharm.* 2004;287:27–37.

72. Fassihi RA, McPhillips AM, Uraizee SA, Sakr AM. Potential use of magnesium stearate and talc as dissolution retardants in the development of controlled drug delivery systems. *Pharm Ind.* 1994;56(6):579–83.

73. Maejima T, McGinity JW. Influence of film additives on stabilizing drug release rates from pellets coated with acrylic polymers. *Pharm Dev Technol.* 2001;6(2):211–21.
74. Banker GS. Film coating theory and practice. *J Pharm Sci.* 1966;55(1):81–9.
75. Knop K, Matthee K. Influence of surfactants of different charge and concentration on drug release from pellets coated with an aqueous dispersion of quarternary acrylic polymers. *STP Pharma Sci.* 1997;7(6):507–12.
76. Bodmeier R, Paeratakul O. Dry and wet strengths of polymeric films prepared from an aqueous colloidal polymer dispersion, Eudragit RS30D. *Int J Pharm.* 1993;96:129–38.
77. Chan LW, Ong KT, Heng PWS. Novel film modifiers to alter the physical properties of composite ethylcellulose films. *Pharm Res.* 2005;22(3):476–89.
78. Bodmeier R, Paeratakul O. Mechanical properties of dry and wet cellulosic and acrylic films prepared from aqueous colloidal polymer dispersions used in the coating of solid dosage forms. *Pharm Res.* 1994;11(6):882–8.
79. Felton LA, Baca ML. Influence of curing on the adhesive and mechanical properties of an applied acrylic polymer. *Pharm Dev Technol.* 2001;6(1):1–9.
80. Muschert S, Siepmann F, Leclercq B, Siepmann J. Dynamic and static curing of ethylcellulose: PVA-PEG graft copolymer film coatings. *Eur J Pharm Biopharm.* 2011;78:455–61.
81. Yang QW, Flament MP, Siepmann F, Busignies V, Siepmann J et al. Curing of aqueous polymeric film coatings: Importance of the coating level and type of plasticizer. *Eur J Pharm Biopharm.* 2010;74:362–70.
82. Yang ST, Ghebre-Sellassie I. The effect of product bed temperature on the microstructure of Aquacoat-based controlled-release coatings. *Int J Pharm.* 1990;60:109–24.
83. Pandey P, Bindra DS, Felton LA. Influence of process parameters on tablet bed microenvironmental factors during pan coating. *AAPS PharmSciTech.* 2014;15(2):296–305.
84. Lippold BH, Sutter BK, Lippold BC. Parameters controlling drug release from pellets coated with aqueous ethyl cellulose dispersions. *Int J Pharm.* 1989;54:15–25.
85. Bodmeier R, Paeratakul O. The effect of curing on drug release and morpholoigal properties of ethylcellulose pseudolatex-coated beads. *Drug Dev Ind Pharm.* 1994;20(9):1517–33.
86. Williams RO, III, Liu J. The influence of plasticizer on heat-humidity curing of cellulose acetate phthalate coated beads. *Pharm Dev Technol.* 2001;6(4):607–19.
87. Porter SC. Scale-up of film coating. In: Levin M, editor. Pharmaceutical Process Scale-Up. Boca Raton: Taylor & Francis; 2006.
88. Pearnchob N, Bodmeier R. Coating of pellets with micronized ethylcellulose particles by a dry powder coating technique. *Int J Pharm.* 2003;268:1–11.
89. Guo J-H, Robertson RE, Amidon GL. An investigation into the mechanical and transport properties of aqueous latex films: A new hypothesis for film-forming mechanism of aqueous dispersion system. *Pharm Res.* 1993;10(3):405–10.
90. Omari DM, Sallam A, Abd-Elbary A, El-Samaligy M. Lactic acid-induced modifications in films of Eudragit RL and RS aqueous dispersions. *Int J Pharm.* 2004;274:85–96.

5 A Proactive Approach to Troubleshooting the Application of Film Coatings to Oral Solid Dosage Forms

Stuart C. Porter

CONTENTS

INTRODUCTION

Film coating, in the formal sense, is a process that has been utilized in the pharmaceutical industry for over half a century. In spite of its relative longevity, it remains a complex process in which all technical issues and their impact are not always fully appreciated with the result that potential problems continue to arise with all too frequent regularity. This chapter will focus on identifying the most common problems and adopting a sensible approach to their effective resolution. To the latter end, the importance of implementing the required tenets of QbD will be stressed.

CHARACTERIZING PROBLEMS IN FILM COATING

Although problems associated with the film coating process can fall into a broad range of categories, by far the largest category is that associated with visual defects, often purely cosmetic in nature, which rarely impact the efficacy of the final dosage form, except in the sense that they might lead to reduced patient confidence.

Problems can generally be classified in this manner, namely those affecting the following:

- Visual coated product quality
- Coated product performance and functionality
- Coated product stability
- Processing efficiencies and manufacturing costs

Failure to adopt a proactive approach when developing quality film-coated products typically results in a situation in which problem resolution

- Basically becomes a "reactive" process because it deals with something that has already gone wrong.
- Becomes seriously constrained simply because changes to the final product/process (the final product, almost certainly by this time, will be an approved commercial product for which problems tend to arise during routine manufacturing operations) are limited by regulatory issues.

For the most part, the multiplicity of problems or defects associated with film-coated products tends to arise because of the following:

- The tablet core formulation is not sufficiently robust to withstand the rigors of the film-coating process (which imparts significant mechanical stress and exposes the product being coated to elevated temperatures and humidity).
- The coating formulation being employed has not been optimized either for the product being coated or the coating process being used.
- Coating process conditions are selected in a rather ad hoc manner and, on occasion, without a full understanding of the overall impact of key process conditions on ultimate product quality.
- There has been ineffective technology transfer from the development group to the chosen production site.
- There is little appreciation for the impact of batch-wise variability in raw materials being used or inherent variations in the coating process on ultimate product quality.

Of course, when it comes to dealing with potential problems, the first necessary step is to identify the problem. In this respect, visual defects are much easier to identify than those associated with functional attributes (such as dissolution and stability-related problems).

Identifying appearance-related problems is relatively easy because they are visually evident, feedback is immediate, and the magnitude of the problem is also often easy to assess.

Visual defects can often be confirmed by reference to tablet samples exemplifying common defects (as shown in Figure 5.1) or by using a decision-tree approach as illustrated in Figure 5.2.

Identifying functionally related problems is more difficult because the problem is often not confirmed until some analytical procedure (which occurs after the conclusion of the coating process) has been carried out. Such problems also become more challenging to identify because the sample size tested (relative to the size of the batch produced) is often quite small, making the true relevance of the problem to the batch in question much more difficult to assess. In this case,

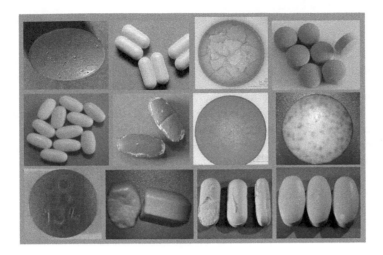

FIGURE 5.1 **(See color insert.)** Common film coating defects.

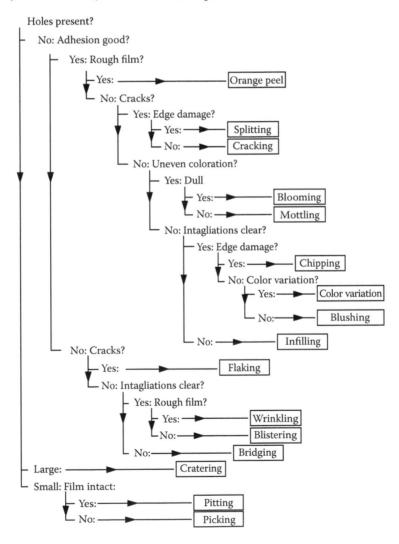

FIGURE 5.2 Example of a decision tree suitable for identifying film coating defects.

application of a proactive approach, employing the principles of QbD, clearly becomes much more relevant.

RESOLVING PROBLEMS IN BOTH A REACTIVE AND PROACTIVE MANNER

Many of the common problems relating to film-coating processes, and ultimately film-coated products, often relate to a lack of understanding of key aspects of the coating process and what happens to the product being coated when it is exposed to such a process.

One of the important factors is stress. Film coating is, by nature, a stressful process because the product being coated is subjected to the following:

- Constant motion, which imparts mechanical stress
- Elevated temperatures, which impart thermal stress
- In the case of an aqueous process, elevated humidities

In addition to these obvious stress factors, others that may be less obvious, such as internal stresses that develop within film coatings as they form on the surface of the product, can contribute to the problem. Rowe [1–3] has gone to great lengths to describe the impact of these internal stresses as well as their origins. Suffice it to say, as the coating forms and solvent is removed, the volume of the coating shrinks and, at a critical point, often called the solidification point of the coating, additional solvent loss cannot be accommodated by further contraction of the coating, and so stresses develop within the film structure. The magnitude of these stresses can be impacted by both coating process conditions as well as by the nature of the coating formulation employed in a given application. Two key negative outcomes of the development of these internal stresses are logo bridging and film cracking as shown in Figure 5.3.

Another key aspect of the film-coating process that needs to be well understood is associated with process thermodynamics. In this case, we are concerned about the balance that is achieved between all of the inputs to the coating process as well as the predominant outputs. Failure to understand this situation means that the ability to control solvent removal from the coating as the coating is being applied will be compromised, resulting in a potentially negative impact on final tablet quality. Film-coating process thermodynamics, while not receiving a great deal of attention, has been discussed by both Ebey [4] and Choi [5].

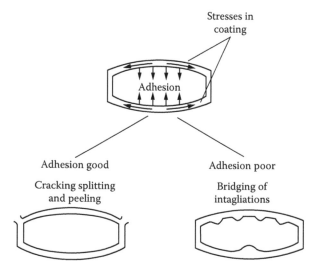

FIGURE 5.3 Schematic diagram of impact of internal stresses.

The key inputs into any typical aqueous film coating process are the following:

- Process air
  - Quantity of air
  - Temperature of air
  - Moisture content of air
- Coating liquid application
  - Spray rate
  - Solids content of coating liquid
  - Enthalpy of coating liquid
- Atomizing/pattern air
  - Air volume
  - Enthalpy of air

On the other side of the equation are these two output elements:

- Process exhaust air
  - Quantity of air
  - Temperature of air
  - Moisture content of air
- Heat loss from the coating process equipment to the surrounding environment

Last, another crucial process element that receives scant attention is spray dynamics. Currently, most coating operations involve the spray application of the coating liquid onto the surface of the product. Today, this process involves the use of air spray guns with which, as coating liquid emerges from the opening of the spray nozzle, the liquid interacts with a steady stream of compressed air required to break the coating liquid up into small droplets. In addition, in most coating pan setups, additional compressed air is used (often called pattern air) to spread the atomized liquid out across the surface of the tablet bed.

Spray dynamics define the following:

- The size (and size distribution) of the droplets created
- The velocity (and hence momentum at impact with the tablet surface) of the droplet stream
- The solvent content of the droplets as they make impact with the surface of the product being coated

The major factors that will ultimately have an influence on final droplet characteristics include the following:

- The pressure and volume of the compressed air used
- The ratio of primary atomizing air to pattern air (when used)
- The rate at which the liquid emerges from the spray nozzle
- The balance between liquid emergence rates at each nozzle for a multi-gun setup
- The orifice diameter of the nozzle from which the liquid emerges
- The viscosity of the coating liquid
- The surface tension of the coating liquid
- The temperature of the coating liquid

Unfortunately, the setup and operation of spray nozzles is often arbitrary (sometimes based on personal experience) without consideration being given to the factors described above. Several fundamental analyses of film-coating operations and their potential impact on coated tablet quality have been undertaken, including those by Suzzi et al. [6] and Muliadi and Sojke [7].

LOGO BRIDGING

Logo bridging of pharmaceutical tablets results directly from the development of those internal stresses that were described earlier, which, on film shrinkage during the drying of the coating, cause the coating to pull away from the tablet surface within the logo and "bridge across" it as shown in Figures 5.4a and 5.4b. Because these stresses increase with increasing coating thickness, logo bridging, when it occurs, becomes progressively worse as more coating is applied.

As this condition is influenced by many factors, including tablet core and coating formulation as well as coating process conditions used, retroactive resolution of the problem has very few options. To some extent, internal stresses can be managed by adjusting process conditions to increase product (and, hence, coating) temperatures as described by Kim et al. [8] and Rowe and Forse [9] and as summarized in Table 5.1.

(a)

(b)

FIGURE 5.4 (a) Typical example of film-coated tablets illustrating logo bridging. (b) SEM cross section illustrating logo bridging.

TABLE 5.1

Impact of Product Temperature on Logo Bridging with Film-Coated Tablets

Inlet Air Temperature (°C)	Spray Rate (ml min⁻¹)	Tablet Bed Temperature (°C)	Incidence of Bridging (%)
60	50	42	49.3
60	40	54	38.5
70	50	50	21.2

Proactively, there are many options that can be followed to eliminate logo bridging. For example,

- Although the impact of the API cannot be ignored, careful selection of appropriate excipients as described by Rowe [2] can be used to improve the overall adhesion characteristics of applied film coatings, decreasing the likelihood that logo bridging will occur.
- In a similar vein, increasing tablet core porosity (as shown in Figure 5.5) can improve film adhesion.
- Judicious selection of an appropriate logo design can increase the area available within the logo to improve adhesion of the applied coating [10].

Adhesion related problems can also be addressed in a proactive manner by careful selection of the coating formulation to be used. Rowe and Forse [11] have described how the appropriate choice of plasticizers can be an effective tool in the management of internal stresses, thus reducing problems such as logo bridging. The data shown in Figure 5.6 can effectively eliminate logo bridging problems.

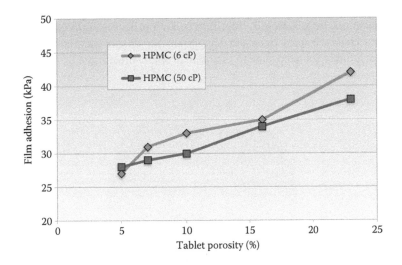

FIGURE 5.5   Example of how increased tablet core porosity can improve film adhesion.

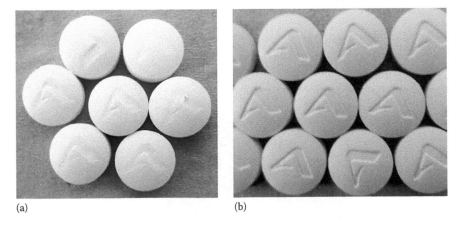

FIGURE 5.6   **(See color insert.)** (a) Standard HPMC. (b) HPMC/copovidone composite. Illustration of how choice of an appropriate coating formulation can eliminate logo bridging problems.

Finally, the importance of designing an optimized film-coating process cannot be ignored. Earlier, it was shown how adjustments in process temperatures can impact the potential for logo bridging to occur. Okutgen et al. [12] have demonstrated how a combination of tablet core characteristics and coating process conditions can be managed in order to reduce the internal stresses within film coatings as a precursor to managing problems such as logo bridging.

INFILLING OF LOGOS

This condition (see example shown in Figure 5.7) produces a similar result to logo bridging in that any intagliation or break line becomes visually indistinct as the result of the application of a film coating. However, unlike logo bridging, infilling is caused by a buildup of dried coating material within the logo. This may be as a result of partially spray-dried material gradually building up within the logo or, as described by Down [13], the buildup of dried foamy material within the logo (see Figure 5.8). Normally, dried material that is deposited on the tablet surfaces would be swept away as tablets slide over each other, but this will not occur for material deposited within the logo.

The general buildup of prematurely dried material can be corrected by optimizing the atomization process to give a more uniform distribution of droplet sizes. Once this has been achieved, further improvement can be achieved by adjusting coating process conditions (from the thermodynamic standpoint to balance spray rates and drying rates) to ensure that most droplets arrive at the tablet surface in a suitably solvated state to facilitate droplet spreading.

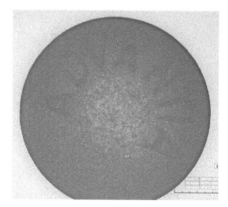

FIGURE 5.7   **(See color insert.)** Example of infilling of tablet logos/break lines.

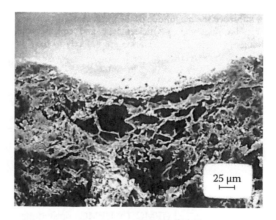

FIGURE 5.8   Illustration of how infilling can occur as a result of build-up of foamy material within a logo.

In the case in which foamy material tends to build up within the logo, this can often be traced back to the preparation of the coating suspension. Because in most cases today coating suspensions are prepared from ready-made dry powder coating formulations, the procedure for making coating suspensions is often well documented. For example, the usual approach is to use a vessel in which the diameter is such that the height of the liquid in the vessel and the vessel diameter are very similar, and in which a vertically mounted mixer (having a mixing head about one third the diameter of the vessel) is inserted into the liquid. The mixer is set into motion at a suitable speed so that a vortex is created without a "gurgling" sound (which is evidence that too much air is being drawn into the liquid); the powdered coating material is then introduced into the vortex steadily but quickly (if the powder addition rate is too fast, large lumps will form that will take some time to break down; if the powder addition rate is too slow, then viscosity will build before all of the powder is added, so it will then become necessary to increase the stirring rate to enable all of the remaining powder to be dispersed). Once all of the powder has been incorporated into the liquid, the stirring speed should then be reduced to a level that allows the whole mass of liquid to be gently agitated for about 45 minutes to allow the polymers to solvate (hydrate) fully.

As most polymers used in coating formulations tend to be surface active in aqueous media, using excessively high stirring rates tends to generate a lot of foam, which, if not allowed to dissipate before coating commences, will ultimately result in foam buildup within the logos.

Unfortunately, all too often, when it comes to the production setting, the equipment available for preparing coating suspensions may be less than ideal. Often, the mixers are not located vertically above the tanks but rather are clamped on the side of the vessel so that the mixing blade is at an angle. Oftentimes, the size of the mixer blade itself is not ideal, so that higher stirring speeds need to be employed; finally, because of the larger quantities of coating powder involved, process operators have a tendency to pour the powder in far too slowly and thus over a much longer period of time. The downside of all of this is that the chances for air entrapment are increased.

Although the use of preformulated coating powders is commonplace, some pharmaceutical companies still prefer to make up coating suspensions from individual ingredients. In this case, some of the challenges to be faced include the following:

- The need to disperse individual polymers into the solvent (water in this case). This process requires great care, and overnight standing may well be required to allow foam to dissipate more effectively.
- Addition of pigments, which can be challenging. This process requires (unless premade pigment dispersions are employed) the use of high-shear dispersing equipment (such as a Silverson or UltraTurrax mixer). Such high-shear dispersing equipment can generate a lot of foam, so the most sensible approach is to make a separate pigment dispersion in water, and then add it to an already prepared polymer solution using gentle agitation.

TABLET EDGE CHIPPING

A classic example of edge chipping is shown in Figure 5.9. As stated earlier, most film-coating processes produce a lot of mechanical stress while the coating is being applied with the consequence that exposed areas of the tablet (particularly the edges) are prone to chipping. Problems with tablet core robustness, mechanical strength of the applied film coating, and excessive mechanical stress imparted in the coating process all contribute to this overall problem.

Reactively addressing this problem typically involves modifying the coating process (within the limits of what has been approved regulatorily). Some useful approaches include the following:

- Reducing the pan speed throughout the coating process (or at least until approximately one third of the coating has been applied so that the tablets receive some protection from the applied coating).

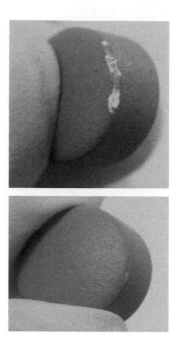

FIGURE 5.9   Example of film-coated tablet edge chipping. (© 2015 AstraZeneca, all rights reserved. Image reproduced with permission of AstraZeneca; image is copyright of AstraZeneca and may not be reproduced without written permission.)

- Instead of prewarming the tablets (a process that usually involves some tablet motion without any coating being applied), commence spraying immediately after the required inlet temperature has been reached.
- Prewarming the tablets to a higher temperature so that a faster spray rate can be employed during the early stages of the coating process, thereby providing faster protection for the tablet edges from the applied coating.

From a more proactive standpoint, several approaches can be used to prevent edge chipping.

The first involves addressing the issue of tablet core robustness. A key approach is to design a formulation in which the tablets have suitable breaking force values for the size of the tablets being prepared. For example, for larger tablets (> 500 mg), breaking force values in the range of 180 to 200 N would be an appropriate target; similarly, for intermediately sized tablets (typically in the range of 150 to 500 mg), appropriate breaking force values might range from 130 to 160 N; and finally, for small-sized tablets (< 150 mg), the target might be 80–100 N. However, of greater importance, in terms of tablet robustness, are the resultant friability values, and ideally tablets that will be film-coated should have friability values of less than 0.1%. The data shown in Figure 5.10 provide a good illustration of how addition of a suitable binder to the tablet formulation can play a significant role in improving tablet robustness.

Another tablet core deficiency that should be addressed involves eliminating the tendencies for capping to occur. In many cases, although tablet cores might not initially show any evidence of capping tendency, when they are introduced into the coating pan and tumbling begins, the applied stress often exposes deficiencies in this regard as indicated in Figure 5.11.

Finally, edge chipping may simply result from inadequate tablet design (shape). If the tablet edges are too sharp (or the land area too pronounced), the risk of edge chipping is increased. This risk can often be mediated by a change in tablet surface profile. For example, switching from a

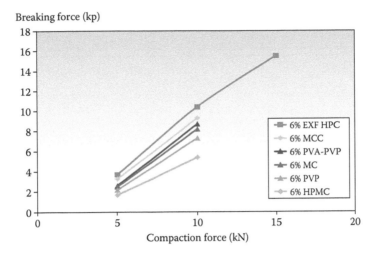

FIGURE 5.10   Indication as to how the addition of suitable binders to a Metformin (60% API content) tablet formulation can improve tablet robustness.

FIGURE 5.11   Example of edge chipping induced by tablet capping tendencies. (© 2015 AstraZeneca, all rights reserved. Image reproduced with permission of AstraZeneca; image is copyright of AstraZeneca and may not be reproduced without written permission.)

conventional convex tablet shape to one that employs a compound radius design (as shown in Figures 5.12 and 5.13) may be all that is required.

With respect to the coating formulation, designing one that is either more flexible (for example, adding more plasticizer) or has greater mechanical strength (using polymers that exhibit higher tensile strength values) can both be effective approaches to eliminate edge chipping (see Figure 5.14). Additionally, employing a coating formulation that can be sprayed at higher solids (say, 20% w/w) instead of the more traditional 12%–15% w/w, will allow the coating to be deposited faster, thus providing greater edge protection.

Generally, to assess the suitability of both tablets and coating formulations in terms of resistance to edge defects, a useful approach involves using a friability test in which the rotational speed is increased significantly (say, to 50 or even 100 rpm), and the test period is increased to at least 30 minutes. In this way, both uncoated tablet cores and those that have also been film coated, can be assessed to minimize robustness problems once they are transferred to a film-coating process, especially on the production scale.

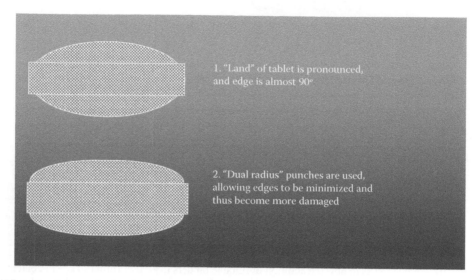

FIGURE 5.12   Schematic showing how a change in tablet profile can reduce the sharpness of tablet edges.

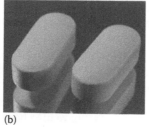

FIGURE 5.13   **(See color insert.)** Example of how changing from standard concave punches to compound radius punches can eliminate tablet edge chipping.

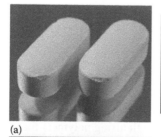

(a)

Property	Value
Film strength (MPa)	30.69
Elongation (%)	21.96
Film adhesion (N m^{-1})	22.51

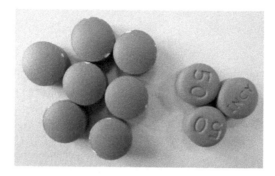

(b)

Property	Value
Film strength (MPa)	53.86
Elongation (%)	16.77
Film adhesion (N m^{-1})	17.16

FIGURE 5.14   **(See color insert.)** Example of how improving coating mechanical properties can eliminate edge chipping problems. (a) The problem and (b) the solution.

TABLET SURFACE EROSION

Tablet surface erosion is another mechanical defect with which conventional reactive remedies can be achieved by addressing coating process issues as described earlier under edge chipping. In addition, because core surface erosion can be facilitated by over-wetting (which causes the tablet surface to soften and hence become less resistant to attritional effects), addressing over-wetting conditions (reducing spray rates, increasing processing temperatures, or both) can produce a positive result.

Proactively, when resolving surface erosion issues, approaches dealing with improving tablet robustness as described in the previous section (dealing with edge chipping) can also be of value. However, additional solutions might include the following:

- Because surface erosion can be linked to the curvature of the upper and lower faces of the tablets, changing punch design. The data shown in Figure 5.15 indicate how tablet surface hardness profiles (not to be confused with tablet hardness or breaking force) change with changing face curvatures.
- Because surface erosion most commonly occurs on the crown of the tablet, the presence of a centrally placed logo can exacerbate this problem (as shown in Figure 5.16). Thus, changing the location of the logo from the center of the tablet faces to the periphery (as shown in Figure 5.17) can be beneficial.

Last, during the development of film-coating formulations, creating one that can be sprayed at higher solids minimizes time in the coating process, reducing the risk that moisture will penetrate into the tablet surface, softening it and making it more prone to erosion.

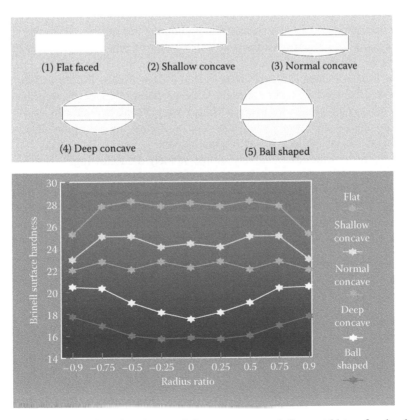

FIGURE 5.15   Data illustrating how changing punch face curvature can influence tablet surface hardness profiles.

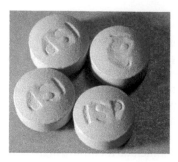

FIGURE 5.16　Photograph illustrating how the presence of a logo in the center of the tablet can have a negative impact on tablet surface erosion.

FIGURE 5.17　Schematic diagram illustrating options for tablet logo placement.

TABLET PITTING

This problem (see Figure 5.18 for a typical illustration of the problem) was previously described by Rowe and Forse [14]. In their case, the problem was caused when agglomerates of the lubricant, stearic acid, melted under the influence of the coating process conditions used, leaving small pits in the tablet surface. Although these pits were eventually covered by the applied film coating, they remained visible afterward. In a similar vein, large particles of super-disintegrant present in the tablet surface on contact with water from the film coating being applied can swell to an extent at which they become loosened from the tablet surface, again leaving pits in the tablet surface so that, after the coating has been applied, they still remain visible.

In both cases, the remedies are relatively simple, namely the following:

- Ensure that the blending process (prior to tableting) is optimized so that the chance of producing agglomerates is minimized, and the offending ingredient is more uniformly mixed.
- Optimize coating process conditions to minimize use of excessively high temperatures (the key factor here will be tablet bed temperatures) when low melting point ingredients are

FIGURE 5.18　**(See color insert.)** Example of tablet pitting.

used in the tablet core and prevent over-wetting conditions (potentially offset by reducing spray rates and/or increasing process temperatures) when the core contains very hydrophilic materials, especially super-disintegrants.

TABLET BREAKAGE

As mentioned previously, perhaps one of the most troublesome defects caused by mechanical stress is tablet breakage, primarily because it is not one that can easily be resolved by minor adjustment to coating process conditions. Prime examples of tablet breakage are shown in Figure 5.19 (in one case, the issue is related to inadequate tablet strength, and in the other case, breakage occurs as a direct result of tablet capping tendencies).

The primary causes of tablet breakage typically relate to the following:

- Mechanical deficiencies in the tablet cores.
- Variations in die fill uniformity during the tableting process, possibly due to variations in hopper fill. Underweight tablets will not possess the same mechanical strengths as those tablets compacted to the required weights and thus will break more easily in a coating process.
- Mishandling of tablets during loading of coating pans, attritional conditions used during application of the applied coating, and unloading of coating pans.

As stated earlier, when designing robust tablets capable of withstanding the rigors of a typical coating process, close attention has to be paid to designing tablets that have sufficient mechanical strength. This has to be considered in the whole context of creating a suitable tablet core formulation, and the challenge is that much greater when the tablet core contains an API that is present in high concentrations in the core formulation and is itself poorly compactable (in which case, a direct compression tableting approach may not be an option, and use of a wet granulation process in which optimal binder attributes can be introduced almost certainly will be required). Creating a tablet core formulation that shows no capping tendencies is also important as tablets that exhibit this defect will readily break apart in a coating process.

In terms of variable tablet breaking strength, especially when related to variable tablet weights, this issue should not really occur with modern tableting processes, but when tablet breakage is observed (particularly as shown by the green tablets in Figure 5.19), the possibility that significant tablet weight variation exists should not be discounted.

When it comes to tablet breakage, careful inspection of defect tablets is critical because what is observed can provide important information leading to a positive resolution. For example, if the broken tablets show all surfaces completely coated, this suggests that breakage occurs during handling, including pan loading, up to the point at which the application of the coating commences. If, on the other hand, broken tablets show completely uncoated surfaces at the conclusion of the coating process, this suggests that breakage occurs during pan unloading. In either case, pan loading and unloading procedures need to be carefully examined. Finally, tablets that exhibit various stages of

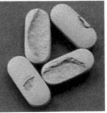

FIGURE 5.19 Common examples of tablet breakage.

coverage by the coating material indicate a distinct possibility that breakage occurs throughout the coating process; in which case, a combination of factors likely need to be taken into account, namely the following:

- Improving tablet core robustness, either by addressing the core formulation or examining the tableting process (slowing down press speeds to increase dwell time, and hence mechanical strength, and possible use of precompression)
- Ensuring that the coating process is optimized with respect to the following:
  - Pan loading (underloaded pans potentially create more attritional problems, especially because tablets can strike the exposed edges of baffles)
  - Pan speed, which can be reduced to minimize mechanical stress

Film Peeling and Flaking

These defects (as illustrated in Figure 5.20) primarily relate to cohesive and adhesive failure of the coating. Initially, the continuity of the coating is lost (typically exacerbated by film cracking), and then the internal stresses within the coating cause it to peel back. The problem may also be compounded by a less than robust tablet in which the tablet surface begins to erode, consequently negatively impacting coating adhesion.

Problems of this nature may be alleviated by changing process conditions (reducing pan speed to reduce mechanically stressful conditions, thus reducing the chances of film cracking, and improving process thermodynamics to allow more moisture to be retained within the film, thus improving plasticization and increasing film flexibility). However, a proactive approach will usually be more effective.

For example, designing the tablet core formulation to improve film adhesion by appropriate excipient selection (see Table 5.2), improving tablet core robustness by appropriate binder selection, as shown earlier in Figure 5.10, to reduce surface erosion issues, and optimizing the coating formulation either to improve robustness (see Figure 5.14) or improve film adhesion (see Figure 5.6).

Tablet-to-Tablet Color Variability

Although variation in the amount of coating applied on each tablet within the batch will, on a weight basis, always exist, the challenge is to ensure that enough coating is applied to prevent any differences in amount of coating applied becoming visually apparent, as shown in Figure 5.21, or even affecting final product functionality.

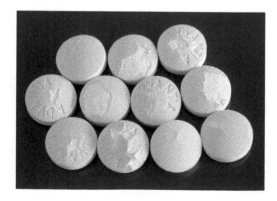

FIGURE 5.20 (**See color insert.**) Example of film peeling and flaking.

TABLE 5.2
Influence of Tablet Core Ingredients on Film Adhesion

Excipient	Measured Adhesion (kPa)
Microcrystalline cellulose	65.4 ± 6.3
Sucrose	44.5 ± 7.9
Anhydrous lactose	51.2 ± 4.3
Spray-dried lactose	24.5 ± 2.7
Dextrose	33.5 ± 9.5
Dicalcium phosphate (dihydrate)	29.9 ± 3.4

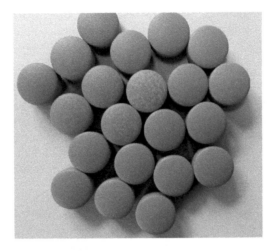

FIGURE 5.21   (See color insert.) Example of tablet-to-tablet color variation. (© 2015 AstraZeneca, all rights reserved. Image reproduced with permission of AstraZeneca; image is copyright of AstraZeneca and may not be reproduced without written permission.)

Tablet color variation may be attributed to many factors, including the following:

- Insufficient amount of coating applied (which becomes more readily apparent when coating tablets of different sizes).
- Coating process efficiency (a measure of how much actual coating is deposited onto the tablets compared to the amount of coating that is actually sprayed onto the batch) is low, thus directly influencing the amount of coating actually deposited.
- The application of the coating, in terms of uniformity of distribution of the coating fluid, is compromised so that each tablet is prevented from being presented to the spray zone an adequate number of times.
- The hiding power of the coating is inadequate (usually certain colors, including darker colors, are more problematic in this regard).

The problem of coating-process efficiency can be dealt with by making sure that overspray onto the pan walls and baffles (typically a combination of gun placement and overuse of pattern air) is minimized and that process conditions that lead to more spray drying (a thermodynamic issue involving spray application rate, airflow, and drying air temperature) are effectively dealt with. If, ultimately, achieving good coating process efficiency (typically > 90% and ideally > 95%) for the equipment being used is difficult, then allowances for this inadequacy can be made by applying an overage of coating solution or suspension to take into account the amount of coating material that will not end up deposited onto the tablet surfaces.

TABLE 5.3

Illustration of the Impact of Tablet Size on Batch Surface Area

Mean Tablet Core Weight (mg)	Individual Tablet Surface Area (mm²)	Total Surface Area of Batch (mm²)[a]	Target Amount of Coating Required per Unit Surface Area (mg mm⁻²)[b]	Coating Weight Gain Required (% w/w)[c]
300	1.954	65,135	4.61	3.0
150	1.242	83,800	4.61	3.8
75	0.780	104,000	4.61	4.8

[a] For a 10 kg batch of tablets.

[b] Calculated based on a 3.0% weight gain for the 300 mg tablets and assuming all tablet sizes require same amount of coating per unit surface area.

[c] Weight gain required to achieve target amount of coating per unit surface area.

Process factors that can lead to improvements in color uniformity involve increasing pan speed to increase the number of presentations to the spray during the coating process and applying the coating in a more dilute form, which effectively reduces the amount of coating deposited during each pass in the spray zone while extending the coating process and hence increasing the number of passes.

Defining the correct amount of coating that needs to be applied to ensure good color uniformity is often an issue that is not effectively managed. Being creatures of habit, most pharmaceutical scientists will make a selection based on a target percentage weight gain (usually 3%). This common habit has two fundamental flaws, namely the following:

- The amount of coating required is actually defined by coating thickness per tablet (often expressed on a batch-wise basis of weight quantity of coating per unit area of tablet surface in the batch); thus, the amount of coating required needs to be adjusted to reflect a change in surface area (for the batch) that occurs when tablets of different sizes and shapes are to be coated. Data shown in Table 5.3 indicate how surface area per batch changes as tablet size changes.
- Additionally, the amount of coating required is impacted by the "hiding power" of the coating. This characteristic will vary from color to color, depending on colorant concentration in the actual coating formulation, and the relative opacity of pigments used (lighter colors tend to exhibit better "hiding power" than darker ones) as well as the light absorption characteristics of those pigments. A useful way of determining the impact of pigments is in terms of their contrast ratios (see Table 5.4), where pigments exhibiting higher contrast ratio values exhibit better hiding power. Unfortunately, color selection is often not the province of the formulator, but rather that of the commercial department, so color selection may not be ideal in terms of covering power, thus requiring some adjustment in the amount of coating that needs to be applied to achieve good visual color uniformity.

TABLET TWINNING

During the application of film coatings and at a critical stage during the drying process, polymer-based coatings can become quite adhesive (a rheological phenomenon). If, at this point, two similarly adhesive surfaces come into contact, the tablets can become stuck together. For regularly shaped tablets, this often results in a problem known as picking (see later discussion); however, when tablets have flat surfaces (especially with capsule-shaped tablets that have deep side walls), the tablets will likely remain "glued" together (see Figure 5.22), creating an issue known as twinning.

While there are certain coating process changes that can be made to obviate this problem, including increasing pan speed to shorten the dwell time in the spray zone (thus reducing over-wetting),

TABLE 5.4

Examples of the Contrast Ratios of Pigment Commonly Used in Film Coating Formulations (Pigment Content in Coating, 16% w/w)

Pigment	Refractive Index[a]	Contrast Ratio (%)
None	1.48	33.3
Titanium dioxide	2.49–2.55	91.6
Red iron oxide	2.94–3.22	99.5
Yellow iron oxide	1.90–2.50	98.4
Black iron oxide	2.40	99.6
FD&C Blue # 2 lake	1.50–1.54	97.5
FD&C Yellow # 6 lake	1.50–1.54	70.1
FD&C Yellow # 5 lake	1.50–1.54	62.9

[a] Pigments with a high refractive index tend to have high contrast ratios because of their opacifying effects; pigments with refractive indices that match those of the polymer film exhibit their hiding power through light absorption effects.

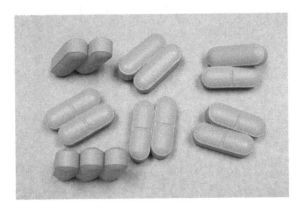

FIGURE 5.22   Example of tablet twinning.

improving the drying conditions (such as increasing drying air temperatures or volumes and/or reducing spray rates) or expanding the spray zone (by increasing the number of spray guns or increasing pattern air) to reduce the risk of localized over-wetting (see Figure 5.23), proactively dealing with the issue during product development provides a more effective solution.

One approach that can be very effective deals with core design. For example, if a capsule-shaped tablet is required, changing from the more traditional design to one in which there is a small degree of curvature on the side walls of the tablets (see Figure 5.24) can reduce the area of contact between tablets and greatly minimize the risk of twinning.

Additionally, while all film coatings will develop some degree of adhesiveness during the drying process, not all coating formulations are equal in this regard. For example, film coatings based on poly (vinyl) alcohol as well as delayed-release coating based on acrylic polymers are well known for their increased tackiness behaviors. Thus, when these kinds of polymers are required, the use of anti-adhesive agents (such as lecithin, talc, glyceryl monostearate, magnesium stearate) in the coating formulation can be quite effective in reducing their tackiness.

TABLET PICKING

Picking, as suggested in the previous section, is essentially a companion problem to twinning, but in this case, tablets tend to break apart as a result of the constant motion within the coating pan,

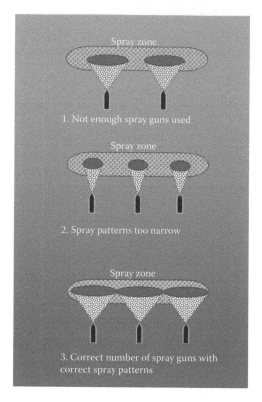

FIGURE 5.23   Illustration of how effective spray gun setup might reduce localized overwetting.

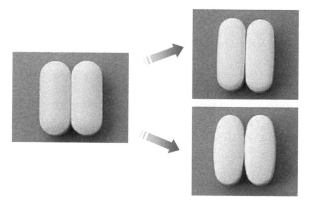

FIGURE 5.24   Example of how small modifications to tablet shape can minimize the risk of twinning.

leaving some tablets exhibiting a discontinuity in the film coating (see Figure 5.25), and other tablets on which a piece of film coating (that has been removed from another tablet) can be clearly seen sticking to the tablet surface. Picking is primarily an over-wetting issue [15] that results in increasing adhesiveness of the coating being applied.

Quite simply, the remedies for resolving this particular problem are essentially the same as those discussed under tablet twinning although additional resolution with respect to changing tablet shape may also include increasing face curvatures to minimize points of contact.

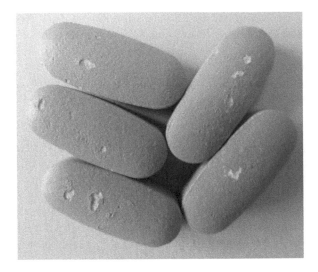

FIGURE 5.25   **(See color insert.)** Example of tablets exhibiting picking.

FILM CRACKING

Cracking of applied film coatings represents a catastrophic condition in which the continuity of the film is disrupted. Cracking defects can be macroscopic or microscopic in nature (see Figure 5.26). Although the condition is clearly unacceptable even for aesthetic film coatings, the situation becomes even more critical when the applied film coating has an important function, such as modifying drug release or protecting the tablet core from detrimental environmental conditions, such as moisture or oxygen. Rowe and Forse [16] have provided a fundamental explanation of this problem, looking at its practical implications.

The source of this problem can be found in all aspects of film coating, namely, the tablet core, the coating formulation, and the coating process. For example, a tablet core that expands as a result of exposure to coating process conditions, a film coating that is insufficiently robust either to withstand the stresses created through such expansion or flexible enough to contain that expansion, and a coating process that either exacerbates core expansion or negatively impacts the mechanical properties of the applied coating will all play a role in creating this defect.

Thus, adjusting coating process conditions to resolve or, at least, minimize these kinds of problems has an important role to play. For example, positive improvements can be achieved when

- Reducing processing temperatures to alleviate core expansion due to thermal effects. Additionally, such a course of action may benefit film robustness by preventing the coating from becoming too dry, which could make the coating more brittle.

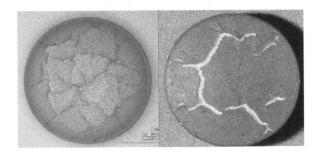

FIGURE 5.26   **(See color insert.)** Examples of cracking of film coatings.

- Improving the drying conditions within the coating process, and thus mediating against further core expansion as a result of swelling due to moisture uptake.
- Reducing pan speeds to minimize stress that could impact a coating the robustness of which may already be suspect.
- Ensuring that pigments are well dispersed during preparation of the coating suspension, so they do not become stress concentrators in the coating, thus initiating the cracking process (see Figure 5.27).

Okutgen et al. [17] provided interesting insight into how coating process conditions might impact issues relating to tablet core expansion, detailing how the characteristics of the coating formulation used and the coating process conditions employed might be managed in order to minimize these kinds of problems.

Because of the catastrophic nature of film cracking, the most effective approaches involve designing tablet core formulations that minimize the risk of significant expansion behavior occurring and create film-coating formulations that are suitably robust to accommodate any core expansion that might occur.

For example, when preparing tablet formulations, avoiding the use of ingredients that are more likely to be associated with core expansion issues, especially when the tablets are exposed to typical coating process conditions, is a very proactive approach. Rowe [18] has described how inorganic fillers (such as calcium carbonate) can increase film cracking as a result of core thermal expansion. Conversely, tablets that are formulated with large quantities of hydrophilic materials (such as microcrystalline cellulose) are likely to undergo significant swelling if moisture penetrates into the core as a result of inadequate drying in the process.

When it comes to designing suitable film-coating formulations, producing coatings that have high film strength [19] or have greater flexibility as a result of effective plasticization [11] will ultimately help to eliminate the risk of film cracking. Both of these approaches certainly combine synergistically with the adoption of appropriate process conditions to eliminate any possibility that film cracking will be a problem.

It may also be preferable to utilize a coating formulation that can be sprayed at higher solids (say 20% w/w) because this will enable the water from the coating suspension to be removed more effectively by the drying environment, thus minimizing penetration into the tablet cores and subsequently inducing swelling.

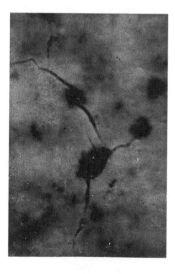

FIGURE 5.27   Example of cracking initiated by poorly dispersed pigments.

TABLET ROUGHNESS

The process of taking a relatively viscous coating suspension and spray atomizing it under conditions in which drying has to be completed in an extremely short time inherently leads to a result in which some droplets of coating liquid are so viscous at the point of impact with a tablet surface that they are unable to coalesce into a smooth film. The result is that the applied coating exhibits many of the characteristics of orange peel. Hence the term "orange peel" is often applied to the surface characteristics of pharmaceutical film coated tablets.

Thus, some degree of roughness is always evident with film-coated tablets. The real question is "When does the level of roughness become unacceptable?" The photograph shown in Figure 5.28 exemplifies what might be considered unacceptable roughness.

Although rough tablets are commonly associated with ineffective atomization of the coating liquid and too rapid drying of the atomized droplets, excessive roughness may also be caused by ineffective drying whereby tablets that are tacky, instead of exhibiting the picking problems described earlier, may well exhibit a "pulling" problem (see Figure 5.29) causing the tablets also to become rough. This kind of tackiness issue, while typically being related to poor drying, may also be influence by the nature of the coating polymer used.

FIGURE 5.28   **(See color insert.)** Example of unacceptably rough film-coated tablets.

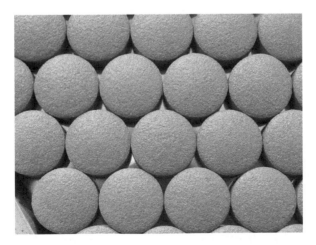

FIGURE 5.29   **(See color insert.)** Example of tablets where roughness of the coating is caused by overwetting. (© 2015 AstraZeneca, all rights reserved. Image reproduced with permission of AstraZeneca, image is copyright of AstraZeneca and may not be reproduced without written permission.)

To reduce coated tablet roughness, coating process adjustments that might be considered are the following:

For over-wetting conditions
- Increasing pan speed to reduce dwell time in the spray zone
- Reducing spray rate and/or increase drying air temperatures and volumes to improve drying
- Ensuring that all spray guns (in a multi-gun setup) are spraying at the same rates and that atomization is not impeded (by, for example, material buildup on the exterior surfaces of the spray nozzles) to the extent coating can accumulate and drip onto the tablets.

For over-drying conditions
- Increasing spray rates or reducing drying air temperatures
- Reducing coating solids (to reduce viscosity and thus improve atomization)
- Readjusting the balance between spray rate and atomizing air to achieve better atomization of the coating liquid

Of course, in each of these cases, the adjustments recommended may fall outside compliance with the registered process, so they likely should be considered to be part of the original process design process.

Creating a film-coating formulation to mitigate against many of the processing limitations described earlier can be an effective course of action. Examples include the following:

- Designing a coating formulation with low viscosity characteristics (see Figure 5.30) to achieve better atomization of the coating liquid, and at the same time, enabling that liquid to be sprayed at higher solids to minimize over-wetting conditions when the coating processes available may exhibit less than optimal drying characteristics.
- Electing to use a coating formulation that is less tacky (either through a reduction in plasticizer or restricting the use of polymers known to be inherently more tacky) are both approaches that can reduce roughness (see Figure 5.31).

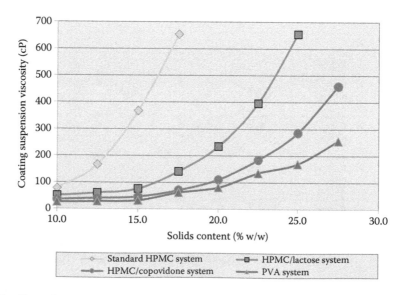

FIGURE 5.30 (See color insert.) Illustration of how adjustment of the coating formulation can be used to reduce the viscosity of the coating liquid.

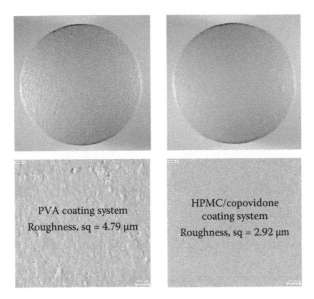

FIGURE 5.31 **(See color insert.)** Example of how changing from a PVA-based coating system to an HPMC/copovidone hybrid system can reduce surface roughness.

TABLET DISCOLORATION

This defect may be expressed in many ways. For example, interaction of ingredients in the core, often in the presence of moisture derived from the coating process, can create dark spots that remain visible through the coating (often seen with multivitamin tablets as shown in Figure 5.32). In a similar vein, bleeding of ingredients from the tablet core, through the coating (as shown in Figure 5.33), again an issue with nutraceutical products, can also be a factor. Finally, poor dispersion of pigment ingredients within the coating formulation can also be a source of tablet discoloration; in which case, this must be dealt with at the time of preparation of the coating suspension to make sure pigments are effectively dispersed into the final coating suspension. For coating formulations prepared from powdered premixes made by a third party vendor, the issue of ineffective pigment dispersion must be dealt with by those vendors.

Minor processing remedies, such as reducing spray rates, increasing processing temperatures, or increasing pan speed to reduce dwell times in the spray zone can reduce over-wetting and thus prevent interaction between tablet core ingredients (such as that between iron salts and ascorbic acid).

FIGURE 5.32 **(See color insert.)** Tablet discoloration caused by interaction of ingredients in the tablet core.

FIGURE 5.33 Tablet discoloration caused by bleed–through of ingredients from tablet core.

Similarly, when ingredients that become liquid and bleed through the coating do so as a result of high processing temperatures, an increase in spray rate and/or reduction in processing temperature can be helpful.

However, potential discoloration issues are more effectively dealt with during product development, especially when past experience suggest that this kind of problem is likely to occur in the current product being developed. For example, designing coating formulations (such as high solids coating formulations) that can minimize moisture penetration into the tablet core during application of the coating (as shown in Figure 5.34) can effectively reduce or eliminate the problem. Alternatively, if discoloration occurs on storage, use of a moisture barrier coating (see Figure 5.35) can also be effective.

TABLET SCUFFING

Tablet scuffing (see Figure 5.36) is generally manifested as the appearance of gray marks on the surfaces of film-coated tablets, especially when using white or light-colored (pastel) coating formulations. Although the problem has been attributed to many causes, it is generally accepted that the abrasive nature and levels of titanium dioxide used in such coating formulations cause metallic material to be removed from the surfaces of the stainless steel coating equipment. Rosoff and Sheen [20] have commented on this phenomenon, linking it to polymorphism that exists with different forms of this pigment. Rowley [21] conducted an investigation into the various causes of such scuffing problems and confirmed that the main issue is related to the level of titanium dioxide used.

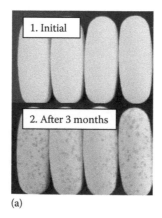
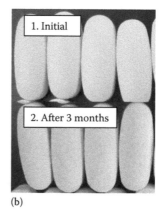

(a)   (b)

FIGURE 5.34 (a) Standard HPMC coating (12.5% w/w solids) and (b) HPMC/copovidone coating (20.0% w/w solids). Example of how a high solids coating can be used to minimize discoloration problems.

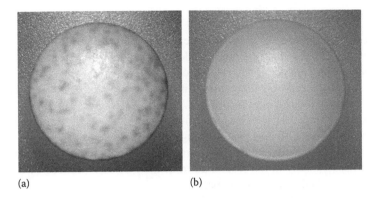

(a)   (b)

FIGURE 5.35   (a) Standard HPMC coating and (b) moisture barrier coating. Reducing discoloration through use of a film coating with moisture barrier properties.

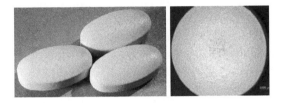

FIGURE 5.36   **(See color insert.)** Examples of surface scuffing of film-coated tablets.

While ensuring that the surfaces of the coating equipment are perfectly clean, an approach that can be augmented by pan passivation, is a relatively simple approach that can help reduce scuffing problems, a more proactive formulation approach can provide a long-term solution. Ogasawara et al. [22] demonstrated how reducing the levels of titanium dioxide used in the coating formulations can be quite effective; however, this approach will also reduce the opacity of the coating formulation, necessitating the use of higher levels of applied coating to achieve complete tablet core coverage.

Scuffing has, at various times, been associated with the use of coating formulations based on poly (vinyl alcohol) (PVA). As can be seen in the example shown in Figure 5.37, the use of hybrid polymer film coatings based on combinations of HPMC and copovidone can provide an effective solution.

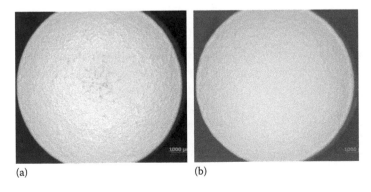

(a)   (b)

FIGURE 5.37   **(See color insert.)** (a) PVA coating and (b) HPMC/copovidone coating. Example of how changing the polymers used in the coating formulation can eliminate scuffing problems.

COATED TABLET DISSOLUTION ISSUES

While film coatings are applied to pharmaceutical oral solid dosage forms for various reasons, most (often designated IR coatings) are not intended to have any influence on drug release. Others (including delayed release and extended release coatings), however, have a specific intent in that regard. Irrespective of the type of film coating involved, none are expected to have any adverse effect on drug release from the dosage form.

Even so, unexpected drug release issues after film coating can occur. Rarely, however, are these issues directly related to the film coat itself (through chemical interaction with an ingredient in the core); rather, the problem usually stems from some aspect of the coating process that influences either drug solubility (typically in the case of an immediate-release product) or film coating structure (especially for modified-release products).

For example, for an immediate-release product, ibuprofen tablets (because of the relatively low melting point of this API) are well known to exhibit dissolution problems when exposed to high processing temperatures. The data shown in Figure 5.38 illustrate how processing conditions (and, to some extent, type of coating system used) can influence drug dissolution from film-coated ibuprofen tablets. All coating systems shown were sprayed at 20% w/w solids except the standard HPMC-based system, which was sprayed at 12.5% w/w solids. The temperatures shown in this figure refer to the product temperatures recorded for each coating process used. As indicated, using a high solids coating system that facilitated the use of lower processing temperatures helped to offset the impact of processing conditions on drug release.

In the case of modified-release products, many coating formulations employ polymer mixtures (a water-insoluble polymer as the primary release modifier, and a water-soluble pore-forming polymer to moderate drug release). Marucci et al. [23] have shown that when using ethylcellulose as the primary insoluble polymer and hydroxypropylcellulose as the water-soluble pore former, coating process conditions can influence the degree to which phase separation can occur with a resulting impact on the size of pores created within the film structure and hence drug release. Similar behavior has been reported by Beissner et al. [24].

Haaser et al. [25] have described how Terahertz Pulsed Imaging can be used to identify variations in film porosity and correlated this to differences in drug release from batches of tablets coated on different processing scales. In each case, the amount of coating applied was the same with differences in drug release being attributed to differences in film density as a result of variations in coating process conditions employed.

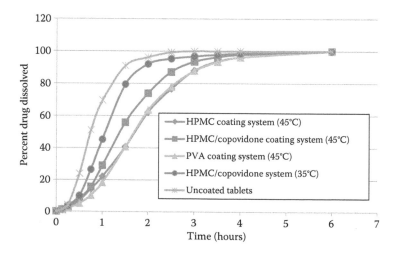

FIGURE 5.38 Data showing the influence of both coating system type and processing temperatures on the dissolution behavior of ibuprofen tablets.

In the case of aqueous coating processes, because modified-release polymers are, by definition, water insoluble, most of the polymers employed are in the form of a latex or polymer dispersion, one characteristic of which is that the final coating is formed through a process of coalescence. The rate and extent to which this occurs are very much impacted by process thermodynamics. Failure to achieve optimal process conditions can result in a process known as "further gradual coalescence" occurring on storage. As this process impacts film structure, drug release behavior can change over the course of time as described by Yang et al. [26].

COATED TABLET STABILITY ISSUES

Film-coating processes, especially those that employ water as the solvent/vehicle for the coating system being used, are inherently stressful processes that expose the product to elevated humidity and temperature for a considerable period of time (typical production scale processes may last from 2 to 4 hours). Understandably, therefore, the risk to product stability is ever present. Even so, stability problems resulting from products being coated with an aqueous-based polymeric film are rare and are likely associated with either moisture penetrating into the tablet cores as the coating is being applied or exposure of those cores to the higher processing temperatures being used.

An example in which moisture penetration into the tablet core during application of the coating material can cause a stability problem was shown earlier in Figure 5.32. In this example, use of a high solids film coating system was shown to obviate this particular problem.

Although this is an example of a chemical stability problem, physical stability issues may also arise; for instance, tablet formulations containing an API in the form of a solid dispersion (designed to enhance drug solubility) may find that failure to control moisture ingress into the tablet core during film coating can increase the risk of recrystallization of the API (thus reducing solubility) on storage. Using a high solids coating formulation to reduce the amount of water present, while improving drying effectiveness, can help to manage this kind of problem.

When the API is heat sensitive, use of coating formulations that enable the coating process to be operated at lower processing temperatures (as depicted in Table 5.5) can effectively be used to minimize stability problems.

TABLE 5.5
Example of How Use of a High Solids Film Coating Can Facilitate the Use of Processing Temperatures Significantly Below those Typically Used in Aqueous Film-Coating Processes

Parameter	Trial A	Trial B	Trial C	Trial D
Inlet temperature (°C)	40.0	40.0	45.0	53.0
Spray rate (g min^{-1})	35.0	25.0	20.0	40.0
Coating solids (% w/w)	25.0	20.0	20.0	20.0
Inlet dew point (°C)	10.0	10.0	10.0	10.0
Bed temperature (°C)[a]	26.0	29.8	35.6	35.3
Bed humidity (% RH)[a]	46.4	39.3	28.2	33.3
Coated tablet photographs				

[a] Measured using Pyrobuttons.

In terms of reducing the exposure times of products to the stressful conditions typically imposed by aqueous film-coating processes, using high solids coating systems (at, say, 20%–25% w/w solids) instead of more conventional coating systems (typically applied at 10%–15% w/w solids), can reduce coating process times and hence exposure times by up to 50%. Alternatively, using continuous film-coating processes can also minimize exposure times because typical residence times in continuous processes are on the order of 15–20 minutes, compared to batch processes where residence times in excess of 2 hours (depending on batch size) are quite common.

Overall, potential tablet stability issues can be avoided by effectively addressing process thermodynamics and selecting an appropriate coating system during product and process development that facilitates more effective moisture removal through enhanced drying characteristics.

EFFECTIVE TOOLS AND APPROACHES FOR DEALING WITH PROBLEMS IN FILM COATING

So far in this chapter, all of the discussion has centered on problems that can occur when oral solid dosage forms are film-coated and how to deal with them. Unfortunately, many such problems tend to show up on the production scale when real opportunities for correction are severely limited by regulatory constraints. Clearly, a proactive approach—essentially dealing with problems by not creating them in the first place—provides the best means for troubleshooting aqueous film-coating processes. The core elements of film-coating processes that impact product quality are the following:

- Process thermodynamics (essentially, the balance between the rate at which the coating liquid is applied and the rate at which the solvent is removed from the processing environment) when two critical factors, heat and humidity, have the greatest impact on product quality
- Mechanical stress, the least predictive factor moving from the laboratory into the production settings
- Uniformity of distribution of the coating material
- Coating structure (especially for functional film coatings) and the impact of the coating formulations and coating processes used on this important attribute

As regulatory agencies impose, to an ever growing degree, the need to implement the principles of quality by design (QbD), this forces greater attention to be placed on the product and process development processes, ensuring that a truly proactive approach is taken to avoiding common problems that have plagued us in the past.

Still, film-coating processes often remain a challenge simply because we still lack a complete understanding of critical elements of this process and how these ultimately impact product quality. One such challenge has involved our ability for accurately determining when the required amount of coating has been applied. Classic methods involving weighing tablets as the coating is being applied (inaccurate because the impact of changing moisture content and the statistical invalidity of such weight methods because of the small sample size) and applying a fixed amount of coating suspension (a technique negatively impacted by variable process efficiencies) are clearly inadequate.

That said, in recent years, many predictive modeling techniques have been used to gain a more concrete understanding of the film-coating process. For example,

- Gaining a better understanding of process thermodynamics, as suggested by Page et al. [27], and Choi [5]
- Gaining insight into the uniformity of distribution of applied film coatings using Monte Carlo simulation as suggested by Pandey et al. [28] and inter-tablet coating variability using a discrete element method (DEM) as described by Kalbag et al. [29]

- Understanding the impact of atomization conditions on pharmaceutical film coating liquids as discussed by Muliadi and Sojke [7]
- Using so-called expert systems as outlined by Rowe and Upjohn [30], which have been applied, *inter alia*, to identify and resolve defects associated with pharmaceutical film-coated tablets
- Employing modeling techniques to predict tablet content uniformity (for an API applied in the coating) and determination of critical process parameters associated with the same [31]

A prime element of any QbD process involves identifying critical process variables (CPV) and establishing their impact on critical quality attributes (CQA). This can be a challenge when it comes to the film-coating process, but it is a challenge that can be overcome by removing the "art" from the process and looking more at the underlying scientific principles governing this process. A systematic approach, as outlined in Figure 5.39, can provide an effective means to achieve a suitable outcome.

Of course, another key element of any QbD approach, once a suitable design space has been created, is to implement an effective means of process monitoring to ensure that the process is operated within the constraints defined by the established design space. This typically means the use of suitable process analytical technology techniques (PAT). While suitable instrumentation for this purpose is available, its application to film-coating processes can be challenging for these reasons:

- The environment inside a typical film coating process is "dirty," meaning a lot of moisture and dust is potentially present that can interfere with measurements being taken by sophisticated instrumentation.
- The movement of product within the process is rapid, making effective data acquisition more challenging.

Nonetheless, several suitable PAT techniques have been successfully applied to film-coating processes as described in a review by Knop and Kleinebudde [32], including use of the following:

- FT nIR techniques [33] to monitor film-coating processes on a manufacturing scale. In one case, the technique was used to monitor tablet moisture content and intra- and inter-tablet coating variability.
- Pyrobuttons to achieve a real-time monitoring of the thermodynamic environment in a pan coater as described by Pandey and Bindra [34] and Porter et al. [35].

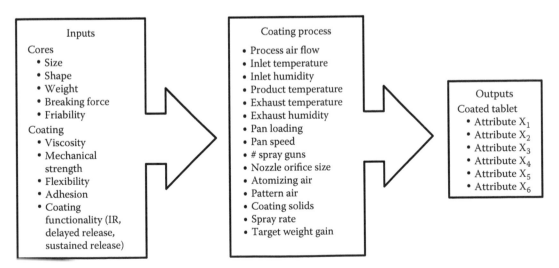

FIGURE 5.39   Film coating: analysis of critical process steps.

- Terahertz pulsed imaging as an in-line technique to monitor the weight gain of an applied film coating to tablets in a side-vented pan coater [36]. In one example, the impact of the environment within the coating process was avoided by taking measurements outside of the coating pan, gaining access to the tablet surfaces through the perforated parts of the pan wall. In this case, the technique was found to be more suitable when dealing with coating processes that involve application of high levels of coating materials (as in the case of modified-release film coatings).
- Raman spectroscopy to monitor the amount of coating being applied and to determine the process end point [37].

In summary, aqueous film-coating processes continue to produce many challenges, and the formulation scientist must determine suitable approaches to resolve them. The implementation of a proactive approach that acknowledges such problems focuses on problem avoidance during product and process development and provides the most effective means of dealing with a subject that for so long has plagued the production of film-coated products.

REFERENCES

1. R. C. Rowe. The cracking of film coatings on film coated tablets—A theoretical approach with practical implications. *J. Pharm. Pharmacol.* 33, 423 (1981).
2. R. C. Rowe. A reappraisal of the equations used to predict internal stress of film coatings applied to tablet substrates. *J. Pharm. Pharmacol.* 35, 112, (1983).
3. R. C. Rowe & S. F. Forse. Bridging of intagliations on film coated tablets. *J. Pharm. Pharmacol.* 34, 277 (1982).
4. G. C. Ebey. A thermodynamic model for aqueous film coating. *Pharm. Technol.* 7 (11), 40. (1987).
5. M. Choi. Application of process thermodynamics in pharmaceutical coating. *Tablets & Capsules.* April (2007).
6. D. Suzzi, R. Radl, & J. G. Khinast. Local analysis of the tablet coating process: Impact of operation conditions on film quality. *Chem. Eng. Sci.* 65, 5699 (2010).
7. A. R. Muliadi & P. E. Sojke. Spatially Resolved Characteristics of Pharmaceutical Sprays. 11th International Annual Conference on Liquid Atomization and Spray Systems, Vail, Colorado, July 2009.
8. S. Kim, A. Mankad, & P. Sheen. The effect of application rate of coating suspensions on the incidence of bridging on film coated tablets. *Drug Dev. & Ind. Pharm.* 12, 801 (1986).
9. R. C. Rowe & S. F. Forse. The effect of process conditions on the incidence of bridging of the intagliations and edge splitting and peeling on film coated tablets. *Acta. Pharm. Technol.* 28, 207 (1982).
10. R. C. Rowe & S. F. Forse. The effect of intagliation shape on the incidence of bridging on film coated tablets. *J. Pharm. Pharmacol.* 33, 412 (1981).
11. R. C. Rowe & S. F. Forse. The effect of plasticizer type and concentration on the incidence of bridging of the intagliations of film coated tablets. *J. Pharm. Pharmacol.* 33, 174 (1981).
12. E. Okutgen, J. E. Hogan, and M. E. Aulton. Effects of tablet core dimensional stability on the generation of internal stresses within film coats. III. Exposure to temperatures and humidities which mimic the film coating process. *Drug Dev. & Ind. Pharm.* 17, 2005 (1991).
13. G. R. B. Down. An alternative mechanism responsible for bridging of intagliations on film-coated tablets. *J. Pharm. Pharmacol.* 34, 281 (1982).
14. R. C. Rowe & S. F. Forse. Pitting—A defect on film coated tablets. *Int. J. Pharm.* 17, 347 (1983).
15. S. C. Porter. Tablet coating—Problems with film coating. *Drug Cosmet. Ind.* 129 (9), 50 (1981).
16. R. C. Rowe & S. F. Forse. The cracking of film coatings on film coated tablets—A theoretical approach with practical implications. *J. Pharm. Pharmacol.* 33, 423 (1981).
17. E. Okutgen, J. E. Hogan, & M. E. Aulton. Effects of tablet core dimensional instability on the generation of internal stresses within film coats. III. Exposure to temperatures and relative humidities which mimic the film coating process. *Drug Dev. & Ind. Pharm.* 17, 2005 (1991).
18. R. C. Rowe. The expansion and contraction of tablets during film coating—A possible contributing factor in the creation of stresses within the film? *J. Pharm. Pharmacol.* 32, 851 (1980).
19. R. C. Rowe & S. F. Forse. The effect of polymer molecular weight on the incidence of film cracking and splitting of film coated tablets, *J. Pharm. Pharmacol.* 583, 52 (1980).

20. M. Rosoff & P.-C. Sheen. Pan abrasion and polymorphism of titanium dioxide in coating suspensions. *J. Pharm. Sci.* 72 (12) 1485 (1983).
21. F. A. Rowley. Toward a greater understanding of the scuffing defect observed on film-coated tablets. Tablets & Capsules (April 2012).
22. Y. Ogasawara, R. Steffenino, & C. Cunningham. Scuffing measurement methodology and improved film coating systems. Poster presented at Annual AAPS Convention November 2008.
23. M. Marucci, J. Arnehed, A. Jarke, H. Matic, M. Nicholas, C, Boissier, & C. von Corswant. Effect of the manufacturing conditions on the structure and permeability of polymer films intended for coating undergoing phase separation. *Eur. J. Pharm. & Biopharm.* 83, 301 (2013).
24. B. Beissner, S. C. Porter, R. Hach, K. Karan, & J. Lian. Extended-release venlafaxine pellets: Influence of coating process conditions. Poster presented at AAPS Annual Convention, October 2015.
25. M. Haaser, K. C. Gordon, C. J. Strachan, & T. Rades. Terahertz pulsed imaging as an advanced characterization tool for film coating—A review. *Int. J. Pharm.* 457, 510 (2013).
26. Q. W. Yang, M. P. Flament, F. Siepmann, V. Busignies, B. Leclerc, C. Herry, P. Tchoreloff, & J. Siepmann. Curing of aqueous polymeric film coatings: Importance of coating level and type of plasticizer. *Eur. J. Pharm. & Biopharm.* 74, 362 (2010).
27. S. Page, K.-H. Baumann, & P. Kleinebudde. Mathematical modelling of an aqueous film coating process in a Bohle lab coater: 1 development of the model. *AAPS Pharm SciTech.* 7 (2), E1 (2006).
28. P. Pandey, M. Katakdaundi, & R. Turton. Modelling weight variability in a pan coating process using Monte Carlo simulation. *AAPS Pharm SciTech.* 7 (4), E1 (2006).
29. A. Kalbag, C. Wassgren, S. S. Penumetcha, & J. D. Perez-Ramos. Inter-tablet coating variability; residence times in a horizontal pan coater. *Chem. Eng. Sci.* 63, 2881 (2008).
30. R. C. Rowe & N. G. Upjohn. Formulating pharmaceuticals using expert systems. *Pharm. Technol. Int.* 5 (8), 46 (1993).
31. W. Chen, S.-Y. Chang, S. Kiang, A. Marchut, O. Lyngberg, J. Wang, V. Rao, D. Desai. H. Stamato, & W. Early. Modeling of pan coating processes: Prediction of tablet content uniformity and determination of critical process parameters. *J. Pharm. Sci.* 99, 3213 (2010).
32. K. Knop & P. Kleinebudde. PAT-tools for process control in pharmaceutical film coating applications. *Int. J. Pharm.* 457, 527 (2013).
33. C. V. Möltgen, T. Puchert, J. C. Menzies, D. Lochmann, & G. Reich. A novel in-line NIR spectroscopy application for monitoring of tablet film coating in an industrial scale process. *Talanta.* 92, 26 (2012).
34. P. Pandey & D. Bindra. Real-time monitoring of thermodynamic microenvironment in a pan coater. *J. Pharm. Sci.* 102 (2), 336 (2013).
35. S. C. Porter, L. Terzian, P. J. Hadfield, L. Pryce, C. Byers, & F. Lajko. Examination of the benefits of the use of high-solids film coating systems as a means of reducing the costs of aqueous coating processes. Poster presented at AAPS Annual Meeting November (2011).
36. R. K. May, M. J. Evans, S. Zhong, I. Warr, L. F. Galdden, Y. Shen, & J. A. Zeitler. Terahertz in-line sensor for direct coating thickness measurement of individual tablets during film coating in real time. *J. Pharm. Sci.* 100, 1535 (2011).
37. A. El Hagrasy, S.-Y. Chang, D. Desai, & S. Kiang. Application of Raman spectroscopy for quantitative in-line monitoring of tablet coating. *Am. Pharm. Rev.* 1, Jan/Feb (2006).

6 Adhesion of Polymeric Films

Linda A. Felton and James W. McGinity

CONTENTS

INTRODUCTION

Adhesion between a polymer and the surface of a solid is a major prerequisite for the film coating of pharmaceutical dosage forms [1–3]. Loss of adhesion may lead to an accumulation of moisture at the film–tablet interface, potentially affecting the stability of drugs susceptible to hydrolytic degradation [4]. Poor adhesion may also compromise the mechanical protection that the coating provides to the substrate [5]. In addition, experiments on adhesion are useful to the pharmaceutical scientist during formulation development to investigate the relationship between tablet excipients and polymeric film-coating formulations [6].

MAJOR FORCES AFFECTING FILM–SUBSTRATE ADHESION

The two major forces that have been found to affect polymer–substrate adhesion are (a) the strength of the interfacial bonds and (b) the internal stresses in the film. Hydrogen bond formation is the primary type of interfacial bonding between the tablet surface and polymer for pharmaceutical products [7]. To a lesser extent, dipole–dipole and dipole-induced dipole interactions also occur. Factors that affect either the type or the number of bonds formed between the polymer and the solid surface will influence film adhesion.

When a polymeric solution or dispersion is applied to a substrate, internal stresses inevitably develop within the film [8]. These stresses include stress due to shrinkage of the film as the solvent evaporates, thermal stress due to the difference in thermal expansion of the film and the substrate, and volumetric stress due to the change in volume when the substrate swells during storage. The total stress within a film is the sum of all the stresses acting on the polymer, and several researchers have developed equations to estimate total stress [8–11]. Equation 6.1, developed by Okutgen et al. [12], includes contributions of volumetric changes of the tablet core in addition to the other well-established mechanisms:

$$P = \frac{E}{3(1-v)}\left[\frac{\Phi_s - \Phi_r}{1 - \Phi_r} + \Delta\alpha_{cubic}\Delta T + \frac{\Delta V}{V}\right]$$

(6.1)

where P is the total internal stress in the film, E is the elastic modulus of the film, v is the Poisson's ratio of the polymer, Φ_s represents the volume fraction of the solvent at the solidification point of the film, Φ_r is the volume fraction of solvent remaining in the dry film at ambient conditions, $\Delta\alpha_{(cubic)}$ is the difference between the cubical coefficient of thermal expansion of the film coat and the substrate, ΔT represents the difference between the glass transition temperature of the polymer and the temperature of the film during manufacturing and storage, ΔV is the volumetric change of the tablet core, and V denotes the original volume of the tablet core. Although this equation has been derived for polymeric solutions, the theory is applicable to polymeric dispersions as well. It is apparent from Equation 6.1 that the total stress within a film is directly proportional to the elasticity of the polymer. Factors that influence the elastic modulus of the polymer will, therefore, affect internal stress and film–tablet adhesion.

METHODS USED TO ASSESS POLYMER ADHESION

A distinction must be made between "fundamental" and "practical" adhesion. Fundamental or "true" adhesion refers to the intermolecular interactions between the polymer and the substrate [13]. Practical or "measured" adhesion refers to the numerical value that results from a variety of testing methods, including shear and tensile tests. In addition to the interfacial interactions, other factors, such as stresses in the film and the adhesion measurement technique will influence measured adhesion [11]. No methods used to quantify polymer adhesion, however, can be directly used to measure fundamental adhesion.

The small size of the tablet and the non-uniform surface roughness of the substrate have presented significant challenges to the pharmaceutical scientist in determining the adhesive properties of a polymer [14,15]. The earliest method for assessing adhesion of thin polymeric films to surfaces was the "Scotch tape" test [16], where a piece of adhesive tape was applied to the film surface and then peeled off. The film either adhered to the solid surface or was removed with the adhesive tape. This method was obviously qualitative in nature and did not provide an accurate measurement of polymer adhesion.

Another method that has been used to provide qualitative information regarding adhesion of polymers to pharmaceutical solids is diametral compression of coated substrates [17]. During compression experiments, the total load will be distributed between the film coating and the solid substrate [5]. The simultaneous fracture of the coating and the substrate is indicative of good adhesion between the polymer and the solid [17,18].

The first quantitative adhesion test was developed by Heavens in 1950 and was known as the "scratch test" [19]. In this technique, the tip of a hard stylus is drawn across the surface of the film. The critical load required to completely detach the film from the substrate along the track of the scratch is determined and related to polymer adhesion. Although it is used extensively to study the adhesion of films cast onto metal surfaces, this method is unsuitable for pharmaceutical systems due to the relative rough surface of tablet compacts [20].

In the 1970s, the peel test was a popular method for the determination of film adhesion to tablets. The peel test uses a modified tensile tester to peel the film from the surface of the tablet at a 90° angle [21]. The primary deficiency of this method is that the peel angle measured at the tablet surface is dependent on the elasticity of the film and the uniformity of adhesion, both of which can produce significant deviations in the data [15].

Several variations of the butt adhesion technique have been reported in the pharmaceutical literature over the past 20 years [22–26]. This method is similar to the peel test. However, the entire film is removed normal to the surface of the tablet rather than sections of the film being peeled. The butt adhesion technique eliminates variations due to the elasticity of the film and is less influenced by the uniformity of adhesion. The experimental setup requires that the film coating around the edge of the tablet be removed using a scalpel. Next, the tablet is affixed to a lower, stationary platen. Double-sided adhesive tape is placed between the tablet surface and the upper platen. Rubber backing may be used to ensure adequate contact. A uniform displacement rate should be used to remove the film from the substrate [27].

In 1980, Rowe [27] investigated the rate effects on measured adhesion of film coatings. Small increases in measured adhesion were found when the crosshead speed was increased from 0.1 to 1.5 mm/second whereas decreased adhesion resulted when the deformation rates were increased above 1.5 mm/second as shown in Figure 6.1. The rates of deformation influence the rheological behavior of different components in the system, including the adhesive tape as well as the polymer itself, and therefore, they affect how the applied stress is transmitted and distributed at the film–tablet interface [27]. Higher rates of deformation resulted in an uneven stress distribution, thus lowering the measured adhesion.

Felton and McGinity [23] used a Chatillon digital force gauge and motorized test stand to conduct butt adhesion experiments. The apparatus was connected to a personal computer, and force–deflection diagrams were constructed from the data, which permitted the visualization of the development of the force within the sample during the adhesion experiments. An example of a force–deflection diagram generated from this equipment is shown in Figure 6.2.

The profile is similar to the stress–strain diagram commonly generated in the tensile testing of free films. From the force–deflection diagrams, the elongation at adhesive failure, the modulus of adhesion, and the adhesive toughness of the polymer in addition to the force of adhesion can be determined. The elongation at adhesive failure, analogous to the elongation at break obtained from tensile testing of free films, reflects the ductility of the polymer. The adhesive toughness is calculated as the area under the force–deflection diagram and is equal to the work required to remove the film from the surface of the solid.

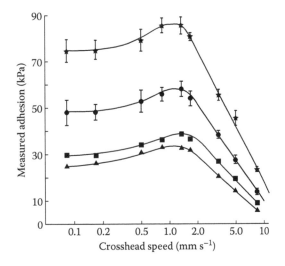

FIGURE 6.1 The effect of cross-head speed on measured adhesion of an organic-based cellulosic film. (★) Microcrystalline cellulose tablet core, 18 μm film thickness (Pharmacoat® 606); (●) microcrystalline cellulose tablet core, 70 μm film thickness (Pharmacoat 606); (■) lactose tablet core, 35 μm film thickness (Pharmacoat 606); (▲) lactose tablet core, 35 μm film thickness (Methocel® E 50). (From R. C. Rowe, *J Pharm Pharmacol*, 1980, 32, 214–215.)

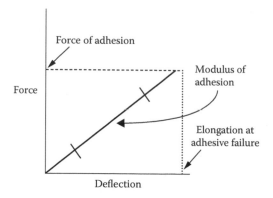

FIGURE 6.2 Example of a force–deflection profile generated using a Chatillon digital force gauge and motorized test stand to quantitate polymer adhesion by employing a butt adhesion technique. (From L. A. Felton, J. W. McGinity, *Pharm Dev Technol*, 1996, 1, 4, 381–389.)

An important factor to consider in the experimental design for investigating polymer adhesion is the shape of the tablet. In 1977, Rowe [28] compared the adhesive force between organic-based cellulosic films and either flat-faced or biconvex tablets. The force required to remove the film from the surface of the biconvex tablets was lower than the same films coated onto flat-faced tablets. A direct relationship between the force of adhesion and the square of the diameter of flat-faced tablets was found whereas a maximum force was reached with biconvex tablets, and no such correlation occurred. Interestingly, a direct relationship between the work required to remove the film from the tablet surface and the square of the diameter of the tablet was found for both flat-faced and biconvex tablets. These findings suggest that the work done to remove the film from the tablet surface provides a more accurate and quantitative measure of film–tablet adhesion for biconvex tablets than the direct force measurement whereas investigation of either the adhesive force or the adhesive toughness would be useful in the study of adhesion involving flat-faced tablets.

The majority of published studies investigating adhesion of polymeric films to pharmaceutical solids involve flat-faced tablets [15,22,23]. Flat-faced tablets, however, may agglomerate in the coating pan apparatus during the coating process. Non-uniform adhesion of the polymer at the edge of the tablets has also been reported due to the high internal stresses within the film at the tablet edge [11,29]. In a study conducted by Felton and McGinity [23], flat-faced punches with a beveled edge were used to achieve a more uniform adhesion of the polymeric film. The bevel decreased the sharp angle at the edge of the tablet and lowered the internal stresses within the film.

Film Thickness

Theoretically, film thickness should not affect the intrinsic adhesion at the film–tablet interface with no influence on adhesion expected after the initial coverage of the substrate. Researchers, however, have found that polymeric film thickness will influence the measured force of adhesion. Rowe [14], for example, showed that for films up to a thickness of 35 μm, increased film thickness resulted in decreased adhesion of an organic-based cellulosic polymer, and films greater than 35 μm in thickness exhibited increased adhesion with increased film thickness. Similar results were reported for aqueous- and organic-based hydroxypropyl cellulose (HPC) [24] and aqueous-based acrylic polymeric films [23].

The effect of film thickness on measured adhesion is thought to be a property of the test method and associated with changes in the stress distribution within the film during the adhesion experiment [14]. During the adhesion test, these stresses will either augment or oppose the applied stress and, therefore, influence measured adhesion. Extrapolation of the force of adhesion to a zero film thickness has been suggested by Reegen and Ilkka [30] as a method of minimizing the effects of residual stresses within a film. In most cases, however, a linear relationship between polymer adhesion and film thickness does not occur, and extrapolation of the force of adhesion to zero film thickness, therefore, would be difficult [14,24]. Furthermore, measured film thickness is a mean value and does not account for variations in thickness that occur when the polymer is applied to the tablet [14].

SUBSTRATE VARIABLES

The physical and chemical characteristics of the substrate can significantly influence the adhesive properties of polymeric films. As mentioned previously, the measured force of adhesion has been shown to be directly related to the square of the diameter of the tablet for flat-faced tablet compacts [15,28]. In addition, the size of the substrate may also affect the error in the data, with higher coefficients of variation in the adhesive force occurring when testing small tablets, due to the difficulties involved in removing the film from the edge of the tablets [15]. The following section describes some of the major substrate variables that impact polymer adhesion.

SURFACE ROUGHNESS

Surface roughness of a tablet and the force of compression used during the tableting process will affect polymer adhesion by altering the effective area of contact between the film coating and the surface of the solid. Above a critical compression force, increased compression pressure during tableting generally results in decreased adhesion as a smoother tablet is produced. Below a critical compression pressure, cohesive failure of the tablet will occur, when the tablet laminates rather than the film being separated from the tablet surface. This type of failure occurs when the intermolecular bonding forces between the film and the tablet surface are stronger than the bonds between the powdered particles within the tablet [23]. In contrast, adhesive failure of film-coated tablets will result in the coating being completely removed from the tablet surface with a minimal amount of powdered particles attached. In order to study film–tablet adhesion, the experimental parameters should be designed such that failure of the film is adhesive in nature [15,23]. Data from cohesive failure should not be compared to data from adhesive failure, due to the different forces that are involved in these processes.

   In a study involving an aqueous-based acrylic polymeric dispersion, Felton and McGinity [23] demonstrated a relationship between tablet hardness and polymer adhesion. Force–deflection profiles, as seen in Figure 6.3, show that as the tablet hardness increased, the force of adhesion, elongation at adhesive failure, and the adhesive toughness of the acrylic polymer decreased.

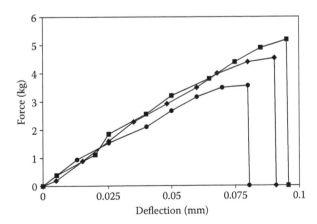

FIGURE 6.3 Force–deflection profiles obtained from butt adhesion experiments of an aqueous-based acrylic resin copolymer as a function of tablet hardness: (■) 7 kg; (♦) 10 kg; (●) 14 kg. (From L. A. Felton, J. W. McGinity, *Pharm Dev Technol*, 1996, 1, 4, 381–389.)

The softer tablets possessed a relatively rougher surface as evidenced by a higher arithmetic mean and root-mean-square roughness. The rougher surfaces of the tablet provided greater interfacial contact with the polymeric film, thus resulting in stronger polymer adhesion. Using a peel test, Nadkarni et al. [1] also found that the compressional force used during tableting influenced the adhesion of poly(methyl vinyl ether/maleic anhydride). Using contact angles between the polymeric solution and the tablet surface, these researchers showed that rougher tablets were more readily wetted by the polymeric solution.

In addition to surface roughness, tablet porosity can influence polymer adhesion. Polymeric films are generally applied to solid dosage forms using a spray atomization technique, and the water in the atomized droplets causes dissolution of the outermost surface of the tablet [26,31]. The rate and depth of polymer solution/dispersion penetration will influence the interfacial contact between the polymer and the tablet with the more porous tablet allowing faster penetration of the polymeric solution [15]. Moreover, drugs and excipients from the tablet can physically mix with the coating [26,31] and affect the adhesive, mechanical, and drug-release properties of the polymer [23,32,33].

TABLET EXCIPIENTS

Adhesion between a polymer and a substrate is due to the intermolecular bonding forces. For pharmaceutical products, hydrogen bond formation is the primary type of interfacial contact between the film and the tablet surface [7]. Excipients used in tablet formulations can alter the chemical properties of the tablet surface, thus influencing polymer adhesion. Sustained-release wax matrix tablets, for example, are generally difficult to coat with aqueous polymeric dispersions due to the poor wettability of the hydrophobic tablet surface [34].

The influence of direct-compression filler excipients on adhesion of organic-based hydroxypropyl methylcellulose (HPMC) films was investigated by Rowe [2]. Polymer adhesion was found to be strongest when microcrystalline cellulose (MCC) was used in the tablet compacts. The interaction between the primary and secondary hydroxyl groups of HPMC and MCC was greater than with other excipients studied, including sucrose, lactose, and dextrose, due to the saturation of the tablet surface with hydroxyl groups [35]. Lehtola et al. [22] found similar results with aqueous-based HPMC. HPMC phthalate was also found to adhere more strongly to MCC tablet compacts than tablets containing lactose or calcium phosphate [26].

Lubricating agents used in tablet formulations may influence polymer adhesion by presenting surfaces consisting of mainly nonpolar hydrocarbon groups, and the extent of the effect is dependent on the nature and concentration of the lubricant. Rowe [2] showed that increased concentrations of stearic acid, a commonly used lubricating agent that has a free polar carboxyl group, improved adhesion of an organic-based cellulosic polymeric film, as shown in Figure 6.4a.

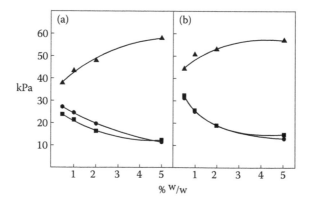

FIGURE 6.4   The effect of lubricant concentration (% w/w) on the measured adhesion (kPa) of hydroxypropyl methylcellulose films: (a) Pharmacoat® 606; (b) Methocel® 60HG viscosity 50; (▲) stearic acid; (●) magnesium stearate; (■) calcium stearate. (From R. C. Rowe, *J Pharm Pharmacol*, 1977, 29, 723–726.)

When this group was combined with glycerol to form the glyceryl esters present in hydrogenated castor oil and vegetable stearin, polymer adhesion decreased, as seen in Figure 6.4b. Similar results were reported by Lehtola et al. for aqueous-based HPMC films [22]. Felton and McGinity [23], investigating an aqueous-based acrylic polymer, found that adhesion decreased when the concentration of the hydrophobic filler hydrogenated castor oil was increased in tablet compacts.

Adhesion to Capsules

Difficulties reported in the film coating of hard gelatin capsules have been attributed to the physical properties of the gelatin and the dosage form itself [36]. In addition to the capsule shell softening and becoming sticky during the coating process due to solubilization of the gelatin, poor adhesion of the polymer to the walls of the hard gelatin capsule may occur. Insufficient adhesion may result in splintering of the film coating. The capsule shell is relatively smooth and generally provides less surface area for interfacial contact between the polymer and the surface of the gelatin than tablet compacts [37,38]. The addition of polyethylene glycol (PEG) 400 and PEG 6000 to the coating formulation has been used to improve adhesion of polymeric films to the gelatin shell [36]. An aqueous–alcoholic solution has also been shown to enhance polymer adhesion to capsule shells [38]. Several studies suggest that hard-shell cellulosic capsules have a relatively rougher surface than the gelatin capsule and thus can provide better film adhesion [39,40].

Felton et al. [17] conducted diametral compression experiments on film-coated soft gelatin capsules and found that adhesion of an aqueous-based acrylic polymer was dependent on the fill liquid of the capsule in conjunction with the plasticizer used in the coating formulation. Good polymer adhesion resulted as evidenced by single-point failure during compression of the coated capsules [5,18], when triethyl citrate (TEC) was incorporated into the coating formulation, regardless of the fill liquid. When the more hydrophobic plasticizer tributyl citrate (TBC) was added to the coating formulation, polymer adhesion was dependent on the fill liquid of the soft gelatin capsule, with better adhesion occurring with the hydrophobic Miglyol® 812 (Sasol Germany GmbH, Witten, Germany) fill liquid compared to the hydrophilic PEG 400.

COATING VARIABLES

Because the strength of adhesion between the film and substrate surface is dependent on the number and type of interfacial interactions, different polymers will exhibit different adhesive properties, depending on their chemical structures. In addition to the polymer itself, film-coating formulations generally include a solvent, a plasticizing agent, an anti-adherent, and pigments, all of which may also influence polymer adhesion. The following section describes some of the major coating formulation components that impact polymer adhesion.

Solvents

The solvent used in a film-coating formulation will interact with the polymer and affect the random coil structure of the polymer chains. It is generally accepted that the greater the polymer–solvent interaction, the greater the end-to-end distance, thus exposing more of the polymer that is capable of interacting with and binding to the surface of the solid. Nadkarni et al. [1] suggested that the solubility parameter of the solvent be used as a qualitative measure of the extent of polymer solvation, with greater polymer solvation resulting in greater film–tablet adhesion. A good correlation between the cohesive energy density of the solvent and the peel strength of methyl methacrylate films coated on a tin substrate was found by Engel and Fitz-water [41]. In 1988, Rowe [42] developed equations using solubility parameters of tablet excipients and polymers to predict trends in film–tablet adhesion.

Early research on film–tablet adhesion focused primarily on organic-based cellulosic films, and several studies have been published on the effects of solvent systems used in the coating formulation on polymer adhesion. Wood and Harder [21] used contact angle measurements as an indication of surface wettability to predict polymer adhesion. Fung and Parrott [6] compared the force of adhesion of HPC films prepared from several solvent systems and found that the force of adhesion varied twofold. Adhesion of films prepared from an aqueous-based system was one fourth to one half that of the organic-based films. These results further emphasize the importance of polymer–solvent interaction because it is the polymer that must interact with and bind to the substrate.

ADDITIVES IN THE COATING FORMULATION

Plasticizers

Plasticizers are included in film-coating formulations to improve the mechanical and film-forming properties of the polymers [43–45]. Several studies have focused on the effects of plasticizing agents on the adhesive properties of polymers. Felton and McGinity [46] investigated the influence of plasticizers on the adhesive properties of an acrylic resin copolymer to both hydrophilic and hydrophobic tablet compacts. Increasing the concentration of the hydrophilic plasticizer TEC in the coating formulation from 20% to 30% caused a slight, insignificant decrease in the force of adhesion. These results are in agreement with those of Fisher and Rowe [15], who found only slight, insignificant decreases in the measured force of adhesion between organic-based HPMC films and tablet compacts when the concentration of propylene glycol was increased from 10% to 20%. Felton and McGinity [46] showed that the plasticizer concentration also influences the elongation at adhesive failure. Moreover, these researchers demonstrated a relationship between the adhesive and mechanical properties of the acrylic polymer and suggested that the elongation at adhesive failure and the adhesive toughness of the polymer in conjunction with the force of adhesion provided a more complete understanding of the mechanisms involved in polymer adhesion.

Felton and McGinity [46] further investigated the effects of hydrophilic and hydrophobic plasticizers on polymer adhesion and found a relationship between adhesion and the glass transition temperature (T_g) of the film, with stronger adhesion occurring when the T_g of the film was lower as shown in Table 6.1.

The water-soluble plasticizers TEC and PEG 6000 lowered the T_g of the films to a greater degree than the hydrophobic plasticizers TBC and dibutyl sebecate, and films containing the hydrophilic plasticizers exhibited stronger adhesion. The researchers attributed these findings to the extent of the polymer–plasticizer interactions and the effectiveness of the plasticizing agent in lowering the internal stresses within the film coating. The addition of plasticizing agents to coating formulations

---

TABLE 6.1

Influence of the Plasticizer in the Coating Formulation on the Force of Adhesion and the Glass Transition Temperature (T_g) of an Acrylic Resin Copolymer to Lactose-Containing Tablets

Plasticizer	Force of Adhesion (SD)	T_g (SD)
Triethyl citrate	4.85 kg (0.27)	36.5°C (1.1)
Polyethylene glycol 6000	4.32 kg (0.25)	38.6°C (2.5)
Tributyl citrate	3.81 kg (0.30)	51.2°C (2.2)
Dibutyl sebecate	3.48 kg (0.33)	62.0°C (3.6)

Source: L. A. Felton, J. W. McGinity, *Eur J Pharm Biopharm*, 1999, 47, 1, 1–14.

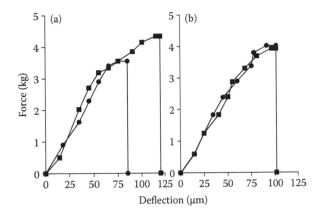

FIGURE 6.5 Force–deflection profiles obtained from butt adhesion experiments of an aqueous-based acrylic resin copolymer as a function of plasticizer type and tablet hydrophobicity: (a) 20% (w/w) polyethylene glycol 6000; (b) 20% (w/w) tributyl citrate; (■) 0% hydrogenated castor oil in tablet core; (●) 30% hydrogenated castor oil in tablet core. (From L. A. Felton, J. W. McGinity, *Int J Pharm*, 1997, 154, 167–178.)

generally decreases the internal stresses within the film by decreasing both the elastic modulus (E) and the glass transition temperature (T_g) of the film coating [11,47,48].

The influence of plasticizers in the coating on adhesion to hydrophilic and hydrophobic tablet compacts was also investigated [46]. Adhesion of the films plasticized with PEG 6000 was found to be significantly influenced by the hydrophobicity of the tablet surface as shown in Figure 6.5a.

These findings are in agreement with previous research showing that increasing tablet hydrophobicity decreased adhesion of both cellulosic and acrylic polymers [2,23]. Interestingly, when TBC was incorporated into the coating formulation, no significant differences in the adhesive properties of the acrylic film were found as seen in Figure 6.5b. Furthermore, these findings were correlated with thermomechanical data, where the T_g of the films plasticized with PEG 6000 was dependent on tablet hydrophobicity, and the amount of wax in the tablet core was not found to affect the T_g of the TBC-plasticized polymer.

Pigments and Fillers

Conflicting reports have been published on the influence of fillers or pigments on polymer adhesion to various substrates. Adhesion of ethylcellulose films cast on aluminum surfaces decreased with the addition of chalk whereas the incorporation of talc into cellulosic films improved polymer adhesion [49]. The addition of titanium dioxide and ferric oxide to methyl methacrylate films sprayed onto polymeric and tin substrates had no effect on adhesion, and mica and talc were found to decrease adhesion [41]. Okhamafe and York [4] suggested that the effects of additives in coating formulations were dependent on the balance between their influence on the internal stress of the film coating and the strength of the film–tablet interface.

Several studies have investigated the influence of talc in coating formulations on the adhesion of polymers to tablet compacts. Talc is a hydrophobic substance that is generally added to the coating formulation to reduce the tackiness of the lacquer during the coating process. Talc has been found to decrease the adhesion of polymers to tablet compacts [4]. The hydrophobic particles become embedded within the polymeric film and interfere with hydrogen bond formation between the tablet surface and the film coating. In addition, talc causes a stiffening of the film and increases the internal stresses within the polymer as evidencedg by an increase in the T_g of the polymer [50,51].

Pigments commonly used in pharmaceutical systems include aluminum lakes of water-soluble dyes, opacifiers such as titanium dioxide, and various inorganic materials, including the iron oxides. Pigments differ significantly in their physical properties, including density, particle shape, particle

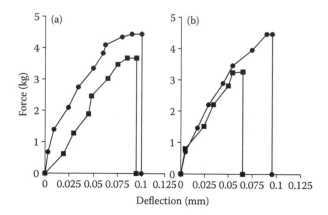

FIGURE 6.6 Force–deflection profiles obtained from butt adhesion experiments of aqueous-based Opadry® and Opadry® II as a function of tablet hydrophobicity: (a) 0% hydrogenated castor oil in tablet core; (b) 30% hydrogenated castor oil in tablet core; (■) Opadry; (●) Opadry II. (From L. A. Felton, J. W. McGinity, *Eur J Pharm Biopharm*, 1999, 47, 1, 1–14.)

size, and morphology, and these differences contribute to the complex relationship with aqueous film coatings [52–54]. In addition to affecting the mechanical properties of films, the incorporation of pigments into coating formulations has also been found to influence polymer adhesion. Fisher and Rowe [15], for example, found a 45% reduction in the force of adhesion of HPMC films with the addition of 10% titanium dioxide to the coating formulation. Okhamafe and York [50] showed that increased concentrations of titanium dioxide produced an increase in the T_g of HPMC films, which the authors attributed to the restriction in the mobility of the polymer chains by the presence of the additives.

Felton and McGinity [55] conducted a study that compared the adhesive properties of Opadry® and Opadry® II, two complete HPMC film-coating systems commercially available from Colorcon (West Point, Pennsylvania, PA). The Opadry II product was formulated with maltodextrins to achieve better adhesion, especially to hydrophobic substrates. Indeed, the addition of the maltodextrins to the cellulosic coating system enhanced polymer adhesion to both hydrophilic and hydrophobic tablet compacts as shown in Figure 6.6.

Surfactants

Previous researchers have used the wettability of a tablet by a polymeric solution as a tool to predict the strength of film–tablet adhesion [1,56]. A polymer solution that spreads more readily across the tablet surface allows for more interactions with the polymer chains and the formation of a greater number of bonds. Many of the polymeric materials commercially available today, however, are formulated as aqueous-based dispersions. Because it is the polymer, not the solvent, that interacts with and adheres to the tablet surface, wettability by polymeric dispersions may not be a valid indicator of film–tablet adhesion.

Surfactants have been incorporated into polymeric solutions to improve the spreadability of the coating across the tablet [57], emulsify water-insoluble plasticizers in aqueous dispersions [47,58], and modulate drug release [59,60]. Felton et al. used surfactants to alter tablet wettability by polymeric dispersions [61]. While the contact angle between the polymer dispersion and the tablet surface was dependent on the type and concentration of the surfactants added to the coating formulation, no correlation between tablet wettability and polymer adhesion was found.

PROCESSING PARAMETERS AND COATING CONDITIONS

The magnitude of internal stresses that inevitably develop during the coating process is dependent upon the interrelationship between many parameters involving both the polymeric coating material and the core substrate [12]. These stresses include stress due to shrinkage of the film upon solvent evaporation, thermal stress due to the difference in the coefficient of thermal expansion of the substrate and polymer, and volumetric stress due to the swelling or contraction of the substrate [8]. Processing parameters may influence the development of these stresses. Okutgen et al. [62], for example, determined the dimensional changes in tablet cores as a function of temperature, simulating temperature variations that tablets generally undergo during the coating process. Tablets containing Avicel® (FMC Biopolymer, Philadelphia, Pennsylvania) maize starch, and Starch® 1500 (Colorcon, West Point, Pennsylvania) all contracted when exposed to elevated temperatures and expanded during the cooling phase while Emcompress® (JRS Pharma, Patterson, New York) tablets exhibited the opposite behavior. These dimensional changes in the tablet core will influence the internal stresses within the films of the final coated products and may ultimately affect polymer adhesion. Selection of tableting excipients and polymeric coating materials with similar coefficients of thermal expansion should minimize internal stresses within the film and could improve polymer adhesion [11].

The process of film formation from polymeric dispersions requires the initial deposition of the atomized polymer droplets onto the substrate surface, followed by evaporation of the water and subsequent coalescence of the polymer chains. The time necessary to form a completely coalesced film has been shown to be dependent on the temperature used during the coating process, the nature and concentration of the plasticizer incorporated into the coating formulation, and the post-coating storage temperature [63,64]. Many commercially available polymeric materials for pharmaceutical film-coating operations require a post-coating thermal treatment or curing step to obtain a fully coalesced film, and this post-coating drying has also been shown to influence adhesion as well as the thermo-mechanical properties of the film [65]. Storage at elevated temperatures was found to increase the force required to separate an acrylic film from the tablet surface, with adhesion equilibrated within four hours of storage at 40°C or 60°C [65]. These findings were attributed to an increased number of polymer–substrate interactions resulting from the coalescence of the film. As the solvent evaporates during curing, the polymer droplets coalesce, and the number of potential polymer–substrate binding sites increases as shown in Figure 6.7.

In addition to processing temperature and post-coating curing, the spray rate will influence the extent of surface dissolution of the substrate and subsequent interfacial mixing at the film–tablet interface [31]. As mentioned previously, surface dissolution and physical mixing at the interface allows for drugs or excipients in the tablet to migrate into the film [31], which can influence internal stresses and thus affect polymer adhesion.

INFLUENCE OF AGING AND STORAGE CONDITIONS ON POLYMER ADHESION

Exposure of coated solids to various temperatures or relative humidities can influence the internal stresses within a film coating and thus affect polymer adhesion. Okhamafe and York [4], for example, showed that adhesion of pigmented and nonpigmented cellulosic films decreased during storage at 37°C and 75% relative humidity (RH). In another study, two weeks of storage at high RH (93%) caused a decrease in adhesion of an acrylic polymer to lactose tablets [46]. These findings were attributed to increased internal stresses in the polymeric films due to differences in the expansion coefficient of the polymer and tablet and volumetric stresses due to the swelling of the tablet core. Although previous researchers have demonstrated that water functions to plasticize polymers [66,67], the swelling of the film and the tablet as water diffuses through the coating during storage weakened the film–tablet interfacial bonding and created new stresses within the polymer.

Felton and McGinity [46] also reported decreased film–tablet adhesion after three months of storage at 0% RH. These findings were attributed to increased internal stresses within the coating due to evaporation of residual water in the polymeric film. Three months of storage at 40°C resulted in no significant change in the measured force of adhesion with only small decreases in the elongation at adhesive failure and adhesive toughness. The authors suggested that, because the tablets

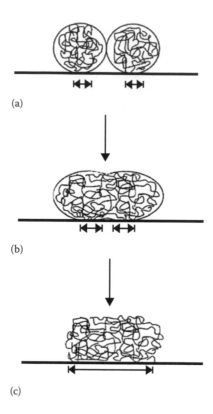

FIGURE 6.7   Schematic of the increase in potential polymer–substrate interactions as film formation proceeds: (a) closely packed polymer spheres due to water evaporation; (b) initiation of coalescence and polymer chain inter-diffusion due to additional water evaporation; (c) completed film formation. (From L. A. Felton, M. L. Baca, *Pharm Dev Technol*, 2001, 6, 1, 1–9.)

were stored at a temperature above the T_g of the film, the polymer chains were more mobile [68] and positioned themselves to minimize internal stresses.

Decreased adhesion between a polymeric film and a capsule shell has been reported to occur during the storage of film-coated hard gelatin capsules at high humidity [36]. The film coating and the gelatin swell to varying degrees and affect the internal stresses within the film. In another study involving film-coated soft gelatin capsules [17], storage at high humidity was found to improve adhesion of an acrylic polymer plasticized with TBC to the capsule containing PEG 400 as the fill liquid. The authors theorized that the fill liquid from the capsule might migrate into the film coating, functioning to further plasticize the polymer and lower the internal stresses of the film.

CONCLUSIONS

Good adhesion between a polymer and the surface of a solid is desirable for a pharmaceutical product. The two major forces that influence adhesion are the strength of the interfacial bonds and the internal stresses within the film. Factors that influence interfacial bonding or internal stresses will therefore affect polymer adhesion. Rougher, more irregular surfaces provide greater interfacial contact between the film and the tablet surface and generally provide for better adhesion. Excipients used in the substrate can also influence the extent of interfacial bonding between the polymeric film and the solid. Additives in the coating formulation, including the solvent system, plasticizer, and pigments, influence internal stresses and thus alter polymer adhesion. Processing parameters used during coating may also affect adhesion. Although many variables have been found to influence polymer adhesion, and direct comparison of the numerical values from one study to another is not practical, further experimentation involving adhesion of polymeric films to solid substrates will provide the pharmaceutical scientist with a better understanding of the mechanisms involved in polymer adhesion.

REFERENCES

1. Nadkarni PD, Kildsig DO, Kramer PA, Banker GS. Effects of surface roughness and coating solvent on film adhesion to tablets. *J Pharm Sci* 1975; 64:1554–1557.
2. Rowe RC. The adhesion of film coatings to tablet surfaces—The effect of some direct compression excipients and lubricants. *J Pharm Pharmacol* 1977; 29:723–726.
3. Felton LA, McGinity JW. Adhesion of polymeric films to pharmaceutical solids. *Eur J Pharm Biopharm* 1999; 47(1):1–14.
4. Okhamafe AO, York P. The adhesion characteristics of some pigmented and unpigmented aqueous-based film coatings applied to aspirin tablets. *J Pharm Phamacol* 1985; 37:849–853.
5. Stanley P, Rowe RC, Newton JM. Theoretical considerations of the influence of polymer film coatings on the mechanical strength of tablets. *J Pharm Pharmacol* 1981; 33:557–560.
6. Fung RM, Parrott EL. Measurement of film-coating adhesiveness. *J Pharm Sci* 1980; 69(4):439–441.
7. Pritchard WH. In: Alner DJ, ed. Aspects of Adhesion, Vol. 6. University of London, 1971, pp. 11–23.
8. Rowe RC. A reappraisal of the equations used to predict the internal stresses in film coatings applied to tablet substrates. *J Pharm Pharmacol* 1983; 35:112–113.
9. Croll SG. The origin of residual internal stress in solvent-cast thermoplastic coatings. *J Appl Polym Sci* 1979; 23:847–858.
10. Sato K. The internal stress of coating films. *Prog Org Coating* 1980; 8:143–160.
11. Rowe RC. The adhesion of film coatings to tablet surfaces—A problem of stress distribution. *J Pharm Pharmacol* 1981; 33:610–612.
12. Okutgen E, Hogan JE, Aulton ME. Quantitative estimation of internal stress development in aqueous HPMC tablet film coats. *Int J Pharm* 1995; 119:193–202.
13. Mittal KL. Interfacial chemistry and adhesion: Developments and prospects. *Pure Appl Chem* 1980; 52:1295–1305.
14. Rowe RC. The measurement of the adhesion of film coatings to tablet surfaces: The effect of tablet porosity, surface roughness, and film thickness. *J Pharm Pharmacol* 1978; 30:343–346.
15. Fisher DG, Rowe RC. The adhesion of film coatings to tablet surfaces—Instrumentation and preliminary evaluation. *J Pharm Pharmacol* 1976; 28:886–889.
16. Strong J. On the cleaning of surfaces. *Rev Scient Instrum* 1935; 6:97–98.
17. Felton LA, Shah NH, Zhang, G, Infeld MH, Malick AW, McGinity JW. Physical-mechanical properties of film-coated soft gelatin capsules. *Int J Pharm* 1996; 127: 203–211.
18. Fell JT, Rowe RC, Newton JM. The mechanical strength of film-coated tablets. *J Pharm Pharmacol* 1979; 31:69–72.
19. Heavens OS. Adhesion of metal films produced by vacuum evaporation. *J Phys Radium* 1950; 11:355–360.
20. Brantley RL, Woodward A, Carpenter G. Adhesion of lacquers to nonferrous metals. *Ind Eng Chem* 1952; 44:2346–2389.
21. Wood JA, Harder SW. The adhesion of film coatings to the surfaces of compressed tablet. *Can J Pharm Sci* 1970; 5(1):18–23.
22. Lehtola VM, Heinamaki JT, Nikupaavo P, Yliruusi JK. Effect of some excipients and compression pressure on the adhesion of aqueous-based hydroxypropyl methylcellulose film coatings to tablet surface. *Drug Dev Ind Pharm* 1995; 21(12):1365–1375.
23. Felton LA, McGinity JW. Influence of tablet hardness and hydrophobicity on the adhesive properties of an acrylic resin copolymer. *Pharm Dev Technol* 1996; 1(4):381–389.
24. Johnson BA, Zografi G. Adhesion of hydroxypropyl cellulose films to low energy solid substrates. *J Pharm Sci* 1986; 75(6):529–533.
25. Sarisuta N, Lawanprasert P, Puttipipatkachorn S, Srikummoon K. The influence of drug-excipient and drug-polymer interactions on butt adhesive strength of ranitidine hydrochloride film-coated tablets. *Drug Dev Ind Pharm* 2006; 32:463–471.
26. Missaghi S, Fassihi R. A novel approach in the assessment of polymeric film formation and film adhesion on different pharmaceutical solid substrates. *AAPS PharmSci Tech* 2004; 5(2):Article 29.

27. Rowe RC. Rate effects in the measurement of the adhesion of film coatings to tablet surfaces. *J Pharm Pharmacol* 1980; 32:214–215.

28. Rowe RC. The adhesion of film coatings to tablet surfaces-measurement on biconvex tablets. *J Pharm Pharmacol* 1977; 29:58–59.

29. Rowe RC, Forse SF. The effect of polymer molecular weight on the incidence of film cracking and splitting on film coated tablets. *J Pharm Pharmacol* 1980; 32:583.

30. Reegen SL, Ilkka GA. The adhesion of polyurethanes to metals. In: Weiss P, ed. Adhesion & Cohesion. New York: Elsevier, 1962:159–171.

31. Felton LA, Perry WL. A novel technique to quantify film-tablet interfacial thickness. *Pharm Dev Technol* 2002; 7(1):1–5.

32. Wu C, McGinity JW. Non-traditional plasticization of polymeric films. *Int J Pharm* 1999; 177:15–27.

33. Dansereau R, Brock M, Redman-Furey N. The solubilization of drug and excipient into a hydroxypropyl methylcellulose (HPMC)-based film coating as a function for the coating parameters in a 24″ accela-cota. *Drug Dev Ind Pharm* 1993; 19(7):793–808.

34. Porter SC. Use of Opadry, Sureteric, and Surelease for the aqueous film coating of pharmaceutical oral dosage forms. In: McGinity JW, ed. *Aqueous Polymeric Coatings for Pharmaceutical Dosage Forms.* New York: Marcel Dekker, Inc., 1997, pp. 327–372.

35. Battista OA, Smith PA. Microcrystalline cellulose. *Ind Eng Chem* 1962; 54(9):20–29.

36. Thoma K, Bechtold K. Enteric coated hard gelatin capsules. Capsugel Technical Bulletin 1986.

37. Thoma K, Oschmann R. Investigations of the permeability of enteric coatings. Part 5: Pharmaceutical-technological and analytical studies of enteric-coated preparations. *Pharmazie* 1991; 46:278–282.

38. Osterwald HP. Experience with coating of gelatin capsules with Driacoater and WSG apparatus, especially rotor WSG. *Acta Pharm Technol* 1982; 28:329–337.

39. Felton LA, Friar AL. Enteric coating of gelatin and cellulosic capsules using an aqueous-based acrylic polymer. American Association of Pharmaceutical Scientists Annual Meeting, Toronto, Canada, 2002.

40. Felton LA, Sturtevant S, Birkmire A. A novel capsule coating process for the application of enteric coatings to small batch sizes. American Association of Pharmaceutical Scientists Annual Meeting, San Antonio, TX, 2006.

41. Engel JH, Fitzwater RN. Adhesion of surface coatings as determined by the peel method. In: Weiss P, ed. Adhesion & Cohesion. New York: Elsevier, 1962, pp. 89–100.

42. Rowe RC. Adhesion of film coatings to tablet surfaces—A theoretical approach based on solubility parameters. *Int J Pharm* 1988; 41:219–222.

43. Honary S, Orafai H. The effect of different plasticizer molecular weights and concentrations on mechanical and thermomechanical properties of free films. *Drug Dev Ind Pharm* 2002; 28(6):711–715.

44. Qussi B, Suess WG. The influence of different plasticizers and polymers on the mechanical and thermal properties, porosity and drug permeability of free shellac films. *Drug Dev Ind Pharm* 2006; 32:403–412.

45. Porter SC. The effect of additives on the properties of an aqueous film coating. *Pharm Tech* 1980; 4:67–75.

46. Felton LA, McGinity JW. Influence of plasticizers on the adhesive properties of an acrylic resin copolymer to hydrophilic and hydrophobic tablet compacts. *Int J Pharm* 1997; 154:167–178.

47. Gutierrez-Rocca JC, McGinity JW. Influence of water soluble and insoluble plasticizers on the physical and mechanical properties of acrylic resin copolymers. *Int J Pharm* 1994; 103:293–301.

48. Johnson K, Hathaway R, Leung P, Franz R. Effect of triacetin and polyethylene glycol 400 on some physical properties of hydroxypropyl methycellulose free films. *Int J Pharm* 1991; 73:197–208.

49. Brantley LR. Removal of organic coatings. *Ind Eng Chem* 1961; 53:310.

50. Okhamafe AO, York P. The glass transition in some pigmented polymer systems used for tablet coating. *J Macromol Sci Phys* 1984–85; B23(4–6):373–382.

51. Okhamafe AO, York P. Relationship between stress, interaction and the mechanical properties of some pigmented tablet coating films. *Drug Dev Ind Pharm* 1985; 11(1):131–146.

52. Lippold BH, Sutter BK, Lippold BC. Parameters controlling drug release from pellets coated with aqueous ethyl cellulose dispersions. *Int J Pharm* 1989; 54:15–25.

53. Gibson SHM, Rowe RC, White EFT. The mechanical properties of pigmented tablet coating formulations and their resistance to cracking II. Dynamic mechanical measurement. *Int J Pharm* 1989; 50:163–173.

54. Rowe RC. Modulus enhancement in pigmented tablet film coating formulations. *Int J Pharm* 1983; 14:355–359.

55. Felton LA, McGinity JW. The influence of plasticizers on the adhesive properties of acrylic resin copolymers. 15th Pharmaceutical Technology Conference, Oxford, England, 1996.

56. Khan H, Fell JT, Macleod GS. The influence of additives on the spreading coefficient and adhesion of a film coating formulation to a model tablet surface. *Int J Pharm* 2001; 227:113–119.
57. Banker GS. Film coating theory and practice. *J Pharm Sci* 1966; 55(1):81–89.
58. Bodmeier R, Paeratakul O. The distribution of plasticizers between aqueous and polymer phases in aqueous colloidal polymer dispersions. *Int J Pharm* 1994; 103: 47–54.
59. Buckton G, Efentakis M, Alhmoud H, Rajan Z. The influence of surfactants on drug release from acrylic matrices. *Int J Pharm* 1991; 74:169–174.
60. Knop K, Matthee K. Influence of surfactants of different charge and concentration on drug release from pellets coated with an aqueous dispersion of quarternary acrylic polymers. *STP Pharma Sci* 1997; 7(6):507–512.
61. Felton LA, Austin-Forbes T, Moore TA. Influence of surfactants in aqueous-based polymeric dispersions on the thermo-mechanical and adhesive properties of acrylic films. *Drug Dev Ind Pharm* 2000; 26(2):205–210.
62. Okutgen E, Hogan JE, Aulton ME. Effects of tablet core dimensional instability on the generation of internal stresses within film coats. Part 1: Influence of temperature changes during the film coating process. *Drug Dev Ind Pharm* 1991; 17(9):1177–1189.
63. Gilligan CA, Li Wan Po A. Factors affecting drug release from a pellet system coated with an aqueous colloidal dispersion. *Int J Pharm* 1991; 73:51–68.
64. Hutchings D, Clarson S, Sakr A. Studies of the mechanical properties of free films prepared using an ethylcellulose pseudolatex coating system. *Int J Pharm* 1994; 104:203–213.
65. Felton LA, Baca ML. Influence of curing on the adhesive and mechanical properties of an applied acrylic polymer. *Pharm Dev Technol* 2001; 6(1):1–9.
66. Hancock BC, Zografi G. The relationship between the glass transition temperature and the water content of amorphous pharmaceutical solids. *Pharm Res* 1994; 11(4):471–477.
67. Gutierrez-Rocca JC, McGinity JW. Influence of aging on the physical-mechanical properties of acrylic resin films cast from aqueous dispersions and organic solutions. *Drug Dev Ind Pharm* 1993; 19(3):315–332.
68. Sinko CM, Yee AF, Amidon GL. The effect of physical aging on the dissolution rate of anionic polyelectrolyes. *Pharm Res* 1990; 7(6):648–653.

7 Influence of Coloring Agents on the Properties of Polymeric Coating Systems

Nasser N. Nyamweya, Stephen W. Hoag,
Tanvi M. Deshpande, and Linda A. Felton

CONTENTS

INTRODUCTION

Coloring agents are widely used in the pharmaceutical industry and are an important component of many oral solid dosage forms. A coloring agent or colorant may be defined as an excipient that imparts color when added to a drug product. Colorants may be incorporated into solid dosage forms by adding them directly to the dosage form (e.g., adding a colorant to a tablet granulation), incorporating the colorant into capsule shells, and by adding the colorant to a coating formulation that would be applied onto the surface of a drug product. Coloring agents may be added to pharmaceutical coatings for a number of reasons, including the following:

1. To enhance product appearance, aesthetic appeal, and product elegance
2. To improve or facilitate product identification for the manufacturer, health care professional, and patient
3. To provide protection from light for photosensitive compounds
4. To provide a brand image and help differentiate the drug product from competitive products
5. To help reduce or prevent counterfeiting

Although coloring agents do not provide any direct therapeutic effects by improving the appearance of the drug product, they can contribute to increasing patient compliance. The use of coloring agents helps to provide drug products with a distinct appearance, which makes it easier for the pharmacist and patient to distinguish between different drug products, thereby reducing the possibility of dispensing and medication errors. Patients taking several different medications will find them easier to distinguish if they have different colors. A unique appearance also contributes to enhancing the brand image of a drug product, which may provide a significant marketing advantage over competitive products. In combination with other factors that can be used to increase the uniqueness of the drug product, such as shape and markings, the addition of color can help to make it more difficult to counterfeit drug products.

Because certain terms for colorants are sometimes used interchangeably, it is important to first provide definitions for the terms "dyes" and "pigments" as they appear in this chapter. The term dye applies to colorants that are soluble in water, and the term pigment applies to colorants that are insoluble in water. The same terms can also be applied to colorants added to non-aqueous liquids. The coloring power of a dye results from the dye molecules being dissolved, and that of pigments is due to dispersion of the pigment particles. The most commonly used colorants in film-coating applications are aluminum lakes, iron oxides, and titanium dioxide. Dyes are also used in some cases although this would usually be in combination with an insoluble colorant.

The visual observation of color requires the following components:

1. A light source (e.g., sunlight)
2. An object (e.g., a red film-coated tablet)
3. An observer

Light is a form of electromagnetic radiation and is characterized by its wavelength. The visible light that is observable by the human eye has wavelengths between about 380 and 780 nm [1]. Coloring agents impart color by selectively absorbing and reflecting certain wavelengths of light within this region. For example, in the case of a tablet that appears red under white light, the colorant in the film coating would predominantly absorb the blue and green wavelengths while the red wavelengths would primarily be reflected from the surface of the tablet (Table 7.1). The eye, which contains light-sensitive receptors, would detect the reflected red wavelengths and send a signal to the brain, which would interpret the color of the coated tablet as red. In addition to absorbance and reflection, light may be transmitted through an object as in the case of a transparent film. Films that do not transmit or transmit very little light are opaque. In translucent films, some light is transmitted, and some is reflected by scattering.

The color of organic dyes results from the select absorption of certain wavelengths of light by chromophores. Chromophores are the part of a molecule responsible for light absorption and hence the color is observed. For organic dyes, these moieties include conjugated double bonds common to functional groups, such as carbonyl, azo, and ethenyl [2]. The absorption of light by chromophores can be enhanced or modified by chemical groups called auxochromes. Examples of auxochromes include amino, alkylamino, methoxy, and hydroxyl groups [2].

TABLE 7.1
Absorbed Colors and Complementary Colors at Different Absorption Wavelengths

Wavelength (nm)	Color of Absorbed Light	Complementary Color
400–420	Violet	Yellow–green
420–450	Indigo blue	Yellow
450–490	Blue	Orange
490–510	Blue–green	Red
510–530	Green	Purple
530–545	Yellow–green	Violet
545–580	Yellow	Indigo blue
580–630	Orange	Blue
630–720	Red	Blue–green

Source: R. C. Schiek, M. Fytelson, J. J. Singer. Pigments. In: Mark, H. F., Othmer, D. F., Overberger, C. G., Seaborg, G. T., Grayson, M., Eckroth, D., editors. *Encyclopedia of Chemical Technology.* 3rd ed. New York: Wiley; 1982, 788–889.

With both inorganic and organic pigments, color arises from absorbed light producing electronic transitions. Electronic transitions in inorganic pigments involve the bonding between transition metal ions and the surrounding geometrical arrangement of molecules or ligands [2]. In addition to the chemical composition, the color of pigments may also be influenced by physical properties such as particle size and particle size distribution. Pigment particles can also scatter and reflect light thereby influencing the opacity of the film coating [2].

COLORING AGENTS USED IN FILM COATING

DYES

The dyes typically used in pharmaceutical applications are synthetic compounds, which are more stable and available in a wider range of colors than natural dyes. A range of colors is available including blue, green, orange, red, and yellow. In addition to their common names, the dyes used in oral dosage forms may be labeled as FD & C (certified for use in food, drugs, and cosmetics) or D & C (certified for use in drugs and cosmetics), a designation given by the U.S. Food and Drug Administration (FDA). The certification refers to the testing of a colorant by the FDA to ensure that it meets identity and purity specifications. The stability of a dye may be affected by heat, light, pH, oxidizing agents, and reducing agents although some dyes are more stable than others to the effects of these factors.

PIGMENTS

Aluminum Lakes

Aluminum lakes are produced by precipitating and adsorbing a water-soluble dye onto a water insoluble substrate, typically aluminum hydroxide [3]. The chemical bonding of the dye to the aluminum hydroxide substrate results in improved stability to light and heat. The exact nature of the adsorption process is not well understood but is believed to be a combination of ionic bonding, hydrogen bonding, and van der Waals forces [4]. The nomenclature of lakes includes the name of the dye and the substrate (e.g., Sunset Yellow or FD & C yellow No. 6 aluminum lake). Lakes are insoluble in water within a certain pH range. Lakes are available in several colors and shades depending on the type and amount of dye used. Common lake colors include blue, orange, red, and yellow. Additional colors or shades may be obtained by mixing two or more lakes of differing color. The amount of dye in a lake may range from 3% to 60% [5].

Iron Oxides

Iron oxides are available in black (Fe_3O_4), red (Fe_2O_3), and yellow ($Fe_2O_3 \cdot H_2O$) colors. They are prepared by the precipitation of iron salts (black and yellow iron oxides), calcination (red iron oxide), or by blending different iron oxides (brown iron oxide) [6]. Iron oxides do occur naturally, but synthetic forms of iron oxides are used predominantly due to their higher quality and the difficulties involved in purifying the natural forms [6,7]. Iron oxides have excellent light and heat resistance [8].

Titanium Dioxide

Titanium dioxide is a white pigment widely used to make films opaque or increase their opacity. Opacity is the degree to which a film containing a pigment can obscure the appearance of a substrate to which the film is applied [7]. The opacifying effect of titanium dioxide is due to its high refractive index, which results in the scattering of visible light. The range of light that is scattered can be varied by changing the particle size [9]. The optimal particle size for scattering visible light is 200 to 300 nm, when the particle size is about half the wavelength of visible light. Titanium dioxide is manufactured using naturally occurring minerals such as ilmenite ($FeO\text{-}TiO_2$) and rutile although the latter is less abundant [7]. Titanium dioxide has different polymorphic forms of which rutile and anatase are the most commonly used commercial forms. The rutile form has a higher density and refractive index, but the anatase form is softer and less abrasive [10]. When titanium dioxide is used in combination with iron oxides or lakes, it tends to produce pastel shades due to its extreme whiteness. Titanium dioxide has excellent heat and light stability.

Talc

Talc is commonly used as an anti-adherent or detackifier in film-coating formulations. Although talc may often not be thought of as a coloring agent because of its anti-adherent function in film coating, it does have a white to grayish-white color and is listed as a color additive in the U.S. Code of Federal Regulations [11]. Talc is a natural mineral, and its composition and physical properties may vary depending on the location where the talc is mined and the method by which it is processed. It is a hydrated magnesium silicate [($Mg_3SiO_4O_{10}(OH)_2$], which may contain small amounts of aluminum silicate, aluminum and iron oxides, calcium carbonate, and calcium silicate depending on the country of origin [12]. Lin and Peck characterized several different United States Pharmacopeia (USP) talc grades and found variation in physical properties, such as particle size and surface area [13]. Talc is a relatively soft mineral and is hydrophobic in nature.

Glyceryl Monostearate

Like talc, glyceryl monostearate (GMS) is also used as an anti-adherent or slipping agent in film-coating formulations. GMS is available as white or cream-colored wax-like flakes. The hydrophobicity of GMS is due to the different proportions of glyceryl esters of fatty acids; predominantly consisting of glyceryl monostearate, followed by glyceryl monopalmitate and glyceryl monomyristate [14,15]. Two distinct grades of GMS based on the percentage of monoglycerides present are available: 40%–55% monoglycerides and 90% monoglycerides. GMS has two main polymorphic forms, the α-form and the more stable β-form, both of which are differentiated by their melting points, and a less abundant metastable β_0-form, which exists during conversion from α-form to the β-form [15–17]. The anti-adherent property of GMS comes from its hydrophobic nature and the ability to intersperse between the polymer chains. It is a more effective anti-adherent as compared to talc and hence can be used at lower concentrations [18]. Unlike talc, GMS does not impart any color to the film coating. It is listed as an emulsifier in food additives according to the U.S. Code of Regulations [19]. In the field of cosmetics, glyceryl monostearate is used as a viscosity enhancer to suspend pigment particles [20].

Pearlescent Pigments

Pearlescent pigments are pigments that impart a shiny, pearl-like luster when incorporated into film coatings. These pigments are available in a variety of colors including blue, gold, green, red, and silver [21]. They are prepared by coating mica (potassium aluminum silicate) platelets with titanium dioxide and/or iron oxide creating a multilayered structure. Incident light undergoes multiple reflections and refractions when transmitted through a coating containing these pigments resulting in the pearlescent visual effect.

REGULATION OF COLORING AGENTS

The use of coloring agents in drug products is regulated by local or regional regulatory agencies with many individual countries having lists of approved colorants. Some colorants, such as iron oxides, are widely accepted globally. Daily intake or usage limits may apply depending on the specific type of pigment. Because regulations on the use of colorants can vary considerably between different countries, drug product manufacturers need to be cognizant of these regulations when developing formulations for international markets [22]. Drug product manufacturers should also work with their color suppliers to determine whether the colorant being used will meet the regulatory requirements in the countries where the drug product will be marketed. For further information on the regulation of coloring agents, the reader is referred to Schoneker and Swarbrick [5].

PROPERTIES OF COLORING AGENTS

DYES

The solid-state properties of a dye (e.g., particle size, surface area, and density) are usually not as important in film-coating applications as they are for pigments because dyes are used in coating formulations after dissolving them in liquid media. Dyes can therefore be distributed in film coatings at the molecular level while, in contrast, pigments exist as much larger undissolved particles. Although smaller size is an advantage in terms of producing a more pronounced and intense coloring effect, the use of dyes in film coating has been limited due to a tendency for the dye molecules to migrate with the evaporating solvent during drying. This migration results in an uneven distribution of color and a mottled film coating. Signorino and Meggos developed uniform, non-mottled coating formulations using dyes with the addition of immobilizing agents [23,24]. The use of dyes is also limited by their lower stability compared to pigments with regards to factors such as pH, light, and heat, which can result in changes in the appearance or color fading of coated products.

PIGMENTS

Pigments are preferred in most film-coating applications because they are insoluble and do not migrate with the evaporating solvent during coating, which results in more uniform color distribution and batch-to-batch control of the film-coating color. Pigments may be characterized by a number of physical tests, including particle size, surface area, morphology, density, refractive index, and surface charge in aqueous media. Table 7.2 compares selected physical properties for various pigments.

The particle size of a pigment can affect the distribution of color, hiding power, tinting strength, gloss, and lightening power in the film [25,26]; sedimentation in liquids; and the surface roughness of the coating. As insoluble colorants, the color produced by pigments is dependent on how well they are dispersed. Rowe investigated the effect of particle size of red iron oxide on the appearance of hydroxypropyl methylcellulose films and found that significant changes in color were observed with different pigment particle sizes [27]. Wou and Mulley studied the influence of the particle size of several aluminum lakes and reported significant differences in color depending on the degree

TABLE 7.2

Selected Physical Properties of Pigments Used in Film-Coating Applications

Pigment	Particle Size (μm)	Surface Area (m²/g)	Density (g/cm³)	Morphology
Aluminum lakes (dye content)				
FD & C Blue No. 2 (36.3%)	2.24	19.45	1.94	Irregular
FD & C Red No. 40 (38.0%)	1.96	21.69	1.84	Irregular
FD & C Yellow No. 6 (37.4%)	2.68	16.13	1.83	Irregular
FD & C Yellow No. 10 (17.8%)	3.37	34.67	2.02	Irregular
Iron oxides				
Black	0.5–1	3.84	4.98	Cubical
Red	0.32	7.56	5.41	Spherical
Yellow	0.5–1	13.07	4.32	Acicular
Titanium dioxide	0.2	7.60	3.78	Rounded

Source: N. Nyamweya, K. A. Mehta, S. W. Hoag, *Journal of Pharmaceutical Sciences*, 2001, 90, 12, 1937–1947; S. Gibson, R. Rowe, E. White, *International Journal of Pharmaceutics*, 1988, 48, 1, 113–117.

of pigment dispersion [28]. A reduction in lake particle size resulted in an increased color strength with submicron particles having the greatest effect on color strength. In order to achieve adequate dispersion of pigments in powder form, high-shear mixing equipment is often recommended to ensure that agglomerates are broken down and that a uniform color distribution is achieved. An alternative to using pigments in powder form is to use pre-dispersed pigment concentrates (suspensions) [29,30]. These color concentrates may contain more than one pigment or may use pigments in conjunction with dyes.

The density of a pigment particle affects its sedimentation rate, especially in coating solutions or dispersions that have a low viscosity. In colloidal polymer dispersions, which typically have low viscosity, commonly used pigments will often tend to settle unless the liquid in which they are dispersed is subjected to flow or mechanical agitation. Approaches to minimizing sedimentation include using pigments with small particle sizes and size distributions, using various additives (e.g., suspending agents), ensuring continuous mixing of the coating dispersion, minimizing the length of the tubing used to deliver the coating dispersion to the coating equipment, and using tubing with a smaller internal diameter to ensure higher liquid flow rates. Additives to prevent sedimentation are generally not recommended because they can alter film properties [31].

Insoluble colorants can acquire a surface charge in aqueous media and develop an electrical double layer around their surface in order for the system to maintain electrical neutrality. In this respect, they have the surface characteristics of a charged colloidal particle. Figure 7.1 shows the electrical double layer of a negatively charged colloidal particle.

The charged surface attracts ions of an opposite charge (counter-ions), which are strongly bound to the surface of the colloid forming the Stern layer. Beyond the Stern layer, a more diffuse layer of counter-ions is observed due to the repulsive forces from counter-ions in the Stern layer. At the same time, ions of similar charge to the surface of the colloidal particle are seen in the diffuse layer as the repulsive forces from the surface start to decrease with increasing distance and are increasingly shielded by the presence of counter-ions. At a certain distance from the charged surface, the concentration of both types of ions reaches equilibrium with the ions in the bulk medium. This distribution of ions and charge in the double layer creates a potential that decreases from the surface of the particle to the bulk medium (Figure 7.2).

As the charged particle moves in the medium, it does so with the tightly bound ions of the Stern layer as well as an associated layer of ions. The shear plane is the point beyond which ions in the diffuse layer do not move with the particle. The potential at the shear plane is the zeta potential, an

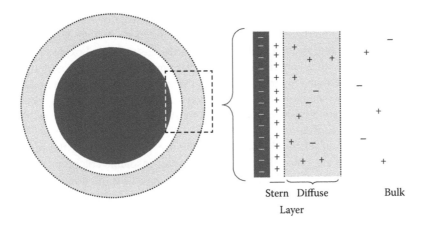

FIGURE 7.1   Electrical double layer of a negatively charged colloidal particle.

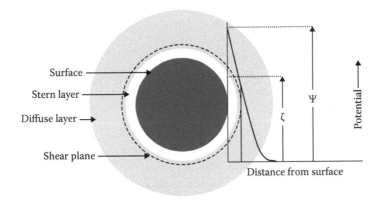

FIGURE 7.2   Change in potential from the surface of a colloidal particle to the bulk medium. *Symbols*: ζ, zeta potential; Ψ, surface potential.

important determinant of the stability of colloidal particles and how they interact with each other in aqueous media. Factors that decrease the zeta potential, such as ions or electrolytes, will allow colloidal particles to approach each other more closely and increase the likelihood of aggregation or coagulation. The effect of a strong electrolyte on the stability of an aluminum lake is shown in Figure 7.3 where the pigment particles are observed to aggregate in a solution of potassium chloride.

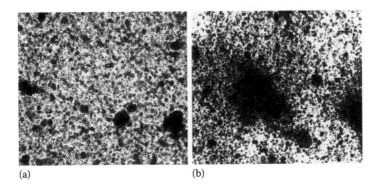

(a)   (b)

FIGURE 7.3   FD & C blue No. 2 lake photomicrographs after 60 minutes in (a) H_2O and (b) 1 M KCl.

For pigments, surface charge may result from and be influenced by the ionization of species on the surface of the pigment and the adsorption of ions from the medium. The surface charge can vary depending on the pH and composition of the system. In the case of metal oxide or hydroxide-based pigments (e.g., iron oxides, titanium dioxide, and aluminum lakes), the surface charge can be influenced by the pH of the medium (Figure 7.4). In fact, the surface charge may be positive or negative depending on the pH of the medium. At a certain intermediate pH value, the pigment particles will have no net charge (referred to as the isoelectric point). At this pH, the particles will be least stable due to absence of repulsive forces.

In the case of aluminum lakes, pH may also influence the stability of the pigment particles. Although aluminum lakes are generally referred to as water-insoluble colorants, dissociation of the lake can occur at low or high pH values. Desai et al. characterized the effect of pH on the stability of aluminum lakes in aqueous media [32]. The study found that although aluminum lakes were stable in aqueous media at intermediate pH values, at more acidic or basic pH values, the aluminum hydroxide substrate of the lake dissolved resulting in the release of the adsorbed dye into solution. Similar findings were reported by Nyamweya and Hoag where the optimal pH stability range was observed to be 4–7 (Figure 7.5) [33].

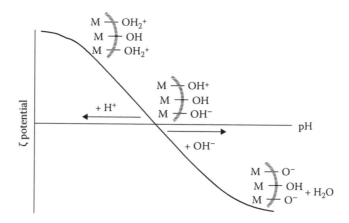

FIGURE 7.4   Influence of pH on the zeta potential of a pigment [metal (M) oxide or hydroxide] in aqueous media.

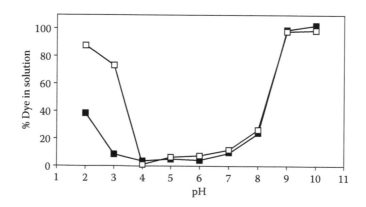

FIGURE 7.5   Effect of pH on the dissolution of an FD & C blue No. 2 aluminum lake after (■) 1 hour and (□) 24 hours. (From N. Nyamweya, K. A. Mehta, S. W. Hoag, *Journal of Pharmaceutical Sciences*, 2001, 90, 12, 1937–1947.)

INFLUENCE OF COLORING AGENTS ON POLYMER COATING DISPERSIONS

SOLUTION AND COLLOIDAL INTERACTIONS

The majority of currently used polymer dispersions for film coating are applied in liquid form in which the film-forming polymer exists in the form of a solution (dissolved polymer molecules) or an aqueous colloidal dispersion. However, there are a number of processes that can be used to apply a coating to a product in a dry state. Examples include compression coating, dry powder coating, and electrostatic coating. It would be expected that in such systems, due to the absence of water or solvents, the potential for interactions between the colorant and the polymer would be much less than for coatings applied in the form of solutions or dispersions.

Film-coating formulations in which the polymer is dissolved may be either aqueous or organic solvent–based. In polymer solutions, incorporation of pigments is usually less of an issue with regards to physical stability. Very few studies have characterized the interaction between pharmaceutical polymer solutions and pigments. Gibson et al. studied the interactions between hydroxypropyl methylcellulose and pigments (an aluminum lake, iron oxides, titanium dioxide, and talc) in aqueous solutions of the polymer using immersion calorimetry [34]. An exothermic reaction was observed following the immersion of pigments into solutions of the polymer. Sawyer and Reed studied the adsorption behavior of hydroxypropyl methylcellulose onto the surface of oxide particles from aqueous suspensions. The more hydrophilic particles (alumina and silica) with highly hydroxylated surfaces did not adsorb the polymer, and the more hydrophobic talc particles showed a significant adsorption of polymer [35]. The authors suggested that the mechanism for the interaction of the polymer with talc was a hydrophobic interaction in which the adsorption of the polymer reduced the free energy of the particle–water interface.

In film-coating formulations in which the polymer is not dissolved but rather exists in the form of colloidal particles, the physical stability of the dispersion needs to be maintained for a successful film-coating process. Colloidal polymer dispersions were introduced in order to enable the aqueous-based coating of functional polymers (e.g., enteric and sustained release polymer coatings). The particle size of the colloidal particles is typically in the submicron range. A colloid in this case may be defined as a system comprising of a dispersed phase (polymer particles) and a dispersion medium (water). Stable colloidal dispersions can be defined as systems in which the original constituent particles retain discreteness with no aggregation or agglomeration. The stability of colloidal systems may be affected by a number of factors including the following:

1. The particle size of the colloidal particles
2. The surface charge of the colloidal particles
3. The pH of the dispersion
4. The viscosity of the dispersion
5. The composition of the dispersion (e.g., the presence and concentration of electrolytes, pigments, and water soluble polymers)

Additionally, the stability of colloidal dispersions may be adversely affected when such systems are subjected to factors such as high shear forces, warm temperatures, and freezing. Instability may be manifested in the form of aggregation or coagulation of the colloidal polymer particles. Colloidal particles have a relatively large surface area and hence a high free energy. Equation 7.1 shows the relationship between the surface area and the free energy in a latex colloidal system.

$$dG = \gamma dA \qquad (7.1)$$

where G is the free energy, γ is the interfacial tension, and A is the surface area. Due to the small particle size and the relatively large surface area of colloidal particles, in the absence of a stabilization mechanism, aggregation would be favored as it would decrease the surface area and consequently the free energy of the system. Therefore, colloidal systems need to have a stabilization mechanism to prevent aggregation. Colloids can be stabilized by:

1. Electrostatic repulsion: arising from like particles having a similar surface charge leading to mutual repulsion
2. Steric stabilization: arising from adsorbed species (e.g., surfactants), which prevent the particles from aggregating
3. Electrosteric stabilization: a combination of both electrostatic repulsion and steric stabilization

With colloidal systems, there may be a potential for interaction with colorants, which can lead to coagulation of the dispersion rendering it unusable for film coating. Aggregated material will not only lead to clogging of the dispersion delivery and spray systems, but it will also hinder the coalescence of colloidal polymer particles and the formation of a uniform film coating.

Depending on the manufacturing process, the polymer, and the composition of the dispersion used for film coating, differential commercial coating products may have different pH values. For example, some film-coating dispersions used in enteric coatings have low pH values due to the acidic nature of their functional groups. The addition of lakes to the acidic dispersions can therefore result in dissociation of aluminum lake pigments. One approach to preventing this type of interaction is to increase the pH of the acidic dispersion to a pH range in which aluminum lakes would not dissociate.

In colloidal systems, the surface charge acquired by the polymer particles plays an important role in the physical stability of the system. Factors that affect the surface charge of the colloidal polymer particles can lead to aggregation or coagulation of the system. The addition of electrolytes or ions to a colloid in sufficient concentrations can affect the surface charge and increase the tendency for coagulation to occur. Nyamweya and Hoag investigated the effect of adding several dyes to anionic, cationic, and nonionic Eudragit® colloidal polymer dispersions [33]. Figure 7.6 shows the minimum amount of dye required to cause coagulation of the colloidal dispersions (critical coagulation concentration). The dyes that were studied were all anionic molecules. Differences in the critical coagulation concentrations were observed based on the surface charge of the colloidal dispersion. The authors related these differences to the zeta potential of the polymer dispersions, shown in Figure 7.7.

The positively charged Eudragit® RS 30 D [(Poly(ethyl acrylate-co-methyl methacrylate-cotrimethylammonioethyl methacrylate chloride), 1:2:0.1)] colloidal dispersion had the lowest critical dye coagulation concentration. This polymer dispersion is stabilized by electrostatic repulsion

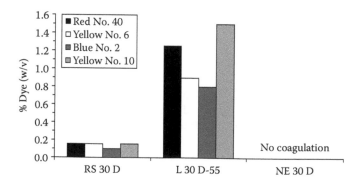

FIGURE 7.6   Dye critical coagulation concentrations for Eudragit® polymer dispersions. (From N. Nyamweya, K. A. Mehta, S. W. Hoag, *Journal of Pharmaceutical Sciences*, 2001, 90, 12, 1937–1947.)

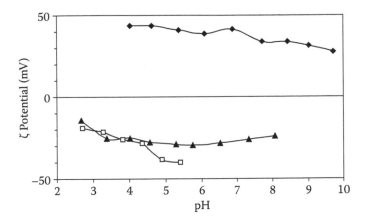

FIGURE 7.7   Zeta potential (ζ) as a function of pH for Eudragit® L 30 D55 (□), Eudragit RS 30 D (♦), and Eudragit NE 30 D (▲). (From N. Nyamweya, K. A. Mehta, S. W. Hoag, *Journal of Pharmaceutical Sciences*, 2001, 90, 12, 1937–1947.)

arising from cationic quaternary ammonium groups in its structure, which give rise to a positive zeta potential [33]. The addition of negatively charged anionic dyes results in neutralization of the stabilizing positive surface charges, leading to coagulation of the polymer dispersion at relatively low dye concentrations. In comparison, higher dye concentrations were required to cause coagulation of the anionic Eudragit® L 30 D-55 [(Poly(methacrylic acid-co-ethyl acrylate), 1:1)] colloidal dispersion because, like the dyes, it is negatively charged. High dye concentrations eventually resulted in coagulation of this polymeric dispersion due to compression of the stabilizing electrical double layer. It was observed that increasing the pH of the polymer dispersion to a range of 5.0–5.2 enhanced dispersion stability, and stable dispersions with dyes that did not coagulate were prepared. Increasing the pH results in an increased absolute value of the zeta potential (Figure 7.7), which makes the polymer more stable in the presence of dyes.

In contrast, the nonionic polymer dispersion Eudragit® NE 30 D [(Poly(ethyl acrylate-co-methyl methacrylate), 2:1)] was stable, and coagulation in the presence of dyes was not observed. Eudragit® NE 30 D is a chemically neutral polymer and has a nonionic emulsifier, nonoxynol 100. The high stability of Eudragit® NE 30 D in the presence of dyes was attributed to the steric stabilization of the polymer dispersion.

Interactions between colloidal polymer dispersions with pigments have been reported in the literature. Dangel et al. observed coagulation in methacrylic acid copolymer dispersions following the addition of red iron oxide–based pigment suspensions [36]. Similar findings for the same copolymer dispersion were reported by Flößer et al. [37]. In both these studies, coagulation was also dependent on the type of plasticizer used. In the former study, the presence of polyvinylpyrrolidone was a factor in dispersion stability, and the authors reported that the coagulation tendency disappeared when this excipient was not included in the dispersion. In another study, the addition of red iron oxide and iron oxide–pearl luster pigments was also observed to cause coagulation of methacrylic acid copolymer dispersions, which was prevented by the addition of sodium carboxymethylcellulose [38]. The authors attributed the enhanced dispersion stability to steric stabilization of the pigments and an increased viscosity from the dissolved sodium carboxymethylcellulose molecules.

Nyamweya and Hoag studied the interactions between Eudragit® RS 30 D, Eudragit® L 30 D-55, and Eudragit® NE 30 D polymeric aqueous dispersions and four aluminum lakes [33]. The stability of the polymer–pigment dispersions was studied by microscopy and particle size measurements. The stability of the polymer dispersions in the presence of lakes was found to be dependent on the pH and surface charge of the components. Eudragit® RS 30 D dispersions were stable in the presence of all the lakes. The addition of aluminum lakes to Eudragit® L 30 D-55 resulted in coagulation.

The authors attributed the coagulation to the low pH of the dispersion, which resulted in dissociation of the lakes and release of electrolytes, which, in turn, affected the stabilizing surface charges of the polymer. Increasing the pH of the polymer dispersion to a pH at which the lakes were stable prevented coagulation.

Although the nonionic polymer Eudragit® NE 30 D was stable in the presence of FD & C red No. 40 or yellow No. 6 lakes, aggregation was observed following the addition of FD & C blue No. 2 or D & C yellow No. 10 lakes. The differences in stability between the different lakes were found to be related to their surface charge. Lakes with positively charged surfaces promoted an interaction with the negative surface charges of the polymer. The lakes that did not cause coagulation were found to have a negative surface charge at the experimental pH. It was observed that the unstable dispersions could be stabilized by the addition of surface-active agents to the pigment dispersions prior to adding them to the polymer dispersion.

Ishikawa et al. investigated the colloidal stability of Eudragit® L 30 D-55, Eudragit® RS 30 D, and Eudragit® NE 30 D in the presence of yellow iron oxide and titanium dioxide at different pH values [39]. Eudragit® L 30 D-55 was evaluated over a pH range of 2–5, and the Eudragit® RS 30 D and Eudragit® NE 30 D dispersions were evaluated over the pH range of 2–11. Stable polymer–pigment dispersions were observed for the Eudragit® RS 30 D and Eudragit® NE 30 D at all pH values in the presence of either pigment. The Eudragit® L 30 D-55 dispersion was also stable with either pigment from a pH range of 3–5 but coagulated when the pH was lowered to a value of 2. However, aggregation of the polymer dispersion when the pH was adjusted to a value of 2 also occurred in the absence of pigments due to a reduction of the absolute value of the zeta potential [40].

Kucera et al. investigated the effects of aluminum lake pigments on the coagulation of Eudragit® EPO [(Poly(butyl methacrylate-co-(2-dimethylaminoethyl) methacrylate-comethylmethacrylate), 1:2:1)] aqueous dispersions [41]. In this study, the authors were able to prevent coagulation of the polymer dispersion in the presence of aluminum lakes by the addition of stabilizing excipients, such as povidone or poloxamer.

FORMULATION OF PIGMENTED COATING DISPERSIONS

Depending on the manufacturer, products used in pharmaceutical film coating are commercially available in the form of powders, granules, solutions, or colloidal dispersions. Some products require the addition of other excipients (e.g., plasticizers and anti-adherents) to the polymer, and other products are available as fully formulated or ready-to-use products (with the required excipients already added by the manufacturer). The advantages of fully formulated systems for drug product manufacturers are a reduction in the number of excipients that must be obtained and a reduction in preparation time and processing steps. On the other hand, there is less flexibility to change the composition of the film-coating formulation, which may become necessary in certain cases, such as when there is an excipient compatibility issue with a component in the film coating.

Many colloidal-based systems are shear sensitive, and the use of low-shear type mixers are often recommended for stirring the dispersions. However, because many pigments and anti-adherents are optimally dispersed with high-shear type mixers, a pigment dispersion is typically made separately and then added to the polymer dispersion with low-shear mixing. For some products, it is important to follow the order of addition in which the pigment dispersion is added to the polymer dispersion to reduce the possibility of coagulation. Adding pigments in a diluted state as opposed to directly adding them to colloidal polymer dispersions is preferable because it provides a more gradual change to the medium in which the colloidal polymer particles are dispersed. To prevent settling of pigment particles, continuous low-shear stirring of the coating dispersion is recommended during the coating process.

INFLUENCE OF COLORING AGENTS ON POLYMER FILMS AND COATED PRODUCTS

Although the main reasons for adding coloring agents are to modify the visual characteristics of the dosage form or provide protection from light, the addition of colorants may unintentionally influence the mechanical properties, permeability, and drug release characteristics of a film coating. In some cases, undesirable effects in the appearance of the film coating (e.g., increased surface roughness) may occur with the incorporation of pigments. With the addition of increasing amounts of colorant, pigment particles at some point will start to reduce intermolecular bonding between polymer molecules and affect the properties of the film. The amount of insoluble excipients that can be added to a polymer film without adversely affecting its intended functions or applications (e.g., sustained release properties) is sometimes referred to as the pigment binding capacity although actual quantitative measured values of this term seem to be absent in the pharmaceutical literature. Polymers with a high pigment capacity can be defined as those that can incorporate very high levels of insoluble additives while still retaining their functional characteristics.

A more well-defined concept, in this regard, is the critical pigment volume concentration (CPVC) [42,43]. According to this theory, below the CPVC, the polymer is able to completely bind and surround the pigment particles forming a dense and continuous film (Figure 7.8). The addition of pigments will initially reduce the permeability of the polymeric film below the CPVC due to an increased tortuosity of diffusion pathway while above the CPVC there is incomplete binding of the pigment particles by the polymer, resulting in the formation of voids within the film [44,45]. In the latter case, there is an increase in film permeability and a reduction in the mechanical strength of the film.

APPEARANCE

The appearance that a coloring agent and film coating impart to a drug product plays an important role in the development of a visually aesthetic product. From a therapeutic standpoint, color may play a role in enhancing patient compliance. Furthermore, color is an important attribute by which different drug products may be distinguished and can hence play an important role in reducing dispensing and patient medication errors. Color may also be reflective of the quality of the product and coating process because color is a readily observable feature of the drug product. Non-uniform distribution of color in the coating, mottled coloration, and color fading may be indicative or suggestive of issues with product quality, changes in the stability of the product, and changes or lack of control in the manufacturing and coating process. Several studies have used color to assess the efficiency and uniformity of the coating process [46,47]. Because color is a basic feature of product identification, consistency of drug product color can be an important factor in the quality control of pharmaceutical products [48]. Consequently, if any changes are made that influence the color of a commercial drug product (such as changes in coating composition or level), color matching of the old and new formulations is important for maintaining the appearance and identity of the product.

Color in film-coated products may be assessed by visual comparison to standards with a defined color or more objectively by the use of color measurement instruments, such as spectrophotometers

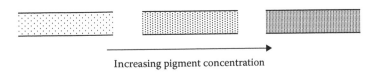

Increasing pigment concentration

FIGURE 7.8 Effect of increasing pigment volume levels in a polymer film.

or tristimulus colorimeters. Color can be measured using color scales, such as the CIE (Commission Internationale l'Eclairage or the International Commission on Illumination) X, Y, Z, or CIE L*a*b* scales [49]. The X, Y, Z functions are based on the average spectral responses (tristimulus values) to red, green, and blue light of human observers. The tristimulus values are based on the three types of cone-shaped receptors in the human eye that are responsible for color vision. Different colors stimulate the cone receptors to different degrees, giving rise to the range of colors visible to the human eye [50]. The X, Y, Z values may be converted into uniform color scales, such as the CIE L*a*b* scale [48]. In the CIE L*a*b* model, color is a function of the values of L*, a*, and b*, which are coordinates in a three-dimensional space (Figure 7.9). L* indicates lightness (ranging from black to white), a* indicates redness–greenness, and b* indicates yellowness–blueness.

In addition to protecting photosensitive products from light, pigments may also serve to mask the appearance of the underlying substrate, which can be important for cores with an unpleasant appearance. The addition of a film coating to cores to obscure their appearance may be one approach to preparing blinded drug products for clinical trials in which the goal is to reduce any bias that may result from observed differences in the drug products being evaluated. Felton and Wiley used over-coating with a hydroxypropyl methylcellulose containing iron oxide and titanium dioxide pigments to visually conceal identifying marks of a sustained release tablet [51]. In such instances, the opacity or hiding power of the film coating is an important factor and is dependent on the type and amount of colorant in the film as well as the thickness of the applied coating. Most polymers used in film coatings will usually form relatively transparent or translucent films and, as such, will not provide much hiding power without the incorporation of pigments having some degree of opacity.

Rowe investigated the effect of several pigments on the opacity of hydroxypropyl methylcellulose–based films [52]. In these studies, opacity measurements of the films were evaluated using a contrast ratio. The contrast ratio (in percentage) was determined by dividing the light reflectance values from a coated black substrate by the light reflectance values from a coated white substrate. With increasing contrast ratio values, the core substrate becomes less visible and increasingly more difficult to see. At equivalent pigment levels, films containing talc had a relatively low contrast ratio (<50%), and titanium dioxide and iron oxides (black, red, and yellow) had high contrast ratios (>90%). The contrast ratio values for aluminum lakes varied from less than 70% to more than 95% and increased in the order yellow < orange < red < blue. Additionally, the contrast ratio values were observed to increase with increasing dye content of the lake. Increasing either the pigment concentration or the film thickness resulted in higher contrast ratio values [53,54].

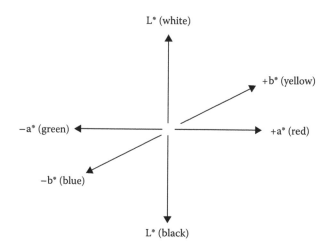

FIGURE 7.9   The coordinates of the CIE L*a*b* color space.

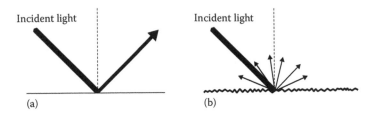

FIGURE 7.10 Specular and diffuse reflection of incident light from (a) smooth and (b) rough surfaces. Dotted line is normal to the surface.

In addition to color, the appearance of a coated product may also be evaluated in terms of gloss and surface roughness. Gloss results from the specular reflection of incident light from a smooth surface as shown in Figure 7.10 when the light is reflected in the opposite direction at an equal angle. In contrast, a rough surface reflects incident light diffusely, scattering it in many directions. Gloss is desirable as it enhances the elegance and aesthetic appeal of the drug product. For example, high gloss contributes to the visual appeal and elegance of pigmented hard gelatin capsules. Rowe studied the effect of pigment particle size on the gloss of film-coated tablets and found that gloss decreased with increasing pigment concentration [55]. Gibson et al. observed a similar relationship in gloss reduction for titanium dioxide and an aluminum lake although black iron oxide exhibited a decrease in gloss at low pigment levels followed by an increase in gloss at higher pigment levels [56]. The authors used the gloss measurements to determine critical pigment volume concentrations for the pigments in hydroxypropyl methylcellulose films.

In contrast to gloss, roughness is an undesirable feature in a coating, especially if it is clearly visible to the naked eye. The effect of pigment concentration on surface roughness was investigated by Rowe in a study that reported that, while low pigment concentrations resulted in a minor increase in the surface roughness of film-coated tablets, at higher pigment concentrations (above the critical pigment volume concentrations), the surface roughness increased considerably [57].

LIGHT PROTECTION

Light can influence the stability of many active pharmaceutical ingredients. For some drug products, light-induced interactions and decomposition may be associated with changes in color. The magnitude of the effects of light on the stability of photolabile compounds can vary considerably from very small amounts of degradation after several weeks of light exposure for some drugs to extensive decomposition in the order of minutes for extremely photosensitive actives [58].

For active pharmaceutical ingredients that are light sensitive, the incorporation of appropriate pigments into the film coating may be an approach to improve drug product stability. The ability of a pigment to provide light protection depends on its ability to reduce the transmittance of light to the substrate or core drug product. Incident light may be reflected, absorbed, or transmitted by a polymeric film coating. Increasing the amount of light that is reflected or scattered will reduce the amount of light that is transmitted to the drug in the core. The amount of light reflected at a polymer–pigment interface can be related to the refractive indices of the components by Equation 7.2 (assuming normal incident light and no absorption) [59].

$$R = \left[\frac{n_1 - n_2}{n_1 + n_2} \right]^2 \tag{7.2}$$

where R is the amount of light reflected at the interface, n_1 is the refractive index of the pigment, and n_2 is the refractive index of the polymer. Increasing the difference between the refractive indices of the pigment and the polymer will increase the amount of light reflected. Pigments that exhibit this behavior are called opacifiers. Rowe and Forse compared the refractive indices of several film-coating polymers and pigments [59]. The refractive indices of the film coating polymers were approximately 1.5, and the refractive indices of pigments ranged from 1.50 to 1.54 for aluminum lakes; 1.54 to 1.59 for talc; 1.94 to 2.51 for yellow oxide; 2.49 to 2.55 for titanium dioxide (anatase form); and 2.94 to 3.22 for red iron oxide. Many pigments are anisotropic and have more than one refractive index, depending on their orientation.

A number of authors have investigated the use of pigmented coatings in stabilizing light-sensitive drugs. Nyqvist and Nicklasson compared the effects of titanium dioxide and yellow iron oxide in hydroxypropyl methylcellulose coatings applied to tablets containing a light-sensitive drug substance [60]. The pigments reduced light-induced changes in the color of the active in the core with a combination of iron oxide and titanium dioxide in the film coating providing the best stability.

Teraoka et al. evaluated the effects of hydroxypropyl methylcellulose–free films containing titanium dioxide or tartrazine (FD & C yellow No. 5) on the stability of the photolabile drug nifedipine [61]. In this study, a sample of the drug was dispersed on a glass plate, which was then covered by a free (unattached to a substrate) polymer film using a special holding device and then exposed to light. Light-transmission measurements indicated that the different colorants transmitted light over different wavelengths with tartrazine having lower light transmission in the visible region and titanium dioxide being more effective in the ultraviolet region. Films containing a mixture of equivalent parts of each colorant had lower light transmittance than either of the individual colorants at equivalent concentrations in the polymer film. When placed on the dispersed drug, films with the binary colorant combination provided better protection against photodegradation of the active films with a single colorant.

Béchard et al. investigated the influence of titanium dioxide concentration and film coating thickness on the photostability of nifedipine tablets coated with hydroxypropyl methylcellulose [62]. The tablets were exposed to fluorescent light for up to three weeks. The authors found that acceptable light protection against drug degradation was obtained for films having contrast ratio values above 98%, which was only achieved by using thick film coatings with high levels of titanium dioxide.

Aman and Thoma evaluated the effectiveness of different formulation approaches in stabilizing light-sensitive molsidomine tablets, including (a) incorporating light-absorbing excipients into the core tablets, (b) incorporating pigments (iron oxides or titanium dioxide) into the core tablets, and (c) coating the tablets with pigmented hydroxypropyl methylcellulose films [63]. Light absorbers and pigments both improved the stability of the tablets when incorporated into the cores with pigments being more effective. However, in both cases, significant drug degradation was still detected upon exposure to light in the time period that was studied. The formulation of a photostable drug product was only achieved by film coating (Table 7.3) or blister packaging. The authors' results indicated

---

TABLE 7.3

Photostability of Various Molsidomine Tablet Formulations after Light Exposure

Formulation	Drug Decomposition (%)			
	After 1 hr	After 3 hr	After 6 hr	After 12 hr
Uncoated tablets	7	20	25	33
Coated tablets				
TiO$_2$, level, film thickness				
4.8% TiO$_2$, 35 μm	2.5	5	9.5	19.5
4.8% TiO$_2$, 73 μm	0	0	0	2
9.9% TiO$_2$, 33 μm	0	0	0	3
TiO$_2$, level. Iron oxide, level, film thickness				
4.8% TiO$_2$, 0.9% red iron oxide, 37 μm	0	0	0	0
4.8% TiO$_2$, 0.9% yellow iron oxide, 39 μm	0	0	0	0

Source: W. Aman, K. Thoma, *Journal of Pharmaceutical Sciences*, 2004, 93, 7, 1860–1866.

that coating thickness, pigment concentration, and pigment type could influence the stability. A combination of iron oxides and titanium dioxide in the film coating provided the most stable tablets.

MECHANICAL PROPERTIES

Film coatings that are applied to solid dosage forms should have sufficient mechanical properties to withstand further processing and handling after the film-coating process, packaging, and transportation of the drug product until it reaches the patient. Brittle film coatings may lead to the formation of cracks, which could compromise the release characteristics of the drug product. The adsorption of polymer molecules on the surface of pigment particles can result in a restriction of polymer mobility [64], which can increase the elastic modulus of the polymer and make the coating more brittle. The addition of pigments to polymer films has been shown to be a factor in increasing coating defects [65]. These effects may be due to insoluble particles acting as stress concentrators, thereby promoting the initiation of cracks in the film and/or the presence of interactions between the additive and the polymer [66]. Poorly dispersed pigments may also play a role by acting as a focus for localized stress in polymer film coatings [67]. Possible formulation approaches for reducing pigment-related coating defects include reducing pigment levels, increasing the amount of plasticizer and using a more flexible or tougher polymer.

High internal stresses in polymer films may lead to a defect known as edge-splitting with which the film breaks and peels back from the edges of a coated tablet. Rowe observed an increase in edge-splitting in tablets coated with hydroxypropyl methylcellulose films with the addition of lakes and iron oxides [65]. In contrast, talc was observed to lower the incidence of edge-splitting. Furthermore, the inclusion of talc or magnesium carbonate in hydroxypropyl methylcellulose films containing a yellow lake pigment was found to lower the incidence of edge-splitting [68]. In contrast to the other pigments studied, the talc and magnesium carbonate particles were both platelet-shaped, suggesting that particle morphology plays an important role in reducing the incidence of edge-splitting in film coatings.

The mechanical properties of polymer films are usually characterized by tensile testing in which a polymer film is subjected to a tensile force until it breaks with the recorded data used to generate a stress–strain curve. The mechanical behavior may be reported in terms of strength at break (the value of stress at which the film breaks), elongation at break (the value of strain at which the film breaks), elastic modulus (the ratio of stress to strain in the initial linear region of the stress–strain curve), and toughness (the energy required to break the polymer film). Chapter 4 of this text provides more details on tensile testing.

Several studies have reported that the addition of pigments can lead to significant changes in the mechanical properties of polymeric films. Porter found that the diametral crushing strength of tablets coated with a hydroxypropyl methylcellulose film decreased when pigments (titanium dioxide and an aluminum lake) were incorporated into the coating [69]. In the same study, tensile tests on free films showed a reduction in strength with the addition of pigments. Okhamafe and York evaluated the effects of different types of titanium dioxide and talc on the mechanical properties of unplasticized and plasticized hydroxypropyl methylcellulose films [70]. In general, the addition of pigments resulted in a reduction in tensile strength and elongation while the elastic modulus increased. In a related study, some of the differences in mechanical behavior of the pigmented polymer films were attributed to differences in pigment particle size, surface area, and morphology [71].

Aulton and Abdul-Razzak studied the effects of three aluminum lakes and titanium dioxide on the mechanical properties of hydroxypropyl methylcellulose films [72]. The inclusion of pigments generally resulted in more brittle polymer films. Aluminum lakes with different colors, but equivalent particle size showed very similar effects on the mechanical properties of the polymer films. Some differences, however, in film toughness were observed when lakes of the same color but different particle size were compared with the finer particle size grades, yielding slightly tougher polymer films. Gibson et al. evaluated the effects of aluminum lakes, iron oxides, and talc on the mechanical properties of hydroxypropyl methylcellulose films [66,73]. Increases in the modulus values were observed in most cases with the addition of pigments to unplasticized polymer films while the effects of pigments on the modulus of plasticized films were noticeably less. In general, the pigments lowered the tensile strength, elongation at break, and toughness of the polymer films although to different degrees, depending on the pigment. The differences in the effects of the different pigments were, in part, related to the shape of the pigment particles. In comparing the data to studies that examined the effects of pigments on the incidence of edge-splitting, the authors reported that all pigments increased the incidence of edge-splitting with the exception of talc, which reduced the occurrence of this defect. Stress-relaxation experiments indicated that talc enhanced the ability of the polymer films to relax in response to applied stress, a finding that the authors attributed to the lamellar shape and orientation of the talc particles facilitating stress relief by slippage of adjacent particles within the polymer film.

The majority of studies that have investigated the effects of pigments on the mechanical properties of polymers have focused on hydroxypropyl methylcellulose. Reports on the effects of pigments on the mechanical properties of other polymers are very limited. Hsu et al. investigated the effect of titanium dioxide on the mechanical properties of polyvinyl alcohol films and observed a reduction in the tensile strength, elongation at break, and film toughness with increasing pigment levels [74]. Nyamweya investigated the influence of aluminum lakes on the mechanical properties on sustained release (Eudragit® RS PO) and enteric (Eudragit® L 100-55) polymethacrylate polymer films [75]. The addition of aluminum lakes increased the rigidity of the polymeric films as evidenced by the increase in the elastic modulus. The elongation at break of the polymer films decreased with the addition of aluminum lakes, indicating a reduction in film flexibility. Both of these findings were in accordance with previous studies that investigated the effects of insoluble additives on hydroxypropyl methylcellulose–based films. However, it was observed that the tensile strength of the films was relatively unchanged (Eudragit® L 100-55 films) or even increased (Eudragit® RS PO films) with the incorporation of lakes. The effects of an aluminum lake on the mechanical properties of the plasticized Eudragit® RS PO films are illustrated in the stress–strain curves generated from tensile testing in Figure 7.11.

The majority of studies that have evaluated the effects of pigments on the mechanical properties of film-coating polymers have been conducted using free (unattached) films. Okhamafe and York investigated the effects of pigments on the mechanical properties of hydroxypropyl methylcellulose film coatings applied to aspirin tablets using an indentation apparatus [76]. The hardness and modulus of the films increased with the addition of talc, and incorporation of titanium dioxide did not

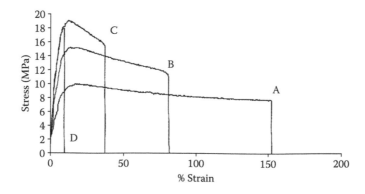

FIGURE 7.11 Effects of FD & C yellow No. 6 aluminum lake on the stress–strain profile of plasticized Eudragit® RS PO films: (A) 0% (w/w) lake; (B) 10% (w/w); (C) 20% (w/w); and (D) 40% (w/w) lake. (From N. Nyamweya, Characterization of the interactions between aluminum lake pigments and Eudragit film coating polymers, PhD Dissertation, University of Maryland, Baltimore, 2001.)

increase either the hardness or modulus. Nyamweya studied the effect of incorporation of aluminum lakes into Eudragit® L 30 D-55 and Eudragit® RS/RL 30 D film coatings on tablet-crushing strength and found that tablets coated with pigmented coatings had slightly higher crushing strength values compared to tablets coated with unpigmented films [75].

THERMAL PROPERTIES

A few studies have investigated the effect of pigments on the glass transition temperature of film-coating polymers. The glass transition temperature, a fundamental property of an amorphous polymer, is the temperature at which a polymer undergoes a change from a hard, brittle, glassy state into a soft, flexible, rubbery state. It is well known that excipients that lower the glass transition temperature, such as plasticizers, make polymers less brittle and more flexible. It is therefore of interest to determine what effect pigments, which have been shown to have significant effects on the mechanical properties of polymers, would have on the glass transition temperatures of these polymers.

Okhamafe and York evaluated the effect of talc and titanium dioxide on the glass transition temperature of hydroxypropyl methylcellulose films [77]. Increases in the glass transition temperature of plasticized and unplasticized pigmented films of up to 15°C to 16°C were observed. Talc had a greater effect on the glass transition temperature of the polymer films than titanium dioxide. The differences between the different pigments on the glass transition temperatures of the films were attributed to differences in their shape, surface area, and interaction with the polymer. Other authors have noted an increase in the glass transition temperature of hydroxypropyl methylcellulose films with the addition of titanium dioxide [66]. Using films applied to tablet compacts, Felton and McGinity also found an increase in the glass transition temperature of an acrylic polymer with increased titanium dioxide concentrations [78].

Nyamweya studied the effects of four aluminum lakes on the glass transition temperature of plasticized Eudragit® RS PO and Eudragit® L 100-55 films (Figure 7.12) [75]. Although the lakes had very little effect on the glass transition temperature of Eudragit® RS PO, significant increases in the glass transition temperature of Eudragit® L 100-55 films were observed, suggesting a greater degree of interaction between the pigments and the latter polymer. In the case of Eudragit® RS PO, changes in the mechanical properties of the films (Figure 7.11) could therefore not be attributed to a change in the glass transition temperature of the polymer. For Eudragit® L 100-55 films, the increases in the glass transition temperature could, in part, account for the changes in mechanical properties of the

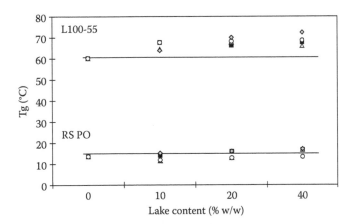

FIGURE 7.12   Effect of aluminum lakes on the glass transition temperature of plasticized Eudragit® L 100-55 and Eudragit RS PO films. *Symbols*: ■, blue No. 2 lake; ◆, red No. 40 lake; ○, yellow No. 6 lake; △, yellow No. 10 lake. (From N. Nyamweya, Characterization of the interactions between aluminum lake pigments and Eudragit film coating polymers, PhD Dissertation, University of Maryland, Baltimore, 2001.)

polymer films; however, for most of the lakes, the glass transition temperature of the polymer did not increase further beyond a level of 10% (w/w) lake while the mechanical properties continued to show a concentration dependency at higher pigment levels.

Permeability

For many solid dosage forms, especially drug products that are sensitive to moisture, it is desirable to have a film coating with low water vapor permeability. Polymeric film coating can be an effective method to reduce the transmission rate of water vapor. The effectiveness of a film coating in this regard will depend on the type of polymer, added excipients, and the thickness of the film coating. An effective moisture-protective coating may also reduce the need for specialized protective product packaging. In addition to reducing the rate of water vapor uptake from the environment, film permeability can also be important in controlling the release of actives. For example, low permeability to salivary fluids is important in coated taste-masking applications, and low permeability of an enteric coating to gastric fluids would be important in protecting an acid-sensitive drug from degradation in the stomach.

A number of studies have investigated the effects of pigments on the water vapor permeability of polymer films. Porter studied the effect of titanium dioxide and an aluminum lake on the moisture permeability of hydroxypropyl methylcellulose and showed the addition of pigments reduced water vapor permeability of the polymer films [67]. These findings were attributed to the pigment particles acting as a barrier to moisture and increased the diffusion path length for permeating water molecules. Other researchers have suggested that the parallel orientation of lamellar or plate-like pigment particles increases the diffusional path to slow the rate of water vapor transmission and hence improve the stability of moisture-sensitive drugs [79]. Porter [67] reported an initial decrease in film permeability followed by an increase in permeability at higher pigment levels. However, pigmented films were still less permeable to moisture than unpigmented films at all pigment concentrations studied, suggesting that the pigment levels were below the critical pigment volume concentrations. Okhamafe and York studied the influence of different grades of talc and titanium dioxide on the moisture permeability of hydroxypropyl methylcellulose–based films [42,45] and also found film permeability first decreased and then increased with increasing pigment levels. Working with polyvinyl alcohol films, Hsu et al. reported slight increases in permeability at low levels of titanium dioxide followed by a sharp increase in permeability at high pigment levels [74].

List and Kassis investigated the effects of talc and titanium dioxide on the water vapor permeability of Eudragit® L 30 D-55 films [80]. They found that the incorporation of talc reduced the water vapor permeability of plasticized Eudragit® L 30 D-55 films while the addition of titanium dioxide resulted in an increase in film permeability to water vapor. The increased permeability of polymer films with titanium dioxide was attributed to the hydrophilic nature of this pigment. Porter and Ridgway investigated the effect of red iron oxide on the permeability of two enteric polymers, cellulose acetate phthalate and polyvinyl acetate phthalate [44]. The addition of iron oxide had little effect on the permeability of polyvinyl acetate phthalate–free films while the permeability of the cellulose acetate phthalate films initially decreased and then increased with increasing pigment concentrations. Similar effects were observed when enteric-coated tablets were tested for permeability to simulated gastric fluid using an acid uptake test.

Drug Release

In functional film coatings, which are designed to control drug release, the addition of colorants to the film coating has been reported to have various effects on the drug release profile. Ghebre-Sellassie et al., for example, investigated the effect of kaolin on the dissolution of pellets coated with Eudragit® NE 30 D and found that increasing levels of this excipient resulted in increasing drug release rates [81]. Chang and Hsiao reported increased dissolution rates with the addition of talc to Eudragit® RS 30 D coated pellets [82].

In another study, Maul and Schmidt investigated the effect of pigments on the release profiles of Eudragit® L 30 D-55 enteric-coated pellets [38]. Platelet-shaped pigments (talc, mica, iron oxide coated mica, and titanium dioxide–coated mica) were found to reduce the drug release rate from the coated pellets, and spherical (titanium dioxide) or needle-shaped pigments (red iron oxide) had little effect or led to an increase in drug release in acidic media. Slower drug release rates were observed for films containing platelet-shaped pigments with either hydrophilic or hydrophobic surfaces, suggesting that particle shape has a greater effect on drug release than surface chemistry. Furthermore, it was observed that platelet-shaped pigments with a larger particle size had a greater effect on reducing drug-release rates compared to platelet-shaped pigments with a smaller particle size. The alignment of platelet-shaped particles, which tend to lie flat and parallel to the surface of the film, may serve as a barrier to the movement of water [38,83]. Similar findings were reported in a later study that investigated the effects of the same pigments on drug-release profiles of pellets coated with Aquacoat® ECD or Eudragit® RS 30 D sustained-release polymers [84].

Nyamweya evaluated the effects on FD & C yellow No. 6 and D & C yellow No. 10 aluminum lakes on the dissolution profiles of enteric (Eudragit® L 30 D-55) and sustained-release (Eudragit® RS/RL 30 D) polymer film–coated tablets [75]. The effects of the aluminum lakes on dissolution profiles of Eudragit® L 30 D-55 coated tablets are shown in Figure 7.13. The Eudragit® L 30 D-55 dispersion was partially neutralized prior to the addition of the lakes to prevent coagulation. In the acid stage of the dissolution test, leaching of the lake dyes from the film coating was observed after the tablets were placed in the dissolution medium due to dissociation of the lake substrate at low pH of the medium. However, the enteric properties of the coating were not adversely affected, and similar dissolution profiles were observed between the pigmented and unpigmented coated tablets. When the aluminum lakes were incorporated into Eudragit® RS/RL 30 D films, a slower initial hydration of film coating was observed compared to unpigmented film-coated tablets. However, subsequent dissolution rates were relatively similar (Figure 7.14). The low pH of the dissolution medium did not lead to faster drug release rates as may have been expected to occur as a result of lake dissociation at low pH values.

Another study showed the importance of the source of talc on drug release from controlled-release coated products. Annamalai et al. studied the effects of talc of various grades from several manufacturers on the dissolution profiles of methacrylate-based enteric and sustained-release coated tablets [85]. The study found that the source and grade of talc could have a significant effect on drug release from

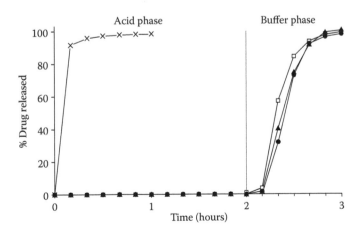

FIGURE 7.13 Dissolution profiles of tablets coated with Eudragit® L 30 D55 (dissolution media: 0.1 N HCl for two hours, followed by pH 6.8). *Symbols*: □, 0% lake; ●, 30% yellow No. 6 lake; ▲, 30% yellow No. 10 lake. (From N. Nyamweya, Characterization of the interactions between aluminum lake pigments and Eudragit film coating polymers, PhD Dissertation, University of Maryland, Baltimore, 2001.)

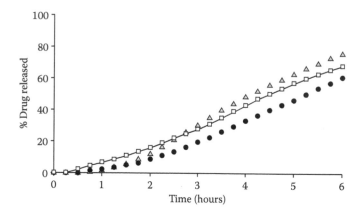

FIGURE 7.14 Dissolution profiles of tablets coated with Eudragit® RS 30 D (dissolution medium, 0.1 N HCl). *Symbols*: □, 0% lake; ●, 30% yellow No. 6 lake; ▲, 30% yellow No. 10 lake. (From N. Nyamweya, Characterization of the interactions between aluminum lake pigments and Eudragit film coating polymers, PhD Dissertation, University of Maryland, Baltimore, 2001.)

tablets coated with either polymer. Because talc is a naturally occurring mineral, the effects of changing to a new talc source (e.g., due to a change in supply from a mine being depleted) need to be considered because this could, in certain cases, result in changes in the dissolution profile of a drug product.

ADHESIVE PROPERTIES

Good adhesion between a polymeric film and the substrate surface is necessary for coated drug products, and the addition of pigments in coating formulations has been reported to influence the adhesive properties of the applied films. Polymer adhesion is dependent on the internal stresses within the film and the strength of the interfacial interactions at the film-substrate interface. Pigments can increase internal stress within the film and also interfere with the interfacial intermolecular bonding, both of which tend to disfavor adhesion. The extent to which pigments affect polymer adhesion is thought to be dependent on the particle size, morphology, and concentration in the coating formulation [86].

However, because pigments differ significantly in their physical properties, such as particle size, morphology, and density, the literature on the relationship between pigments and substrate adhesion is somewhat contradictory. For example, Fisher and Rowe [87] found a 45% reduction in adhesion of hydroxypropyl methylcellulose films when 10% titanium dioxide was added to the coating formulation whereas Lehtola et al. [88] reported increased adhesion of hydroxypropyl methylcellulose films with increasing concentrations of titanium dioxide. Felton and McGinity [78] also found stronger film–tablet adhesion when titanium dioxide was added to an aqueous dispersion of an acrylic polymer, which the authors attributed to the titanium dioxide increasing hydrogen bond formation at the film-tablet interface with minimal increases in the internal stresses in the film.

In contrast to titanium dioxide, which was thought to enhance interfacial interactions, pigment particles can embed themselves at the film–tablet interface and disrupt intermolecular bonding. Okhamafe and York [89] showed that talc adversely affected adhesion and attributed these findings to the hydrophobic particles interfering with hydrogen bond formation between the tablet surface and the film coating. Smaller pigment particles are expected to disrupt intermolecular bonding at the film–substrate interface to a lesser extent than larger particles. Felton and McGinity [78], for example, found that adhesion was strongest when the small yellow iron oxide pigment was incorporated in an acrylic polymer whereas adhesion was compromised when the larger mica particles were incorporated into the coating formulation.

SUMMARY

Coloring agents are important components of coated pharmaceutical drug products and can be used to enhance product elegance, product identification, and product differentiation and to improve the stability of light-sensitive compounds. Coloring agents also play an important role in reducing medication errors and may contribute to the development of drug products with unique visual features making them more difficult to counterfeit. There are a number of coloring agents available, and from a regulatory standpoint, the intended regions and countries where drug products containing the colorant will be marketed should be considered. Pigments may influence the properties of polymer dispersions and the resulting polymer films in a number of ways, which may have a significant effect on the performance of the drug product. An understanding of the physical and chemical properties of colorants as well as how they interact with polymers used in coating applications will enable the formulation of coated products with colorants.

REFERENCES

1. Berns RS. Billmeyer and Saltzman's Principles of Color Technology. New York: Wiley; 2000.
2. Schiek RC, Fytelson M, Singer JJ. Pigments. In: Mark HF, Othmer DF, Overberger CG, Seaborg GT, Grayson M, Eckroth D, editors. *Encyclopedia of Chemical Technology*. 3rd ed. New York: Wiley; 1982. pp. 788–889.
3. Wou LLS, Mulley BA. Microstructure of aluminum hydroxides and the formation of aluminum dye lakes. *Journal of Pharmaceutical Sciences*. 1984;73(12):1738–44.
4. Bell JH, Stevenson NA, Taylor JE. A moisture transfer effect in hard gelatin capsules of sodium cromoglycate. *Journal of Pharmacy and Pharmacology*. 1973;25(Suppl.):96P–103P.
5. Schoneker DR, Swarbrick J. Coloring agents for use in pharmaceuticals. *Encyclopedia of Pharmaceutical Technology*. 3rd ed. New York: Informa Healthcare USA, Inc. 2007:648–70.
6. Marmion DM. Handbook of U.S. Colorants for Foods, Drugs, and Cosmetics. 2nd ed. New York: John Wiley & Sons; 1984.
7. Trojan M, Šolc Z, Novotny M. Pigments, Kirk-Othmer Encyclopedia of Chemical Technology, Vol. 19, New York: John Wiley & Sons; 1996.
8. Bauer KH. Coated Pharmaceutical Dosage Forms: Fundamentals, Manufacturing Techniques, Biopharmaceutical Aspects, Test Methods, and Raw Materials: CRC Press; 1998.
9. Weller P. Titanium dioxide. *Handbook of Pharmaceutical Excipients*. 5th ed. Washington, DC: American Pharmacists Association. 2006:782–4.
10. Battista OA, Smith PA. Microcrystalline cellulose. *Journal of Industrial & Engineering Chemistry*. 1962;54(9):20–9.
11. Barrer RM, Barrie JA. Sorption and diffusion in ethyl cellulose. Part IV. Water in ethyl cellulose. *Journal of Polymer Science*. 1958;28:377–86.
12. Kibbe A. Talc. In: Rowe RC, Sheskey PJ, Owen SC, editors. *Handbook of Pharmaceutical Excipients*. 5th ed. Washington, DC: American Pharmacists Association, 2006, pp. 767–769.
13. Lin K, Peck GE. Characterization of talc samples from different sources. *Drug Development and Industrial Pharmacy*. 1994;20(19):2993–3003.
14. Burdock GA. *Encyclopedia of Food and Color Additives*. CRC Press; 1997.
15. Yajima T, Itai S, Takeuchi H, Kawashima Y. Determination of optimum processing temperature for transformation of glyceryl monostearate. *Chemical and Pharmaceutical Bulletin*. 2002;50(11):1430–3.
16. Howland H, Fahmy R, Hoag SW. Analysis of curing of a sustained release coating formulation by application of NIR spectroscopy to monitor changes associated with glyceryl monostearate. *Drug Development and Industrial Pharmacy*. 2014(0):1–11.
17. Rowe RC, Sheskey PJ, Owen SC. *Handbook of Pharmaceutical Excipients*. 5th ed. London: Pharmaceutical Press and American Pharmacist Assoication; 2006.
18. Kriangkrai W, Puttipipatkhachorn S, Sriamornsak P, Pongjanyakul T, Sungthongjeen S. Impact of anti-tacking agents on properties of gas-entrapped membrane and effervescent floating tablets. *AAPS Pharmscitech*. 2014;15(6):1357–69.
19. Binns JS, Davies MC, Mella CD, editors. A study of polymer hydration and drug distribution within hydrophilic polymer matrices by cryogenic SEM and EDX. International Symposium on Controlled Release of Bioactive Materials; 1990; Reno, Nevada: Controlled Release Society, Inc.

20. Kirk-Othmer. Kirk-Othmer Chemical Technology of Cosmetics: Wiley; 2012.

21. OPADRY® fx™ Product Information. An Immediate Release, Pearlescent, Film Coating System from Colorcon®, available at http://www.colorcon.com/literature/marketing/fc/Opadry%20fx/pi_opadry_fx_prop .pdf (accessed June 24, 2016).

22. Wheatley TA. What are excipients? *Drugs and the Pharmaceutical Sciences.* 2000;103:1–19.

23. Signorino CA, Meggos H. Solutions with opacifiers of titanium dioxide and iron oxide and immobilizer for dyes. Google Patents; 1995.

24. Signorino CA, Meggos H. Dye compositions and methods for film coating tablets and the like. Google Patents; 1997.

25. Tyagi S, Ray A, Sood Y. Surface characteristics of coating layers formed by coating pigments with different particle sizes and size distribution. *Journal of Coatings Technology and Research.* 2010;7(6):747–56.

26. Völz HG. Pigments, Inorganic, 1. General. Ullmann's Encyclopedia of Industrial Chemistry: Wiley-VCH Verlag GmbH & Co. KGaA; 2000.

27. Rowe R. The effect of the particle size of synthetic red iron oxide on the appearance of tablet film coatings. *Pharmaceutica Acta Helvetiae.* 1984;60(5–6):157–61.

28. Wou LLS, Mulley B. Effect of dispersion on the coloring properties of aluminum dye lakes. *Journal of Pharmaceutical Sciences.* 1988;77(10):866–71.

29. Bawa R. Ocular Inserts. In: Mitra AK, editor. Ophthalmic Drug Delivery Systems. 1993 ed. New York: Marcel Dekker, Inc.; 1993. pp. 223–60.

30. Signorino CA, Levine S, Barkley A, Forcellini L. The use of acrylic resins for improved aqueous enteric coating. *Pharmaceutical Technology.* 2004:32–9.

31. Bajdik J, Bölcskei É, Kelemen A, Pintye-Hódi K. Rapid method to study the sedimentation of a pigment suspension prepared for coating fluids. *Journal of Pharmaceutical and Biomedical Analysis.* 2007;44(5):1159–62.

32. Desai A, Peck G, Lovell J, White J, Hem S. The effect of aluminum hydroxide dissolution on the bleeding of aluminum lake dyes. *Pharmaceutical Research.* 1993;10(10):1458–60.

33. Nyamweya N, Mehta KA, Hoag SW. Characterization of the interactions between polymethacrylate-based aqueous polymeric dispersions and aluminum lakes. *Journal of Pharmaceutical Sciences.* 2001; 90(12):1937–47.

34. Gibson S, Rowe R, White E. Quantitative assessment of additive-polymer interaction in pigmented hydroxypropyl methylcellulose formulations using immersion calorimetry. *International Journal of Pharmaceutics.* 1988;48(1):113–7.

35. Sawyer CB, Reed JS. Adsorption of hydroxypropyl methyl cellulose in an aqueous system containing multicomponent oxide particles. *Journal of the American Ceramic Society.* 2001;84(6):1241–9.

36. Dangel C, Schepky G, Reich H-B, Kolter K. Comparative studies with Kollicoat MAE 30 D and Kollicoat MAE 30 DP in aqueous spray dispersions and enteric coatings on highly swellable caffeine cores. *Drug Development and Industrial Pharmacy.* 2000;26(4):415–21.

37. Flößer A, Kolter K, Reich HB et al. Variation of composition of an enteric formulation based on Kollicoat® MAE 30 D. *Drug Development Industrial Pharmacy.* 2000;118:103–12.

38. Maul K, Schmidt P. Influence of different-shaped pigments on bisacodyl release from Eudragit L-30D. *International Journal of Pharmaceutics.* 1995;118(May 1):103–12.

39. Ishikawa Y, Aoki N, Ohshima H. Colloidal stability of aqueous polymeric dispersions: Effect of water insoluble excipients. *Colloids and Surfaces B: Biointerfaces.* 2005;45(1):35–41.

40. Ishikawa Y, Katoh Y, Ohshima H. Colloidal stability of aqueous polymeric dispersions: Effect of pH and salt concentration. *Colloids and Surfaces B: Biointerfaces.* 2005;42(1):53–8.

41. S Kucera, MA. The stabilization of Eudragit® E PO in the presence of aluminum lake pigments. American Association of Pharmaceutical Scientists Annual Meeting; Oct 28–Nov 2, 2006; San Antonio, TX.

42. Okhamafe AO, York P. Effect of solids-polymer interactions on the properties of some aqueous-based tablet film coating formulations. I. Moisture permeability. *International Journal of Pharmaceutics.* 1984;22(2):265–72.

43. Hogan J. Additive effects on aqueous film coatings. *Manufacturing Chemist.* 1983;54:43.

44. Porter S, Ridgway K. The permeability of enteric coatings and the dissolution rates of coated tablets. *Journal of Pharmacy and Pharmacology.* 1982;34(1):5–8.

45. Okhamafe A, York P. Studies on the moisture permeation process in some pigmented aqueous-based tablet film coats. *Pharmaceutica Acta Helvetiae.* 1985;60(3):92.

46. Chan L, Chan W, Heng P. An improved method for the measurement of colour uniformity in pellet coating. *International Journal of Pharmaceutics.* 2001;213(1):63–74.

47. Smith GW, Macleod GS, Fell JT. Mixing efficiency in side-vented coating equipment. *AAPS Pharmscitech*. 2003;4(3):71–5.

48. Hunter R. Tristimulus colour measurement of pharmaceuticals. *Pharmaceutical Technology*. 1981;5: 63–7.

49. Woznicki E, Schoneker D. Coloring agents for use in pharmaceuticals. In: Swarbrick J, Boylan J, editors. *Encyclopedia of Pharmaceutical Technology*. New York: Marcel Dekker; 1990. pp. 65–100.

50. Guyton A, Hall J. The eye: II. Receptor and neural function of the retina. Textbook of Medical Physiology 10th ed, WB Saunders Co, Philadelphia. 2000:578–90.

51. Felton LA, Wiley CJ. Blinding controlled-release tablets for clinical trials. *Drug Development and Industrial Pharmacy*. 2003;29(1):9–18.

52. Rowe R. The opacity of tablet film coatings. *Journal of Pharmacy and Pharmacology*. 1984;36(9):569–72.

53. Rowe R. Quantitative opacity measurements on tablet film coatings containing titanium dioxide. *International Journal of Pharmaceutics*. 1984;22(1):17–23.

54. Rowe R. The measurement of the opacity of tablet film coatings in-situ. *Acta Pharmaceutica Suecica*. 1984;21(3):201–4.

55. Rowe R. Gloss measurement on film coated tablets. *Journal of Pharmacy and Pharmacology*. 1985;37(11):761–5.

56. Gibson S, Rowe R, White E. Determination of the critical pigment volume concentrations of pigmented film coating formulations using gloss measurement. *International Journal of Pharmaceutics*. 1988;45(3):245–8.

57. Rowe R. The effect of some formulation and process variables on the surface roughness of film-coated tablets. *Journal of Pharmacy and Pharmacology*. 1978;30(1):669–72.

58. Tønnesen HH. Formulation and stability testing of photolabile drugs. *International Journal of Pharmaceutics*. 2001;225(1):1–14.

59. Rowe R, Forse S. The refractive indices of polymer film formers, pigments and additives used in tablet film coating: Their significance and practical application. *Journal of Pharmacy and Pharmacology*. 1983;35(4):205–7.

60. Nyqvist H, Nicklasson M, Lundgren P. Studies on the physical properties of tablets and tablet excipients. V. Film coating for protection of a light-sensitive tablet-formulation. *Acta Pharmaceutica Suecica*. 1982;19(3):223.

61. Teraoka R, Matsuda Y, Sugimoto I. Quantitative design for photostabilization of nifedipine by using titanium dioxide and/or tartrazine as colourants in model film coating systems. *Journal of Pharmacy and Pharmacology*. 1989;41(5):293–7.

62. Bechard S, Quraishi O, Kwong E. Film coating: effect of titanium dioxide concentration and film thickness on the photostability of nifedipine. *International Journal of Pharmaceutics*. 1992;87(1):133–9.

63. Aman W, Thoma K. How to photostabilize molsidomine tablets. *Journal of Pharmaceutical Sciences*. 2004;93(7):1860–6.

64. Rowe RC. A comment on the localized cracking around pigment particles in film coatings applied to tablets. *International Journal of Pharmaceutical Technology & Product Manufacture*. 1982;3(2): 67–8.

65. Rowe R. The effect of pigment type and concentration on the incidence of edge splitting on film-coated tablets. *Pharmaceutica Acta Helvetiae*. 1982;57(8):221.

66. Gibson S, Rowe R, White E. Mechanical properties of pigmented tablet coating formulations and their resistance to cracking I. Static mechanical measurement. *International Journal of Pharmaceutics*. 1988;48(1):63–77.

67. Porter S. The practical significance of the permeability and mechanical properties of polymer films used for the coating of pharmaceutical solid dosage forms. *International Journal of Pharmaceutical Technology & Product Manufacture*. 1982;3:21–5.

68. Rowe R. The effect of white extender pigments on the incidence of edge splitting on film coated tablets. *Acta Pharmaceutica Technologica*. 1984;30(3):235–8.

69. Porter S. The effect of additives on the properties of an aqueous film coating. *Pharmaceutical Technology* 1980;4:65–75.

70. Okhamafe AO, York P. Effect of solids-polymer interactions on the properties of some aqueous-based tablet film coating formulations. II. Mechanical characteristics. *International Journal of Pharmaceutics*. 1984;22(2):273–81.

71. Okhamafe AO, York P. Relationship between stress, interaction and the mechanical properties of some pigmented tablet coating films. *Drug Development and Industrial Pharmacy*. 1985;11(1):131–46.

72. Aulton ME, Abdul-Razzak MH. The mechanical properties of hydroxypropyl methycellulose films derived from aqueous systems. Part 2: The Influence of Solid Inclusions. *Drug Development and Industrial Pharmacy.* 1984;10(4):541–61.

73. Gibson S, Rowe R, White E. The mechanical properties of pigmented tablet coating formulations and their resistance to cracking II. Dynamic mechanical measurement. *International Journal of Pharmaceutics.* 1989;50(2):163–73.

74. Hsu E, Gebert M, Becker N, Gaertner A. The effects of plasticizers and titanium dioxide on the properties of poly (vinyl alcohol) coatings. Pharmaceutical Development and Technology. 2001;6(2):277–84.

75. Nyamweya N. Characterization of the interactions between aluminum lake pigments and Eudragit film coating polymers. PhD Dissertation, University of Maryland, Baltimore, 2001.

76. Okhamafe AO, York P. Mechanical properties of some pigmented and unpigmented aqueous-based film coating formulations applied to aspirin tablets. *Journal of Pharmacy and Pharmacology.* 1986; 38(6):414–9.

77. Okhamafe AO, York P. The glass transition in some pigmented polymer systems used for tablet coating. *Journal of Macromolecular Science, Part B: Physics.* 1984;23(4–6):373–82.

78. Felton LA, McGinity JW. Influence of pigment concentration and particle size on adhesion of an acrylic resin copolymer to tablet compacts. *Drug Development and Industrial Pharmacy.* 1999;25(5):597–604.

79. Mathiazhagan A, Joseph R. Nanotechnology—A New Prospective in Organic Coating. 2011.

80. List P, Kassis G. Über die Wasserdampf-und Sauerstoffdurchlässigkeit verschiedener Tablettenüberzüge. *Acta Pharmaceutica Technologica.* 1982;28:21–33.

81. Ghebre-Sellassie I, Gordon RH, Nesbitt RU, Fawzi MB. Evaluation of acrylic-based modified-release film coatings. *International Journal of Pharmaceutics.* 1987;37:211–8.

82. Chang RK, Hsiao C. Eudragit RL and RS Pseudolatices: Properties and performance in pharmaceutical coating as a controlled release membrane for theophylline pellets. *Drug Development and Industrial Pharmacy.* 1989;15(2):187–96.

83. Rowe RC. The orientation and alignment of particles in tablet film coatings. *Journal of Pharmacy and Pharmacology.* 1983;35:43–4.

84. Maul K, Schmidt P. Influence of different-shaped pigments and plasticizers on theophylline release from Eudragit RS30D and Aquacoat ECD30 coated pellets. *STP Pharma Sciences.* 1997;7(6):498–506.

85. Annamalai AM, Bradley R et al. Influence of various talc grades on in vitro dissolution properties of enteric and controlled release coated tablets with various Eudragit® polymers. American Association of Pharmaceutical Scientists Annual Meeting; Nov 7–11, 2004; Baltimore, Maryland, 2004.

86. Felton LA, McGinity JW. Influence of insoluble excipients on film coating systems. *Drug Development and Industrial Pharmacy.* 2002;28(3):225–43.

87. Fisher DG, Rowe RC. The adhesion of film coatings to tablet surfaces—Instrumentation and preliminary evaluation. *Journal of Pharmacy and Pharmacology.* 1976;28(12):886–9.

88. Lehtola VM, Heinamaki JT, Nikupaavo P, Yliruusi JK. The mechanical and adhesion properties of aqueous-based hydroxypropyl methylcellulose coating systems containing polydextrose and titanium dioxide. *Drug Development and Industrial Pharmacy.* 1995;21(6):675–85.

89. Okhamafe AO, York P. The adhesion characteristics of some pigmented and unpigmented aqueous-based film coatings applied to aspirin tablets. *Journal of Pharmacy and Pharmacology.* 1985;37(12):849–53.

8 Process and Formulation Factors Affecting Drug Release from Pellets Coated with the Ethylcellulose-Pseudolatex Aquacoat®

Juergen Siepmann, Florence Siepmann,
Ornlaksana Paeratakul, and Roland Bodmeier

CONTENTS

INTRODUCTION

Ethylcellulose is one of the most widely used water-insoluble polymers for the coating of solid dosage forms (Rekhi and Jambhekar 1995, Siepmann et al. 2008a). Although coating with organic polymer solutions is still widespread, aqueous ethylcellulose dispersions have been developed to overcome problems associated with organic solvents (environmental concerns, high cost, residual solvents, toxicity, and explosion hazards) (Banker and Peck 1981, Chang et al. 1987, Felton 2013).

Aquacoat® (FMC) and Surelease® (Colorcon) are two commercially available aqueous ethylcellulose dispersions. Aquacoat® (30% solids content) is prepared by a direct emulsification–solvent evaporation method (FMC 2006). The pseudolatex is stabilized with sodium lauryl sulfate and cetyl alcohol and requires the addition of plasticizers prior to use. Surelease® (25% solids content) is prepared by a phase inversion–in situ emulsification technique (Colorcon 2006). It contains ammonium oleate as a stabilizer and dibutyl sebacate as a plasticizer. Upon drying and film formation, ammonia evaporates, leaving oleic acid as a plasticizer within the film. Detailed information about the composition and properties of Aquacoat® and Surelease® are presented in other chapters in this book. In addition to the colloidal ethylcellulose dispersions, a micronized ethylcellulose powder with an average particle size of a few micrometers is available in Japan from Shin-Etsu (Shin-Etsu

1991, Nakagami et al. 1991). The polymeric powder is dispersed in water with the addition of relatively large amounts of plasticizer prior to use.

This chapter will summarize our research with the aqueous ethylcellulose dispersion, Aquacoat, and will discuss some important formulation and process variables published in related references.

ADDITIVES IN THE AQUEOUS POLYMER DISPERSIONS

Additives in aqueous colloidal polymer dispersions can be classified into those added during or shortly after the preparation of the polymer dispersion and those added just prior to use. The first group includes surfactants necessary to physically stabilize the dispersion during preparation and storage, preservatives for microbiological stability, and antifoaming agents. Plasticizers, anti-tack agents, or additives modifying the permeability of the ethylcellulose film (e.g., hydrophilic polymers, such as hydroxypropyl methylcellulose [HPMC]) are generally added shortly before the application of the polymer dispersion.

These additives will not only fulfill their task (such as in the case of surfactants of physically stabilizing the dispersion), but will also be present in the final film coating. They can therefore affect various film properties, such as mechanical stability and, in particular, the permeability and hence drug release from coated dosage forms. The effects of various additives on these properties are discussed below.

Surfactants

Surfactants play an important role in the preparation, formulation, and application of colloidal polymer dispersions. The surfactants are used to lower the interfacial tension between the organic polymer solution and the aqueous phase during pseudolatex formation and to prevent agglomeration and coalescence of the dispersed polymer particles during storage. However, the surfactants will also be present in the coating during and after drying, and therefore, they can affect the film-formation process and thus can potentially modify the film structure and release properties.

Cellulose acetate membranes prepared from aqueous dispersions containing sodium lauryl sulfate as a stabilizer underwent phase separation (Bindschaedler et al. 1987, 1989, Zentner et al. 1985). Above a certain concentration, sodium lauryl sulfate altered the structure as well as the mechanical and permeation properties of cellulose acetate films due to the redistribution of the surfactant into small islets during the film formation process. In Aquacoat, sodium lauryl sulfate (4% w/w of total solids), an anionic surfactant, is used in combination with cetyl alcohol (9% w/w of total solids) to stabilize the ethylcellulose dispersion. Ethylcellulose is a nonionic polymer; the drug release is expected to be pH-independent for drugs with pH-independent solubility. However, several studies with Aquacoat-coated beads showed a faster drug release in simulated intestinal fluid when compared to simulated gastric juice. Goodhart attributed the faster drug release to the ionization of sodium lauryl sulfate (Chang et al. 1987, Goodhart 1984) whereas Lippold suggested that the presence of carboxyl groups on the polymer chain was responsible for the pH-dependent effects (Lippold et al. 1989, Sutter 1987).

In order to clarify the effect of surfactant and cosurfactant levels, drug-loaded beads were coated with ethylcellulose pseudolatices containing varying concentrations of the surfactant (sodium lauryl sulfate) and cosurfactant (cetyl alcohol). Chlorpheniramine maleate (CPM) was used as the model drug because of its pH-independent solubility at physiological pH levels (Bodmeier and Paeratakul 1991a).

The chlorpheniramine maleate release from cured Aquacoat-coated beads is shown in Figure 8.1. The drug was released faster in pH 7.4 buffer. Similar results were also observed with beads coated with self-prepared ethylcellulose pseudolatices having a composition identical to that of Aquacoat. To clarify the contributions of sodium lauryl sulfate or of ethylcellulose to the pH-dependent drug release, ethylcellulose pseudolatices with varying surfactant concentrations were prepared. The effect of sodium lauryl sulfate concentration in the coating on the drug release is shown in Figure 8.2.

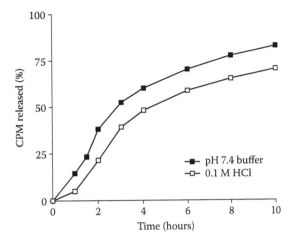

FIGURE 8.1 Chlorpheniramine maleate release from cured Aquacoat-coated beads in pH 7.4 phosphate buffer and 0.1 M HCl. (Reprinted from *Int J Pharm*, 70/1–2, R. Bodmeier and O. Paeratakul, Process and formulation variables affecting the drug release from chlorpheniramine maleate-loaded beads coated with commercial and self-prepared aqueous ethylcellulose pseudolatexes, 59–68, copyright 1991, with permission from Elsevier.)

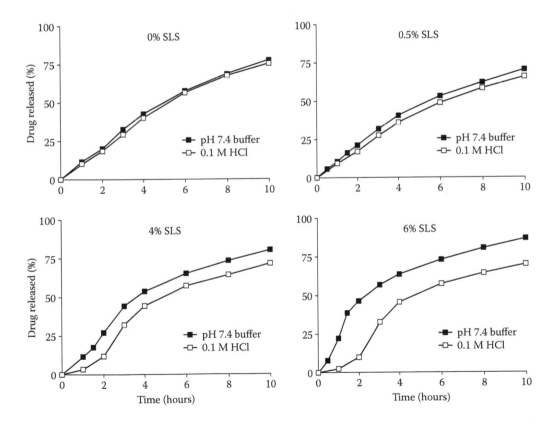

FIGURE 8.2 Effect of sodium lauryl sulfate (SLS) concentration (% w/w of coating) on the chlorpheniramine maleate release in pH 7.4 phosphate buffer and 0.1 M HCl from cured beads. (Reprinted from *Int J Pharm*, 70/1–2, R. Bodmeier and O. Paeratakul, Process and formulation variables affecting the drug release from chlorpheniramine maleate-loaded beads coated with commercial and self-prepared aqueous ethylcellulose pseudolatexes, 59–68, copyright 1991, with permission from Elsevier.)

The difference between the drug release in the two media increased with increasing concentration of sodium lauryl sulfate. By visually comparing the release profiles in the two media at higher sodium lauryl sulfate concentrations, it appeared that the release profiles in 0.1 M HCl were similar to the release profiles in pH 7.4 buffer after a lag time. The faster initial drug release in pH 7.4 buffer may be an indication of better wetting of the beads with this medium when compared to 0.1 M HCl. Sodium lauryl sulfate, an anionic surfactant with a pK_a of 1.9, is surface-active, particularly in the ionized state. It is approximately 10% ionized in 0.1 M HCl when compared to complete ionization in pH 7.4 buffer. The wetting hypothesis was confirmed by measuring the contact angles between pseudolatex-cast ethylcellulose films and the two dissolution media. As shown in Table 8.1, the contact angle was the same on surfactant-free ethylcellulose films. The contact angle decreased with increasing concentrations of sodium lauryl sulfate in the film and was significantly lower on films wetted with pH 7.4 buffer than on the films wetted with 0.1 M HCl. A lower contact angle indicated better wetting and therefore explained the initial faster drug release in pH 7.4 buffer. No dissolution media effects were seen with drug release profiles from beads coated with sodium lauryl sulfate–free ethylcellulose pseudolatices.

In addition to sodium lauryl sulfate, cetyl alcohol, a cosurfactant present in Aquacoat, also had a pronounced effect on the drug release. Cetyl alcohol, a long-chain fatty alcohol, is present in Aquacoat at a concentration of 9% w/w of total solids to stabilize the pseudolatex. Its effect on the drug release was investigated by dissolving different amounts into the organic polymer–plasticizer solution prior to emulsification of the organic phase into the aqueous phase during pseudolatex formation. The drug release decreased with increasing concentration of the cosurfactant (Figure 8.3). The presence of cetyl alcohol rendered the film coat more hydrophobic as indicated by an increased contact angle (Table 8.1). The pH sensitivity of the films decreased with increasing amount of cetyl alcohol.

These results clearly demonstrated that the pH-dependent drug release from Aquacoat-coated beads was caused by the presence of the anionic surfactant, sodium lauryl sulfate, and not the polymer. The surfactant system present in aqueous colloidal polymer dispersions should be taken into consideration when developing a latex- or pseudolatex-coated dosage form because of its potential impact on the film formation and drug release properties. In addition, charged surfactants, such as sodium lauryl sulfate, could form insoluble complexes with cationic drugs present in the core (Kositprapa et al. 1993, Kositprapa and Bodmeier 1994). Cationic drugs (e.g., CPM, propranolol

TABLE 8.1

Contact Angles between Ethylcellulose Pseudolatex–Cast Films and 0.1 M HCl or pH 7.4 Phosphate Buffer

Film	0.1 M HCl	0.1 M pH 7.4 Buffer
Aquacoat	67.9 ± 3.9	40.8 ± 3.8
Sodium Lauryl Sulfate (%)		
0	63.6 ± 2.1	63.1 ± 1.9
2	62.3 ± 0.6	38.6 ± 3.0
4	50.4 ± 3.1	31.0 ± 1.8
6	47.9 ± 1.5	24.1 ± 0.8
Cetyl Alcohol (%)		
0	17.5 ± 1.3	10.4 ± 1.4
9	50.4 ± 3.1	31.0 ± 1.8

Source: Reproduced from R. Bodmeier, O. Paeratakul, *Int J Pharm*, 70, 1–2, 59–68, 1991, with permission from Elsevier.

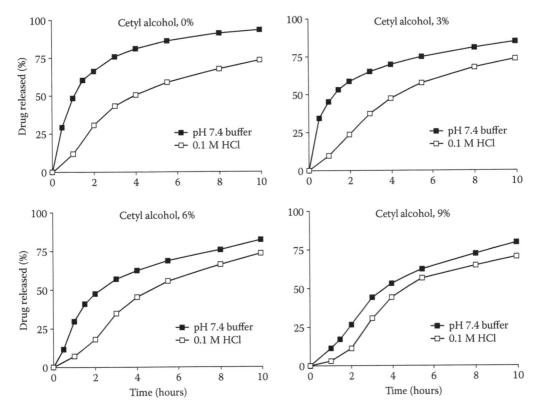

FIGURE 8.3 Effect of cetyl alcohol concentration (% w/w of coating) on the chlorpheniramine maleate release in pH 7.4 phosphate buffer and 0.1 M HCl from cured beads. (Reprinted from *Int J Pharm*, 70/1–2, R. Bodmeier and O. Paeratakul, Process and formulation variables affecting the drug release from chlorpheniramine maleate-loaded beads coated with commercial and self-prepared aqueous ethylcellulose pseudolatexces, 59–68, copyright 1991, with permission from Elsevier.)

HCl, diltiazem HCl, and quinidine sulfate) interacted with the anionic surfactant, sodium lauryl sulfate, and formed a water-insoluble ion–pair complex. This interaction could affect the release of cationic drugs from dosage forms coated with Aquacoat.

WATER-SOLUBLE ADDITIVES

Water-soluble additives have been incorporated into ethylcellulose coatings to modify the drug release. Typical water-soluble additives include (a) low molecular weight materials, including various sugars (e.g., sucrose, lactose, sorbitol), salts (e.g., sodium chloride, calcium phosphate), and surfactants such as sodium lauryl sulfate or (b) hydrophilic polymers, including polyethylene glycol, polyvinylpyrrolidone and, in particular, cellulose ethers, such as HPMC (Goodhart et al. 1984, Harris and Ghebre-Selassie 1997, Kallstrand and Ekman 1983, Li et al. 1990, McAinsh and Rowe 1979, Nesbitt 1994, Rekhi et al. 1989). During dissolution studies, these additives might leach from the coating membrane or hydrate in the coating in the case of high molecular weight polymers, resulting in a more permeable membrane and generally in a faster drug release. However, a decrease in drug release with increasing HPMC concentration was observed with acetaminophen as a model drug; this was attributed to the lower solubility of the drug in the HPMC-containing Aquacoat film (Zhang et al. 1990). Appel and Zentner (1991) used urea as a pore-forming agent with Aquacoat to form microporous films in order to increase the release of drugs from coated osmotic tablets. The drug release could also be increased by incorporating drug powder in the coating formulation

(Li et al. 1990). Theophylline was incorporated in the coating and resulted in faster drug release due to an increase in film porosity after dissolving from the coating.

The water-soluble high molecular weight polymers are usually not considered as true pore-forming agents because they often do not completely leach out from the coating to leave a well-defined pore structure. A critical HPMC concentration was identified below which very little polymer leaches from the coating and no pores are formed. Above 24% HPMC, the polymer leaches from the ethylcellulose films, resulting in pore formation and an increase in drug release (Lindstedt et al. 1991). Using an interesting experimental setup developed by the same research group (a pressurized cell device in which the permeability was measured in dependence on an applied tensile stress), the permeability properties of ethylcellulose/HPMC were shown to increase with increasing HPMC content (Lindstedt et al. 1989).

During coating with the HPMC-containing ethylcellulose dispersion (Aquacoat), we observed the appearance of a sediment in the colloidal dispersion upon standing, indicating destabilization of the colloidal ethylcellulose particles by HPMC (Wong et al. 1991, Wong and Bodmeier 1996). With organic solvent–based coatings, HPMC and the water-insoluble polymer ethylcellulose are co-dissolved in an organic solvent or solvent mixture and therefore applied as a mixed polymer solution. Combining HPMC with the colloidal ethylcellulose dispersion prior to the coating process results in a system of HPMC being in solution and the water-insoluble polymer in colloidal dispersion. It is a well-known fact in colloidal science that water-soluble polymers can cause flocculation of dispersed polymer particles (Feigin and Napper 1980, Ottewill 1990, Sperry et al. 1981). The addition of HPMC to the ethylcellulose pseudolatex results in the flocculation of the colloidal polymer particles above a critical HPMC concentration. Flocculation is indicated by the appearance of a sediment upon standing. Photomicrographs of the dispersions showed no aggregates with HPMC-free polymer dispersions whereas flocculation became visible at concentrations in excess of 3% HPMC E5, with the number of aggregates clearly increasing with increasing concentration of HPMC.

The critical HPMC concentration necessary to cause flocculation moved to lower concentrations with increasing molecular weight of the water-soluble polymer. The higher molecular weight grades were more efficient flocculants. In addition to the type and concentration of HPMC, the flocculation of the colloidal dispersion was also affected by its solids content. The HPMC concentration necessary to cause flocculation decreased with increasing solids content of the polymer dispersion.

The observed flocculation phenomena could interfere with the film formation of the colloidal polymer dispersion upon removal of water and thus could affect the drug release from the polymer-coated dosage forms. The HPMC concentrations commonly used in the coating of solid dosage forms (3%–10% w/w), based on the water-insoluble polymer, often fell into the flocculated region. This may have important implications for the film formation and coating with aqueous colloidal dispersions. A steep increase in drug release was observed at HPMC concentrations above the critical flocculation concentration (Wong 1994).

Also the addition of small amounts of other hydrophilic polymers has been proposed for Aquacoat-based film coatings, namely poly(vinyl alcohol)-poly(ethylene glycol) graft copolymer (PVA-PEG graft copolymer, marketed under the trade name Kollicoat IR by BASF) (Siepmann et al. 2007a, 2008c) and carrageenan (Siepmann et al. 2007b). In contrast to HPMC, these additives do not cause flocculation of the coating dispersion. As an example, Figure 8.4 shows how the resulting theophylline release rate from coated beads can effectively be adjusted by varying the PVA-PEG-graft copolymer content. The significant increase in film coating permeability can primarily be attributed to the following two phenomena:

1. The presence of the hydrophilic copolymer within the macromolecular network leads to a marked increase in water uptake rate and extent (Figure 8.5a, symbols = experimental results). Interestingly, water penetration into free, thin films of identical composition as the film coatings is predominantly diffusion-controlled with constant diffusivities, irrespective of the additive content in the investigated range (Figure 8.5a, curves = fittings of a

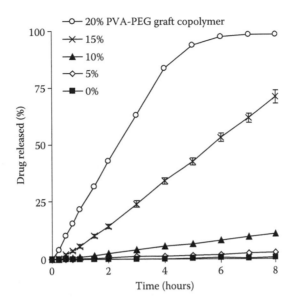

FIGURE 8.4 Effect of the addition of varying amounts of PVA-PEG graft copolymer to Aquacoat on drug release from theophylline beads in phosphate buffer pH 7.4. (Reproduced from *J Control Release*, 119/2, F. Siepmann et al., How to adjust desired drug release patterns from ethylcellulose-coated dosage forms, 182–189, copyright 2007, with permission from Elsevier.)

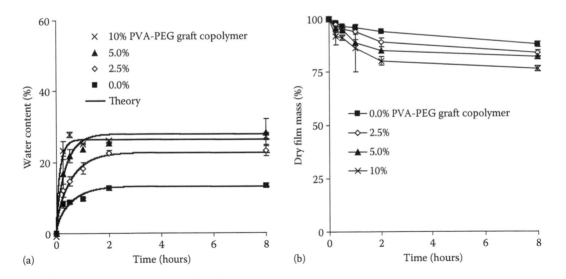

FIGURE 8.5 Effect of the addition of varying amounts of PVA-PEG graft copolymer to Aquacoat on the (a) water uptake behavior of free, thin films upon exposure to 0.1 M HCl (symbols = experiments, curves = mathematical modeling based on Fick's law); (b) dry mass loss of free, thin films upon exposure to phosphate buffer pH 7.4. (Reproduced from *J Control Release*, 119/2, F. Siepmann et al., How to adjust desired drug release patterns from ethylcellulose-coated dosage forms, 182–189, copyright 2007, with permission from Elsevier.)

mathematical theory based on Fick's law). Thus, apparent water diffusion coefficients can be determined and the effects of the film composition on the hydration behavior quantitatively predicted.

2. PVA-PEG graft copolymer (at least partially) leaches out of the films upon exposure to aqueous release media, resulting in increased rates and extents of dry mass loss of the polymeric systems (Figure 8.5b).

Both phenomena, leading to increased water contents and decreased dry film masses (and hence less dense polymeric structures) result in increased drug permeability and release rates.

Interestingly, also drug transport through the polymeric networks is primarily diffusion-controlled with constant diffusivities, irrespective of the PVA-PEG-graft copolymer content. Figure 8.6a shows as an example the release of theophylline from free, thin films in phosphate buffer pH 7.4 (symbols = experiments, curves = mathematical modeling). As variations in the film thickness alter the length of the diffusion pathway, the results have been normalized to this parameter. Based on these calculations, the apparent diffusion coefficient of theophylline in these polymeric systems (being a measure for drug mobility) could be determined (Figure 8.6b). The diffusivity significantly increased with increasing additive PVA-PEG-graft copolymer content, irrespective of the type of release medium. This knowledge and appropriate solutions of Fick's law can help to quantitatively predict the effects of the film coating composition on the resulting drug release kinetics from coated dosage forms (Siepmann and Siepmann 2012). However, it must be pointed out that the drug release mechanisms of coated systems are generally more complex than of free, thin films. Additional phenomena, such as convective water influx at early time points and the creation of significant hydrostatic pressure within bead cores (potentially resulting in the formation of water-filled cracks in the film coatings) can also be of major importance. This is discussed in more detail in the section on drug release mechanisms.

Importantly, the addition of small amounts of PVA-PEG graft copolymer to Aquacoat-based film coatings can also help providing long-term stability during storage of the coated dosage forms (Muschert et al. 2009a, Siepmann et al. 2008c). For example, Figure 8.7 illustrates theophylline

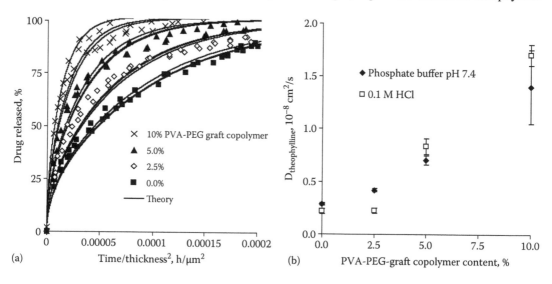

FIGURE 8.6 Effect of the addition of varying amounts of PVA-PEG graft copolymer to Aquacoat on the (a) release of theophylline from free, thin films in phosphate buffer pH 7.4 (symbols = experiments, curves = mathematical modeling based on Fick's law) (the time is normalized to the film thickness to account for slight variations in this parameter) and (b) apparent diffusion coefficient of theophylline in the polymeric systems upon exposure to phosphate buffer pH 7.4 or 0.1 M HCl. (Reproduced from *J Control Release*, 119/2, F. Siepmann et al., How to adjust desired drug release patterns from ethylcellulose-coated dosage forms, 182–189, copyright 2007, with permission from Elsevier.)

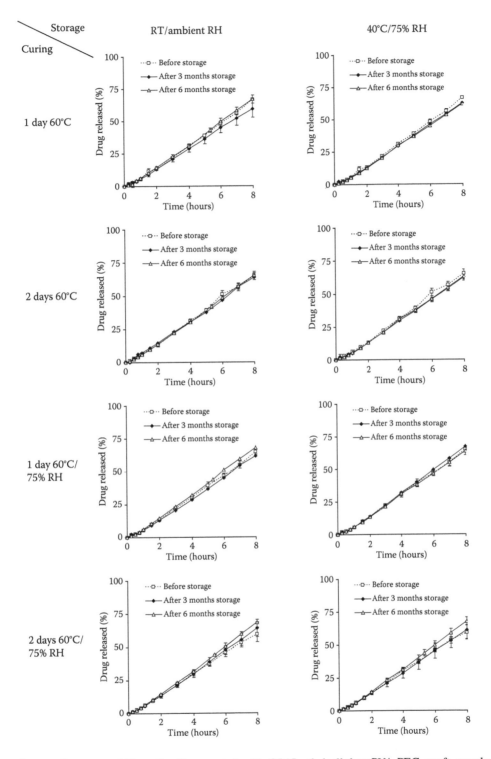

FIGURE 8.7 Storage stability of pellets coated with 85:15 ethylcellulose:PVA-PEG graft copolymer: Theophylline release in 0.1 M HCl before (dotted lines) and after three and six months storage (full lines, as indicated). The curing conditions are shown on the left, the storage conditions at the top (coating level = 20%). (Reproduced from *J Control Release*, 126, F. Siepmann et al., How to improve the storage stability of aqueous polymeric film coatings, 26–33, copyright 2008, with permission from Elsevier.)

release from pellets coated with 85:15 ethylcellulose:PVA–PEG graft copolymer before (dotted lines) and after three and six months storage (full lines) in 0.1 M HCl. The curing conditions are shown on the left, the storage conditions at the top (20% coating level). As can be seen, there were no significant time-, temperature-, or humidity-dependent changes in the drug-release profiles during open storage (no packaging material). This is of great practical importance because it indicates that the presence of small amounts of PVA–PEG graft copolymer in Aquacoat-based film coatings effectively improves film formation during coating or curing and/or hinders further polymer particle coalescence during long-term storage. Importantly, significant structural changes within the polymeric systems during open storage under ambient as well as stress conditions are effectively avoided. Improved film formation during coating or curing might be attributable to the hydrophilic nature of PVA–PEG graft copolymer, trapping water within the polymeric systems during film formation. At this stage, the water content is of crucial importance because (a) water is mandatory for the capillary effects driving the individual polymer particles together and (b) water acts as a plasticizer for ethylcellulose and, thus, increases the mobility of the macromolecules and facilitates polymer particle fusion. Alternatively, the presence of PVA–PEG graft copolymer chains between incompletely fused ethylcellulose particles might sterically hinder further film formation during long-term storage. A further example for a water-soluble, macromolecular additive that does not cause flocculation of the coating dispersion is propylene glycol alginate (Siepmann et al. 2008b). As it contains free carboxylic groups, the resulting film-coating properties (including water uptake and dry mass loss behavior as well as drug permeability) are pH dependent and, thus, triggered by the environment within the gastrointestinal tract (GIT). At nonacidic pH, the carboxylic groups are ionized, leading to a more pronounced hydration and leaching of propylene glycol alginate out of the film coatings. For certain drugs, for example, weak bases with pH-dependent solubility, this type of coating can be advantageous: At low pH (in the stomach) the elevated drug solubility is combined with a relatively low film permeability whereas at higher pH (in the intestine) the decrease in drug solubility might be compensated by a simultaneous increase in film coating permeability. Also combinations of Aquacoat and the enteric polymer Eudragit® L can be used for this purpose (Lecomte et al. 2003).

In addition, the presence of a suitable water-soluble macromolecular additive in Aquacoat-based controlled release film coatings can result in ethanol-resistance (Rosiaux et al. 2013a,b, 2014). It has to be pointed out that the sensitivity of controlled-release dosage forms to the presence of ethanol in the GIT can be highly critical, for example, if the incorporated drug is potent and exhibits severe side effects. This is, for instance, the case for most opioid drugs. The co-ingestion of alcoholic beverages can lead to dose dumping and potentially fatal consequences. Because ethylcellulose is soluble in ethanol, ethylcellulose-based controlled-release film coatings might present safety concerns for the patient. Interestingly, blends of Aquacoat and medium or high viscosity guar gums have been shown to be able to provide ethanol-resistant drug release. The basic principle is that ethylcellulose is insoluble in water but soluble in ethanol whereas guar gum is soluble in water but insoluble in ethanol. Thus, the presence of guar gum can be used to effectively hinder the potential dissolution of ethylcellulose in aqueous media containing high ethanol concentrations. Vice versa, the presence of ethylcellulose can be used to effectively hinder the potential dissolution of guar gum in pure aqueous media. As an example, Figure 8.8 shows theophylline release from pellets coated with Aquacoat containing 10% or 15% medium and high viscosity guar gum. As can be seen, drug release was virtually unaffected by the addition of 40% ethanol to the release medium. This can be of great practical importance.

Furthermore, the presence of particular types of macromolecular additives in Aquacoat-based film coatings might allow for site-specific drug delivery within the GIT. For example, polymeric compounds can be added, which are preferentially degraded by bacterial enzymes present in the colon (Karrout et al. 2009a–c, Maroni et al. 2013). In this case, drug release in the upper GIT can be limited (ethylcellulose being insoluble and the second polymeric compound not being degraded to a noteworthy extent). However, once the dosage forms reaches the colon, the bacterial enzymes start

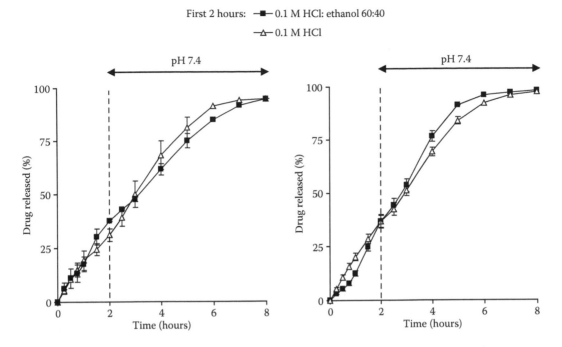

FIGURE 8.8 Ethanol-resistant drug release: Theophylline release from pellets coated with 20% ethylcellulose:high η guar gum 90:10 (left hand side), or 20% ethylcellulose:medium η guar gum 85:15 (right hand side), upon exposure to 0.1 M HCl for two hours and phosphate buffer pH 7.4 for the subsequent six hours (open triangles), or 0.1 M HCl:ethanol 60:40 for two hours and phosphate buffer pH 7.4 for the subsequent six hours (filled squares). (Reproduced from Y. Rosiaux, S. Muschert, R. Chokshi et al., *J Control Release*, 169, 1–9, 2013 with permission from Elsevier.)

to degrade the polymeric additive, resulting in an increase in drug permeability of the film coating and, thus, the onset of drug release (Karrout et al. 2009d, 2010a,b, 2011). This type of "colon targeting" can for instance be very helpful to improve the efficacy of inflammatory bowel disease treatments (Karrout et al. 2015).

WATER-INSOLUBLE ADDITIVES

Insoluble ingredients may be included in the coating formulations for a variety of reasons. One application is to use materials such as magnesium stearate or talc as anti-tack or separating agents that help reduce agglomeration or sticking of coated particles during the coating process (Goodhart et al. 1984). Talc and kaolin are the most commonly used ingredients in aqueous film coating. In general, the separating agent should be inert with respect to the drug and the release characteristics of the film. Surface and morphological properties, including the hydrophilicity of insoluble filler particles have been shown to be important factors that may contribute to the properties of the final film (Horvath and Ormos 1989). The amount of insoluble filler incorporated in the aqueous dispersion must be optimized without exceeding the maximum carrying capacity of the polymer or critical pigment volume concentration (CPVC). The pigment concentration has a strong influence on the final film properties, such as mechanical strength and permeability (Ramig 1970, Rodriguez et al. 2004, Patton 1979). Care must be taken when incorporating coloring agents into an aqueous dispersion of high pH value, such as Surelease, because the basicity of the dispersion can destroy dye–substrate complexes. Colorants such as aluminum lakes should be replaced with inorganic pigments such as titanium dioxide (Porter 1990).

Plasticizers

Plasticizers are usually high-boiling organic solvents used to impart flexibility to otherwise hard or brittle polymeric materials. Plasticizers generally cause a reduction in the cohesive intermolecular forces along the polymer chains, resulting in various changes in the polymer properties, such as a reduction in tensile strength, increases in elongation and flexibility, and reduction in the glass transition or softening temperature of the polymer.

With aqueous colloidal polymer dispersions, the addition of plasticizers is required for systems having a minimum film formation temperature (MFT) above the coating temperature (Bindschaedler et al. 1983). During plasticization, the plasticizer diffuses into and softens the polymeric particles, thus promoting particle deformation and coalescence into a homogeneous film. The effectiveness of a plasticizer for a particular polymer or polymer dispersion depends on the plasticizer–polymer compatibility and the permanence of the plasticizer in the film during coating, storage, and during contact with artificial or biological fluids.

With aqueous polymer dispersions, water-soluble plasticizers dissolve whereas water-insoluble plasticizers have to be emulsified in the aqueous phase of the dispersion. During plasticization, the plasticizer diffuses into the colloidal polymer particles with the rate and extent of diffusion being dependent on its water solubility and affinity for the polymer phase. With insoluble plasticizers, the plasticized polymer dispersion can be visualized as a three-phase system composed of the water phase, polymer particles, and emulsified droplets. During plasticization, the plasticizer diffuses from the emulsion droplets through the water phase and is then absorbed by the polymer (Dillon et al. 1953).

Factors influencing the rate and extent of the plasticizer uptake by the colloidal particles, such as type and concentration of the plasticizer and type and solids content of the polymer dispersion, were investigated with Aquacoat (Bodmeier and Paeratakul 1994a, Paeratakul 1993). The plasticizers were classified into water-soluble (triethyl citrate [TEC] and triacetin [TA]) and water-insoluble plasticizers (acetyltriethyl citrate [ATEC], acetyltributyl citrate [ATBC], dibutyl phthalate [DBP], dibutyl sebacate [DBS], diethyl phthalate [DEP], and tributyl citrate [TBC]).

A separation scheme that allowed the quantification of the amount of plasticizer in the aqueous and polymer phases was developed in order to determine the distribution of the plasticizer in the colloidal polymer dispersion. The plasticizer present in the different phases could be separated by centrifugation and/or ultracentrifugation because of differences in the densities of the plasticizers, water, and the polymer particles. The amount of plasticizer in each phase was determined by a high-performance liquid chromatography (HPLC) method (Bodmeier and Paeratakul 1991b).

The extent of the partitioning of both water-soluble and -insoluble plasticizers in Aquacoat after a plasticization time of 24 hours is shown in Table 8.2. The water-soluble plasticizers, triethyl citrate and triacetin, were dissolved in both the aqueous and polymer phase. The higher amount of triacetin in the aqueous phase, when compared to triethyl citrate, could be explained with its higher solubility in the supernatant. With water-insoluble plasticizers, between 85% and 90% of the incorporated plasticizer partitioned into the colloidal polymer particles or polymer phase after 24 hours. The remaining plasticizer existed in the aqueous phase predominantly in the emulsified (between 7% and 14% of the total amount of plasticizer incorporated) and not dissolved form. This clearly showed that water-insoluble plasticizers were not completely taken up by the colloidal polymer particles within a 24-hour period, a result previously reported with dibutyl sebacate (Sutter 1987). This may have important implications for the coating with polymer dispersions when compared to organic polymer solutions in which the plasticizer is completely dissolved. During coating, in addition to the plasticized polymer particles, the emulsified plasticizer droplets are sprayed onto the solid dosage forms. This could result in an uneven plasticizer distribution within the film, potentially causing changes in the mechanical and especially release properties upon aging. A thermal treatment following the coating (curing step) (Bodmeier and Paeratakul 1991a, Goodhart et al. 1984, Lippold et al. 1989), which is nowadays widely used to promote further coalescence of the colloidal polymer

TABLE 8.2
Extent of Plasticizer Diffusion in Aquacoat (Solids Content, 15% w/w; Level of Plasticizer, 20% w/w of Polymer)

| | Plasticizer Conc. (%) | | | |
| | Aqueous Phase | | | |
Plasticizer	Dissolved	Emulsified	Polymer Phase	Recovery (%)
		Water-Soluble		
TEC	49.87 ± 0.02	–	50.10 ± 0.18	99.97 ± 0.16
TA	63.96 ± 1.12	–	35.41 ± 1.67	99.36 ± 0.55
		Water-Insoluble		
ATEC	7.63 ± 0.27	7.25 ± 0.37	84.72 ± 1.33	99.60 ± 0.69
ATBC	0.44 ± 0.01	12.19 ± 0.58	86.37 ± 1.98	98.99 ± 1.39
DBS	10.77 ± 0.03	1.51 ± 0.01	87.43 ± 0.25	99.73 ± 0.25
DEP	2.46 ± 0.15	10.59 ± 1.23	87.41 ± 1.02	100.46 ± 0.37
DBP	0.37 ± 0.01	13.74 ± 1.46	85.92 ± 0.88	100.03 ± 0.89
TBC	0.81 ± 0.01	9.88 ± 0.69	89.22 ± 0.75	99.91 ± 0.06

Source: Reproduced from R. Bodmeier, O. Paeratakul, *Int J Pharm*, 1994a, 103, 47–54, with permission.
Note: Abbreviations are defined in text.

particles and to overcome stability problems, may also result in a more homogeneous distribution of the plasticizer.

The triethyl citrate uptake into the polymer phase increased with increasing polymer content of the polymer dispersion whereas the fraction dissolved in the aqueous phase decreased. Similar trends were also seen with the water-insoluble plasticizer dibutyl sebacate with the amounts of plasticizer dissolved or emulsified into the aqueous phase decreasing with increasing pseudolatex solids content. Although most polymer dispersions available for the coating of solid dosage forms are obtained with a solids content of 30%, the coating is generally performed after diluting the dispersions to a solids content between 10% and 15%. In order to have most of the plasticizer present in the polymer particles, it is therefore recommended to add the plasticizer to the concentrated dispersions followed by dilution to the desired solids content just prior to coating rather than first diluting the dispersion followed by addition of the plasticizer.

The rate at which the plasticizer diffuses into the colloidal particles determines the amount of plasticizer taken up by the polymer as a function of plasticization time. The diffusion or uptake rate thus affects the film formation process. Plasticizers differ greatly in the rate of the diffusion process (Bodmeier and Paeratakul 1997). With water-soluble plasticizers, such as triethyl citrate and triacetin, the distribution behavior of the plasticizers was virtually not affected by the mixing time or the degree of agitation.

Iyer et al. (1990) determined the uptake of the water-insoluble plasticizer dibutyl sebacate into Aquacoat by using an alkaline partition column to separate the unbound plasticizer and gas chromatography for the plasticizer assay. The uptake of dibutyl sebacate was found to be complete within 30 minutes, irrespective of the amount used, and the uptake rate was faster with increasing solids content of pseudolatex or when smaller quantities of plasticizer were incorporated. However, a previous study reported the presence of visible dibutyl sebacate droplets in Aquacoat after 1 week of mixing, indicating incomplete plasticization even after such a long plasticization time (Sutter 1987).

When emulsified in the aqueous colloidal dispersion, a water-insoluble plasticizer exists mainly in either the polymer or the emulsified phase, whereas a minor portion is present as plasticizer dissolved in water. The rate at which the plasticizer is taken up by the polymer particles corresponds well to the rate at which it is lost from the emulsified phase. Therefore, the rate of uptake can be expressed by the rate at which the emulsified plasticizer disappears from the aqueous phase (into the polymer phase). The effect of the plasticizer (acetyltributyl citrate; ATBC) concentration (% w/w of polymer) on the rate of ATBC uptake by Aquacoat is shown in Figure 8.9. The rate of plasticizer uptake is expressed as the amount of emulsified plasticizer remaining in the aqueous phase as a function of plasticization time. The pseudolatex, having a solids content of 15% w/w, was diluted from the original dispersion (30% w/w) with an equal volume of water. The ATBC uptake was fastest when 10% w/w ATBC was used; it decreased with increasing ATBC concentration. After 24 hours of plasticization, emulsified ATBC droplets were detectable at ATBC concentrations in excess of 10% w/w. The presence of the excess emulsified portion indicated an incomplete plasticizer uptake and therefore possible saturation of the polymer at the plasticizer concentrations used. However, when the dispersions were aged for a longer period, the emulsified plasticizer droplets gradually disappeared and could no longer be detected after 1 week of mixing.

Based on the experimentally determined uptake kinetics of water-insoluble plasticizers in Aquacoat, an appropriate mathematical theory could be developed taking into account all relevant mass transport phenomena (Siepmann et al. 1998). Figure 8.10a shows schematically the processes taken into account: (a) dissolution of the plasticizer droplets in the aqueous phase and (b) diffusion of the plasticizer within the polymer particles. Initially, after adding the plasticizer to the aqueous dispersion, dissolution governs the overall transport kinetics (dissolution rate < diffusion rate). But increasing amounts of plasticizer located within the polymer particles lead to decreasing concentration gradients and subsequent declining diffusion rates. A change of the governing mechanism is observed. During the final portion, diffusion controls the uptake kinetics (dissolution rate > diffusion rate). Dissolution and diffusion are taken into account simultaneously in the theory, which was used to determine the exact composition of the three phase system at any time point (Figure 8.10b) as well as the diffusion coefficients and dissolution rate constants of various types of water-insoluble plasticizers in Aquacoat. Knowing these parameters, the minimum stirring time necessary for sufficient plasticizer uptake (to avoid inhomogeneous coatings) can be calculated.

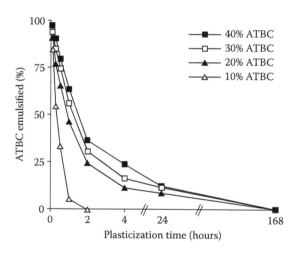

FIGURE 8.9 Effect of acetyltributyl citrate (ATBC) concentration added (% w/w of polymer) on the rate of plasticizer uptake into Aquacoat, expressed as the amount of emulsified plasticizer (% w/w of total ATBC added) remaining in the aqueous phase as a function of plasticization time (solids content, 15% w/w). (Reproduced from *Int J Pharm*, 152/1, R. Bodmeier and O. Paeratakul, Plasticizer uptake by aqueous colloidal polymer dispersions used for the coating of solid dosage form, 17–26, copyright 1997, with permission from Elsevier.)

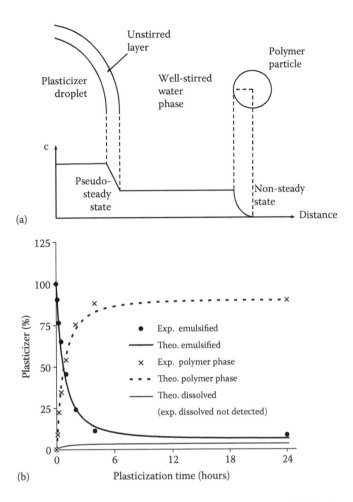

FIGURE 8.10 Mathematical modeling of the uptake of a water-insoluble plasticizer into aqueous ethylcellulose dispersion: (a) schematic presentation of the processes taken into account; (b) experimentally measured (symbols) and theoretically calculated amounts (curves) of ATBC dissolved versus emulsified in the aqueous phase and taken up into the polymer particles (solids content of the polymer dispersion = 15% w/w, plasticizer concentration = 20% w/w based on polymer). (Reproduced from *Int J Pharm*, 165, F. Siepmann et al., Modeling plasticizer uptake in aqueous polymer dispersions, 191–200, copyright, 1998, with permission from Elsevier.)

Not only will plasticizers affect the film formation from colloidal polymer dispersions or the mechanical properties of the resulting films, but their choice will also affect the drug release from the coated dosage form (Banker and Peck 1981, Chang et al. 1987, Goodhart et al. 1984, Guo 1996, Hutchings et al. 1994, Iyer et al. 1990). Increasing the concentration of dibutyl sebacate or triethyl citrate decreased the drug release from Aquacoat-coated dosage forms (Goodhart et al. 1984), probably because of better fusion of the colloidal polymer particles.

The effect of the type of plasticizer on the theophylline release from beads coated with Aquacoat is shown in Figure 8.11. The release from beads containing triacetin as a plasticizer was very rapid, indicating poor film formation even after curing at 60°C for one hour. Triethyl citrate resulted in intermediate release properties, indicating good film formation, with the release being faster when compared to the water-insoluble plasticizers because of its higher water solubility. Triethyl citrate has been reported to be a very effective plasticizer for Aquacoat (Selinger and Brine 1988). As expected, the drug release was slower with the water-insoluble plasticizers.

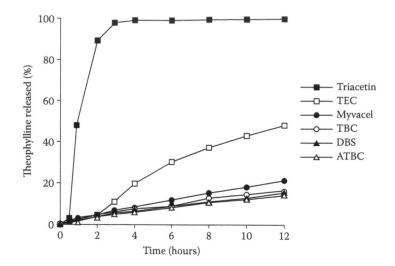

FIGURE 8.11 Effect of the type of plasticizer on the theophylline release from beads coated with Aquacoat in 0.1 M HCl. (Reproduced from X. Guo, Physicochemical and Mechanical Properties Influencing the Drug Release from Coated Dosage Forms, PhD dissertation, University of Texas at Austin, 1996, with permission.)

The appropriate choice of the type of plasticizer is particularly important when macromolecular release modifiers are added to Aquacoat (Lecomte et al. 2004a). The affinity of the plasticizer to ethylcellulose can be different from the affinity to the release modifier, resulting in potential redistributions within the polymeric networks and subsequent changes in the release profiles during storage. Importantly, appropriate preparation (in particular curing) conditions can effectively avoid these phenomena.

PROCESS VARIABLES

The coating process and equipment also have a significant impact on the release behavior of controlled release products as described in various chapters of this book. Process variables, such as spray rate, droplet size, bed temperature, spray mode, and so forth, can strongly influence the drug release (Chang et al. 1987, Coloron 2006, FMC 2006, Gundert-Remy and Moller 1990, Yang et al. 1990).

The coating temperature should be sufficiently high to achieve efficient water removal and subsequent particle coalescence. In general, it should be 10°C–20°C higher than the MFT of the polymer dispersion (Dashevsky et al. 2005, Fukumori et al. 1988). The drug release with Surelease-coated theophylline pellets decreased with increasing the product temperature from 32°C to 48°C because of a more complete film formation (Laicher et al. 1993). However, an excessively high inlet temperature can potentially cause difficulties in processing, such as electrostatic interactions and agglomeration of the beads, because of excessive drying or softening and sticking of the coating (Colorcon 2006). It could also cause premature coalescence resulting in a more porous and inhomogeneous film structure.

A process variable of particular importance is the curing of the coated dosage forms, determining the degree of film formation.

CURING OF AQUACOAT®-COATED SOLID DOSAGE FORMS

The coalescence of the colloidal polymer particles into a homogenous film is often incomplete after coating with aqueous polymer dispersions. As a consequence, changes in the drug release from the coated dosage form caused by further coalescence during storage have been observed as a function of storage temperature and time (Bodmeier and Paeratakul 1991a, Dahl and Sue 1990, FMC 2006, Lippold et al. 1990, Nakagami et al. 1991, Shah et al. 1994, Sutter 1987).

A curing step or thermal treatment (storage of the coated dosage forms at elevated temperatures for short periods) is often recommended to accelerate the coalescence of the polymer particles prior to long-term storage. Changes in drug release during storage can be circumvented. During the curing step, the coated dosage forms are subjected to a heat treatment above the glass transition temperature of the polymer. This is achieved either by storing the coated dosage forms in an oven or through further fluidization in the heated fluidized bed coater for a short time immediately after the completion of the coating process (Gendre et al. 2013, Muschert et al. 2011). The storage temperature should be about 10°C above the MFT (Lippold et al. 1989). Higher curing temperatures could cause excessive tackiness and agglomeration of the solid dosage forms. An HPMC overcoat has been used to overcome tackiness and allow the curing step to be performed in the fluidized bed after coating (Harris et al. 1986).

Chlorpheniramine maleate–containing beads were coated at 40°C with self-prepared ethylcellulose pseudolatexes identical in composition to Aquacoat (Bodmeier and Paeratakul 1991a). In that study, the coated beads were subsequently cured (in an oven) at 40°C, 50°C, and 60°C for periods of 1–24 hours. The drug release in pH 7.4 buffer was strongly affected by the curing conditions (Figure 8.12). Although the bed temperature was above the minimum film formation temperature of the pseudolatex, evaporation of water during the coating process could have resulted in a cooling effect and might have kept the temperature on the bead surface below the minimum MFT (Patton 1979). While curing at 40°C for 24 hours was insufficient, curing at either 50°C or 60°C resulted in a significant reduction in drug release. The limiting drug release pattern was approached after curing the beads for one hour at 60°C. This value was also found by other authors (Gilligan and Li Wan Po 1991). At a curing temperature of 50°C, longer curing times were required to approach the limiting drug release pattern.

As an alternative to oven curing, Aquacoat-coated beads have been cured directly in the fluidized bed after the coated beads have been applied with a thin layer of HPMC (Harris et al. 1986). The hydrophilic overcoat prevented the sticking and agglomeration of the beads without altering the release profiles of the original coated pellets. The same technique has been used in the coating of pellets with other latex and pseudolatex coating systems (Chang et al. 1987, Christensen and Bertelsen 1990).

The effect of thermal treatment was apparent with chlorpheniramine maleate (CPM)-loaded beads coated with Aquacoat plasticized with varying plasticizer concentrations. Figure 8.13 shows the effect of the plasticizer (triethyl citrate) concentration and the curing temperature on CPM release in pH 7.4 buffer. The drug release was very rapid when low concentrations (10% w/w of polymer) of triethyl citrate were used. This concentration range was insufficient for film formation and curing could not further retard the drug release. With an intermediate concentration of triethyl citrate (20% w/w of polymer), curing temperatures of 50°C and 60°C resulted in a significant reduction in the drug release. The drug release approached the limiting pattern as the plasticizer concentration was increased to 25%–35% w/w, and similar release profiles were obtained, irrespective of the curing temperatures. A curing step may therefore not be necessary at higher plasticizer concentrations.

Depending on the physicochemical properties of the drug and polymeric coatings, curing can influence the performance of the pseudolatex-coated dosage forms differently. While curing of the Aquacoat-coated CPM beads produced a retarding effect on drug release, curing of ibuprofen beads coated with a comparable coating system resulted in more complex drug release patterns (Bodmeier and Paeratakul 1994a).

Figure 8.14 shows the effect of curing time on ibuprofen release from beads cured at 50°C. The release was initially rapid with the beads cured for 15 minutes, then decreased with increasing curing time up to a curing period of four hours. The ibuprofen release was characterized by a rapid burst followed by a linear portion indicating a region of constant drug release. At curing times in excess of four hours, the drug release increased. The initial decrease in drug release (up to a curing

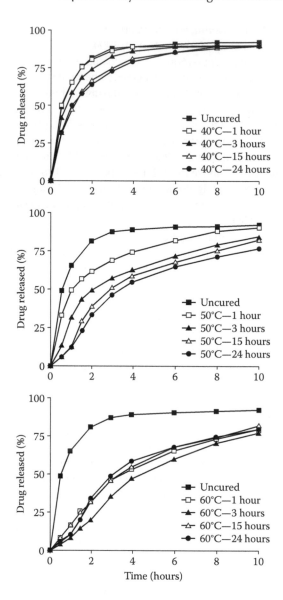

FIGURE 8.12 Effect of curing conditions (curing temperature–curing time) on the chlorpheniramine maleate release in pH 7.4 phosphate buffer. (Reprinted from *Int J Pharm*, 70/1–2, R. Bodmeier and O. Paeratakul, Process and formulation variables affecting the drug release from chlorpheniramine maleate-loaded beads coated with commercial and self-prepared aqueous ethyl cellulose pseudolatexes, 59–68, copyright 1991, with permission from Elsevier.)

period of four hours) was due to the further coalescence of the polymer particles in the ethylcellulose film. The increase in drug release (curing periods in excess of four hours) could be explained with the migration of ibuprofen from the bead interior to the bead surface through the ethylcellulose coating during the curing step. After the application of the Aquacoat layer onto the drug beads, the surface of uncured beads was uniform and smooth. However, after the coated beads were subject to a curing step, large drug crystals could be observed throughout the coated surface by scanning electron microscopy. The formation of drug crystals indicated an outward migration of the drug during thermal treatment that resulted in subsequent drug recrystallization on the film surface. During thermal treatment of ibuprofen pellets, the processes of particle coalescence and migration

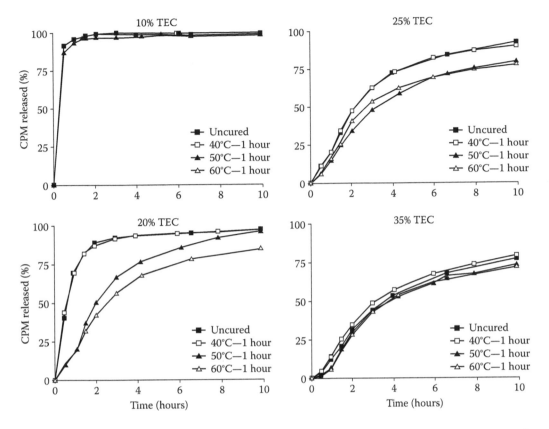

FIGURE 8.13 Effect of triethyl citrate (TEC) concentration and curing conditions (curing temperature–curing time) on the chlorpheniramine maleate release in pH 7.4 phosphate buffer. (Reproduced from R. Bodmeier, O. Paeratakul, *Drug Dev Ind Pharm*, 1994, 20, 9, 1517–1533, with permission.)

of ibuprofen to the bead surface occurred simultaneously. When compared to CPM, ibuprofen has a much lower melting point (CPM: 130°C–135°C; ibuprofen: 75°C–77°C). The drug–polymer affinity coupled with the drug's low melting point could thus serve as an explanation for the phenomenon of drug migration, a process that was accelerated at elevated temperatures. The diffusion of guaifenesin, another low-melting drug, through Aquacoat coatings during storage of coated beads has also been observed (Sutter 1987). In order to retard or avoid the drug migration during the curing step, the drug beads can be seal-coated with a polymer having a low affinity for the drug, thus avoiding direct contact of the drug and ethylcellulose.

Process variables involved in the coating of solid dosage forms and post-coating thermal treatment can significantly affect the drug release from the pseudolatex beads. Curing of the coated dosage forms not only can positively affect the coalescence of the polymeric particles resulting in homogeneous films, but can also enhance the interaction of the drug core with the polymer coating. Both a retardation and an increase in drug release were observed, with the extent being dependent on the drug type and curing conditions. The physicochemical properties of the drug and of the polymeric coatings including their interaction were important factors determining the drug release and the long-term stability of the coated dosage forms (Shawn et al. 2013, Siepmann and Siepmann 2013).

In addition, the coating level and type of plasticizer might alter the impact of the curing conditions on drug release (Yang et al. 2010). For example, the release rate of diltiazem HCl monotonically decreased with increasing harshness of the curing conditions (time, temperature, and relative humidity) from pellets coated with Aquacoat, when plasticized with triethyl citrate (TEC)

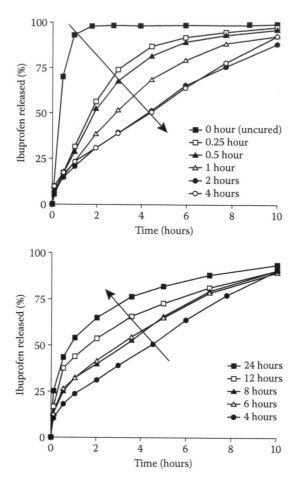

FIGURE 8.14   Effect of curing time on the ibuprofen release in pH 7.4 phosphate buffer from uncured and cured Aquacoat-coated beads (curing temperature 50°C). (Reproduced from R. Bodmeier, O. Paeratakul, *Drug Dev Ind Pharm*, 1994, 20, 9, 1517–1533 with permission.)

(irrespective of the coating level). In contrast, in the case of dibutyl sebacate (DBS), and distilled acetylated monoglycerides (Myvacet), this type of relationship was only observed at low coating levels (5%). At intermediate coating levels (around 7.5%), the curing conditions had virtually no effect on drug release. At higher coating levels (10%), the release rate initially increased and then decreased with increasing harshness of the curing conditions.

MECHANICAL PROPERTIES OF ETHYLCELLULOSE FILMS

Polymer coatings are often characterized with respect to permeability and morphological and mechanical properties. The mechanical properties of dry polymer films are mainly affected by the thermomechanical properties of the polymer, such as glass transition or softening temperature, and by film additives, such as plasticizers and fillers. They are rarely measured to predict the performance of the final coated dosage form under applied stress (e.g., compression, shipment) or in an aqueous environment but primarily to study the effect of certain process or formulation factors on properties such as tensile strength, elongation, and various moduli. However, an important question to be answered relates to the performance of the coated dosage forms in dissolution or biological fluids. With oral drug delivery systems, the drug release process is initiated by diffusion of aqueous fluids across the polymeric coating. The polymer films are hydrated and can contain significant amounts of water. In addition to film hydration, plasticizers or other film additives might leach into the aqueous environment. What are the mechanical properties of these hydrated films and how could they potentially affect the performance of the drug delivery system? The coated dosage form might exposed to significant mechanical stress factors caused internally by the buildup of hydrostatic pressure due to water-soluble core ingredients or externally through peristaltic movements in the gastrointestinal tract. A rupturing of the film coat can result in a loss in protective or sustained-release properties.

The mechanical properties (e.g., puncture strength and percentage of elongation at break) of polymeric films in the dry and wet state can be evaluated and compared using a puncture test (Bodmeier and Paeratakul 1994b). They are strongly affected by the type of polymer dispersion. The ethylcellulose pseudolatexes Aquacoat and Surelease resulted in very brittle films in the dry state and weak and soft films in the wet state with low values for puncture strength and elongation (<5%) in both cases (Table 8.3). The brittle nature of the ethylcellulose films can possibly be explained with the interchain hydrogen bonding and the bulkiness of the glucose subunits. Surelease, which is an ethylcellulose dispersion already plasticized with dibutyl sebacate, had slightly better mechanical properties in the wet state when compared to those of Aquacoat films. Both ethylcellulose pseudolatexes are stabilized with anionic surfactants. However, in the case of Surelease, ammonium oleate converts to oleic acid, which then acts as a plasticizer during drying. With Aquacoat films, the presence of sodium lauryl sulfate might have been responsible for the lower wet strength as well as the higher water uptake when compared to Surelease films.

As curing is often recommended after the coating with colloidal polymer dispersions in order to enhance and complete the coalescence of the colloidal polymer particles in a homogenous film, it was thought that curing might improve the mechanical properties of the Aquacoat films. However, as shown in Table 8.4, the drying temperature and time had only minimal effects on the mechanical properties of films plasticized at two triethyl citrate concentrations. Although the puncture strength increased with both dry and wet films after curing, the elongation was still less than 1%. The water-soluble plasticizer triethyl citrate almost completely leached from the films during exposure to aqueous media. In dry films, the actual triethyl citrate content decreased with increased drying time and temperature, indicating evaporation and/or possible degradation of the plasticizer. In contrast, the mechanical properties of films prepared with another ethylcellulose dispersion, Surelease, were dependent on the coalescence temperature and type of plasticizer (Parikh et al. 1993). No change in mechanical properties was observed at temperatures in excess of 60°C.

Ethylcellulose films when cast from organic solutions were stronger (higher puncture strength) in both the dry and wet state than Aquacoat films (Table 8.5). However, the elongation values were still low. Interestingly, triethyl citrate leached almost completely from the pseudolatex-cast film whereas more than 75% of the original plasticizer was still present in films cast from organic solutions. This can be attributed to the different film coating structures as detailed in the section on drug

---

TABLE 8.3

Mechanical Properties of Dry and Wet Films Prepared from Ethylcellulose Dispersions

Polymer Dispersion (Film Thickness, µm)	Puncture Strength (MPa)		Elongation (%)	
	Dry	Wet	Dry	Wet
Aquacoat[a] (309)	0.34 (0.11)	0.10 (0.02)	0.94 (0.18)	0.13 (0.02)
Surelease (394)	0.23 (0.04)	0.74 (0.10)	0.62 (0.12)	4.89 (0.9)

Source: Reproduced from R. Bodmeier, O. Paeratakul, *Pharm Res*, 1994c, 11, 6, 882–888 with permission.

[a] Plasticized with triethyl citrate (20% w/w).

TABLE 8.4

Effect of Drying Conditions on the Mechanical Properties of Dry and Wet Aquacoat-Triethyl Citrate Films

Triethyl Citrate,% w/w (Film Thickness, μm)	Puncture Strength (MPa)	Elongation (%)	Actual TEC Content (%)
	Dry Films		
Drying temperature and time: 40°C–48 hours			
20 (385)	0.21 (0.01)	0.25 (0.03)	19.89 (0.86)
30 (356)	0.23 (0.02)	0.97 (0.26)	27.86 (0.23)
Drying temperature and time: 40°C–24 hours + 60°C–24 hours			
20 (385)	0.35 (0.02)	0.56 (0.08)	16.78 (0.28)
30 (361)	0.34 (0.05)	1.00 (0.16)	25.77 (0.46)
	Wet Films		
Drying temperature and time: 40°C–48 hours			
20	0.07 (0.00)	0.08 (0.01)	2.61 (0.70)
30	0.08 (0.01)	0.02 (0.00)	0.84 (0.09)
Drying temperature and time: 40°C–24 hours + 60°C–24 hours			
20	0.13 (0.01)	0.13 (0.02)	3.98 (0.14)
30	0.17 (0.00)	0.14 (0.00)	2.81 (0.51)

Source: Reproduced from R. Bodmeier, O. Paeratakul, *Pharm Res*, 1994c, 11, 6, 882–888, with permission.

TABLE 8.5

Mechanical Properties of Solvent- and Pseudolatex-Cast Ethylcellulose–Triethyl Citrate Films

Polymeric Film (Film Thickness, μm)	Puncture Strength (MPa)	Elongation (%)	Triethyl Citrate Content in Films,% w/w	Water Content, g, Water/g, Polymer
	Dry Films			
Ethylcellulose[a] (313)	3.04 (0.00)	2.08 (0.00)	20.02 (0.75)	–
Aquacoat (385)	0.35 (0.02)	0.56 (0.08)	16.78 (0.28)	–
	Wet Films			
Ethylcellulose[a]	0.56 (0.10)	0.45 (0.15)	16.29 (0.81)	0.116 (0.017)
Aquacoat	0.13 (0.01)	0.13 (0.02)	3.98 (0.14)	0.426 (0.005)

Source: Reproduced from R. Bodmeier, O. Paeratakul, *Pharm Res*, 1994c, 11, 6, 882–888, with permission.
[a] Solvent-cast.

release mechanisms. The pseudolatex-cast films took up almost 43% water when compared to only 12% with the solvent-cast films.

The permanence of the plasticizer in the film during coating, storage, and contact with artificial or biological fluids is important to assure the stability of the dosage form and consistent drug release. Bodmeier and Paeratakul (1992) studied the leaching of water-soluble plasticizers (e.g., triethyl citrate) from polymeric films prepared by casting and drying of plasticized Aquacoat dispersion. The leaching was quite rapid from Aquacoat films and increased with increasing level of plasticizer. Although the selection of a leachable plasticizer had no negative effect on the film formation from

TABLE 8.6

Mechanical Properties of Dry and Wet Aquacoat Films Plasticized with Different Plasticizers (30% w/w)

Plasticizer (Film Thickness, μm)	Puncture Strength (MPa)		Elongation (%)	
	Dry	Wet	Dry	Wet
Water-Soluble				
TEC (309)	0.34 (0.11)	0.10 (0.02)	1.34 (0.18)	0.13 (0.02)
Triacetin (302)	0.12 (0.04)	0.03 (0.01)	0.10 (0.05)	0.03 (0.01)
Water-Insoluble				
ATBC (314)	0.16 (0.05)	0.19 (0.02)	0.18 (0.09)	1.69 (0.21)
ATEC (323)	0.18 (0.05)	0.06 (0.00)	0.38 (0.15)	0.31 (0.05)
DBP (327)	0.60 (0.02)	0.22 (0.02)	1.21 (0.07)	2.28 (0.09)
DBS (324)	0.19 (0.04)	0.09 (0.01)	0.25 (0.09)	0.30 (0.06)
DEP (324)	0.18 (0.02)	0.11 (0.02)	0.21 (0.12)	0.28 (0.12)
TBC (319)	0.50 (0.06)	0.16 (0.01)	2.25 (0.45)	1.79 (0.66)

Source: Reproduced from R. Bodmeier, O. Paeratakul, *Pharm Res*, 1994c, 11, 6, 882–888, with permission.

aqueous polymer dispersions, it could have a significant impact on the permeability and mechanical properties of polymeric coatings during dissolution studies or in a biological environment.

Various pharmaceutically acceptable plasticizers have been used with ethylcellulose dispersions. Plasticizers are added to induce and enhance the coalescence of the colloidal polymer particles into a homogeneous film by reducing the glass transition and minimum film formation temperature and to improve the mechanical properties of the dried films. The effect of the water-soluble plasticizers triethyl citrate and triacetin and of the water-insoluble plasticizers tributyl citrate, acetyltributyl citrate, acetyltriethyl citrate, dibutyl sebacate, dibutyl phthalate, and diethyl phthalate on the mechanical properties of dry and wet Aquacoat films are shown in Table 8.6. The mechanical properties of Aquacoat films were similar for all plasticizers. Dry films were very brittle and wet films soft and weak as indicated by a low puncture strength and elongation. The elongation was less than 2% in most cases.

In summary, ethylcellulose films are weak in both the dry and wet state with low puncture strength and elongation values. In contrast, acrylic-based polymeric films are stronger and more flexible (Bodmeier and Paeratakul 1994c).

DRUG RELEASE MECHANISMS

Drug release from polymer-coated dosage forms is often controlled by diffusion through the intact polymeric film, through water-filled channels, or a combination of both. As the mobility of most drugs is much higher in aqueous fluids than in (dense) macromolecular networks, diffusion through water-filled pores is generally more important if both types of diffusion pathways are available. In addition to diffusional mass transport, convection may play an important role, for instance, in the case of highly osmotically active inner bead cores. The water influx at early time points can, for example, hinder drug diffusion in the opposite direction. Furthermore, limited drug solubility, drug partitioning between aqueous and polymeric phases, drug–polymer interactions, (partial) plasticizer leaching into the surrounding bulk fluid as well as bead swelling and rupture of the film coatings can be involved in the overall control of drug release. Thus, not only the permeability of ethylcellulose coatings, but also their mechanical stability (in particular in the wet state) is of outmost importance for the underlying drug-release mechanisms (Lecomte et al. 2005a). As

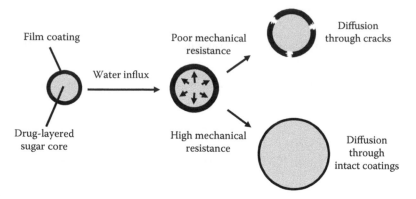

FIGURE 8.15 Schematic presentation of the underlying drug release mechanisms from polymer-coated, drug-layered sugar cores, depending on the mechanical stability of the film coatings. (With kind permission from Springer Science+Business Media: *Pharm Res*, pH-sensitive polymer blends used as coating materials to control drug release from spherical beads: elucidation of the underlying mass transport mechanisms, 22, 2005, 1129–1141, F. Lecomte et al.)

illustrated in Figure 8.15, the water influx into polymer coated, drug-layered sugar cores can generate significant hydrostatic pressure inside the systems acting against the film coatings. In the case of poor mechanical film stability, cracks are created after a certain lag time (as soon as a critical threshold pressure is attained), and drug release occurs primarily via diffusion (and/or convection) through water-filled channels (Lecomte et al. 2005b). In the case of mechanically stable film coatings, drug release is generally controlled by diffusion through the intact polymeric networks. For instance, diltiazem HCl release from pellets coated with Aquacoat containing small amounts of poly(vinyl alcohol)-poly(ethylene glycol) graft copolymer has been found to be primarily controlled by drug diffusion through the intact polymeric membranes, irrespective of the type of starter core (consisting of microcrystalline cellulose or sugar, optionally coated with ethylcellulose) (Muschert et al. 2009a,b). Importantly, the apparent diffusion coefficient of the drug in the macromolecular networks could easily be determined with thin free films and successfully be used to quantitatively predict the release rate from coated pellets.

The mechanical stability of ethylcellulose film coatings can significantly be altered by adding varying amounts of a more flexible polymer, for example, Eudragit L (Lecomte et al. 2009a, Siepmann et al. 2008a). Figure 8.16a shows how the underlying drug release mechanism from Aquacoat:Eudragit® L coated verapamil HCl-layered sugar cores can be shifted from transport through water-filled channels to diffusion through the intact polymeric coatings. The onset of crack formation can be experimentally monitored using different techniques, for example, scanning electron microscopy (however, the risk of artifact creation during sample preparation should not be underestimated). Also the changes in the pellets' diameter during drug release can serve as an indicator for crack formation: As long as the film coating remains intact and water continues to diffuse into the beads, their diameter increases. As soon as crack formation occurs, the increase in bead diameter levels off and the generated hydrostatic pressure within the pellets can eventually squeeze out liquids, resulting in pellet shrinking (this phenomenon is more likely in the case of high hydrostatic pressures built up in the bead core). Figure 8.16b shows, as an example, the experimentally measured changes in pellet diameter of Aquacoat:Eudragit L coated verapamil HCl-layered sugar cores upon exposure to 0.1 M HCl. The composition of the coatings was varied, resulting in different mechanical stabilities (ethylcellulose-rich films being brittle) and, thus, altered onsets of crack formation (indicated by the leveling off/decreasing bead diameter-time-curves). The observed swelling kinetics agree well with the drug release profiles from these systems (Figure 8.16a).

Furthermore, it has been shown that the type of coating technique (using aqueous polymer dispersions vs. organic polymer solutions) significantly affects the inner film coating structure and,

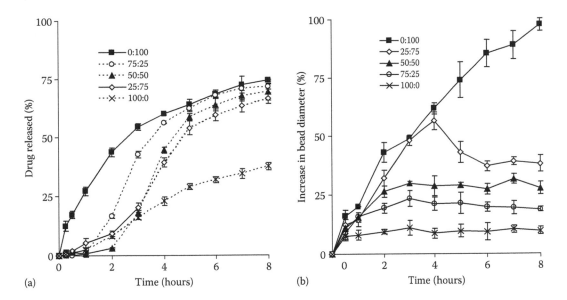

FIGURE 8.16 Behavior of Aquacoat:Eudragit L (blend ratio indicated in the figures) coated, verapamil HCl-layered sugar cores in 0.1 M HCl: (a) drug release; dotted curves indicate diffusion through water-filled cracks, solid curves through the intact film coatings; (b) change in pellet diameter. (With kind permission from Springer Science+Business Media: *Pharm Res*, pH-sensitive polymer blends used as coating materials to control drug release from spherical beads: elucidation of the underlying mass transport mechanisms, 22, 2005, 1129–1141, F. Lecomte et al.)

hence, mechanical stability of the macromolecular networks (Lecomte et al. 2004b). In organic solutions, the mobility of the polymer chains is high, leading to elevated degrees of polymer chain inter-diffusion and thus intensive polymer chain entanglement in the resulting film coatings. Consequently, the polymeric barriers are stable. As ethylcellulose is poorly permeable for most drugs, this generally results in low release rates. In contrast, the mobility of the macromolecules within colloidal polymer particles is highly restricted and polymer chain inter-diffusion is limited during film formation. Hence, the degree of polymer chain entanglement within the resulting films is lower than in systems prepared from organic solutions. Consequently, the coatings are mechanically weaker and cracks can more easily be formed, causing faster drug release (Lecomte et al. 2004b, Wesseling et al. 1999).

As discussed above, the type of plasticizer can also affect the mechanical properties of ethylcellulose film coatings. It has recently been shown with Aquacoat-coated, propranolol HCl-layered sugar cores that different drug release patterns can be achieved depending on the type of plasticizer used (water-insoluble dibutyl sebacate [DBS] vs. water-soluble triethyl citrate [TEC]) (Lecomte et al. 2004a). In contrast to DBS, TEC rapidly leaches out of the coatings, resulting in decreasing mechanical resistances of the films and thus facilitated crack formation. In addition, the hydrophilicity of the plasticizer significantly affects the water uptake behavior of the film coatings and, hence, changes in the coatings' toughness and drug permeability upon exposure to the release media. Thus, the type of plasticizer can significantly affect the underlying drug release mechanisms in Aquacoat-coated dosage forms.

REFERENCES

Appel LE, Zentner GM. Use of modified ethylcellulose lattices for microporous coating of osmotic tablets. *Pharm Res* 1991; 8:600–604.

Banker GS, Peck GE. The new, water-based colloidal dispersions. *Pharm Technol* 1981; 5:55–61.

Bindschaedler C, Gurny R, Doelker E. Theoretical concepts regarding the formation of films from aqueous micro-dispersions and application to coating. *Labo Pharma Probl Tech* 1983; 31:389–394.

Bindschaedler C, Gurny R, Doelker E. Influence of emulsifiers on film formation from cellulose acetate latexes. Experimental study of phase separation phenomena due to sodium dodecyl sulfate. I. *J Appl Polym Sci* 1987; 34(8):2631–2647.

Bindschaedler C, Gurny R, Doelker E. Influence of emulsifiers on film formation from cellulose acetate latexes. Modeling approach to the fate of emulsifiers in highly plasticized films. II. *J Appl Polym Sci* 1989; 37(1):173–182.

Bodmeier R, Paeratakul O. Process and formulation variables affecting the drug release from chlorpheniramine maleate-loaded beads coated with commercial and self-prepared aqueous ethyl cellulose pseudolatexes. *Int J Pharm* 1991a; 70(1–2):59–68.

Bodmeier R, Paeratakul O. Determination of plasticizers commonly used in pharmaceutical dosage forms by high performance liquid chromatography. *J Liq Chromatogr* 1991b; 14:365–375.

Bodmeier R, Paeratakul O. Leaching of water-soluble plasticizers from polymeric films prepared from aqueous colloidal polymer dispersions. *Drug Dev Ind Pharm* 1992; 18(17):1865–1882.

Bodmeier R, Paeratakul O. The distribution of plasticizers between aqueous and polymer phases in aqueous colloidal polymer dispersions. *Int J Pharm* 1994a; 103:47–54.

Bodmeier R, Paeratakul O. The effect of curing on drug release and morphological properties of ethylcellulose pseudolatex-coated beads. *Drug Dev Ind Pharm* 1994b; 20(9):1517–1533.

Bodmeier R, Paeratakul O. Mechanical properties of dry and wet cellulosic and acrylic films prepared from aqueous colloidal polymer dispersions used in the coating of solid dosage forms. *Pharm Res* 1994c; 11(6):882–888.

Bodmeier R, Paeratakul O. Plasticizer uptake by aqueous colloidal polymer dispersions used for the coating of solid dosage form. *Int J Pharm* 1997; 152(1):17–26.

Chang RK, Hsiao CH, Robinson JR. A review of aqueous coating techniques and preliminary data on release from a theophylline product. *Pharm Technol* 1987; 3:56–68.

Christensen FN, Bertelsen P. Abstracts of papers, 17th International Symposium on Controlled Release of Bioactive Materials, Reno, NV, July 22–25, 1990:124–125.

Colorcon Ltd. technical literature. Surelease® ethylcellulose dispersion 2006.

Dahl TC, Sue IT. The effects of heat and desiccation treatment on the controlled release properties of aqueous silicone latex coated tablets. *Drug Dev Ind Pharm* 1990; 16(14):2097–2107.

Dashevsky A, Wagner K, Kolter K et al. Physicochemical and release properties of pellets coated with Kollicoat SR 30 D, a new aqueous polyvinyl acetate dispersion for extended release. *Int J Pharm* 2005; 290(1–2):15–23.

Dillon RE, Bradford EB, Andrews Jr. RD. Plasticizing a synthetic latex. *Ind Eng Chem* 1953; 45(4):728–735.

Feigin RI, Napper DH. Depletion stabilization and depletion flocculation. *J Colloid Interf Sci* 1980; 75: 525–541.

Felton LA. Mechanisms of polymeric film formation. *Int J Pharm* 2013; 457(2):423–427.

FMC BioPolymer technical literature. Aquacoat® Aqueous Coating 2006.

Fukumori Y, Yamaoka Y, Ichikawa H et al. Coating of pharmaceutical powders by fluidized bed process. IV. Softening temperature of acrylic copolymers and its relation to film-formation in aqueous coating. *Chem Pharm Bull* 1988; 36(12):4927–4932.

Gendre C, Genty M, Fayard B et al. Comparative static curing versus dynamic curing on tablet coating structures. *Int J Pharm* 2013; 453:448–453.

Gilligan CA, Li Wan Po A. Factors affecting drug release from a pellet system coated with an aqueous colloidal dispersion. *Int J Pharm* 1991; 73:51–68.

Goodhart FW, Harris MR, Murthy KS et al. An evaluation of aqueous film-forming dispersions for controlled release. *Pharm Technol* 1984; 8(4):64–71.

Gundert-Remy U, Moller H, eds. Oral Controlled Release Products: Therapeutic and Biopharmaceutic Assessment. Stuttgart: Wissenschaftliche Verlagsgesellschaft, 1990.

Guo X. Physicochemical and Mechanical Properties Influencing the Drug Release from Coated Dosage Forms. PhD dissertation, University of Texas at Austin, 1996.

Harris MR, Ghebre-Sellassie I, Nesbitt RU. Water based coating process for sustained release. *Pharm Technol* 1986; 10(9):102–107.

Harris MR, Ghebre-Sellassie I. Aqueous polymeric coating for modified release oral dosage forms. In: McGinity JW, ed. Aqueous Polymeric Coatings for Pharmaceutical Dosage Forms. New York: Marcel Dekker, 1997:81–100.

Horvath E, Ormos Z. Film coating of dragee seeds by fluidized bed spraying methods. *Acta Pharm Technol* 1989; 35(2):90–96.

Hutchings D, Kuzmak B, Sakr A. Processing considerations for an EC latex coating system: Influence of curing time and temperature. *Pharm Res* 1994; 11(10):1474–1478.

Iyer U, Hong WH, Das N et al. Comparative evaluation of three organic solvent and dispersion-based ethylcellulose coating formulations. *Pharm Technol* 1990; 14(9):68–86.

Kallstrand G, Ekman B. Membrane-coated tablets: A system for the controlled release of drugs. *J Pharm Sci* 1983; 72(7):772–775.

Karrout Y, Neut C, Wils D et al. Novel polymeric film coatings for colon targeting: How to adjust desired membrane properties. *Int J Pharm* 2009a; 371:64–70.

Karrout Y, Neut C, Wils D et al. Colon targeting with bacteria-sensitive films adapted to the disease state. *Eur J Pharm Biopharm* 2009b; 73:74–81.

Karrout Y, Neut C, Wils D et al. Characterization of ethylcellulose:starch-based film coatings for colon targeting. *Drug Dev Ind Pharm* 2009c; 35:1190–1200.

Karrout Y, Neut C, Wils D et al. Novel polymeric film coatings for colon targeting: Drug release from coated pellets. *Eur J Pharm Sci* 2009d; 37:427–433.

Karrout Y, Neut C, Wils D et al. Enzymatically activated multiparticulates containing theophylline for colon targeting. *J Drug Deliv Sci Technol* 2010a; 20:193–199.

Karrout Y, Neut C, Siepmann F et al. Enzymatically degraded EC:Eurylon 6 HP-PG film coatings for colon targeting in IBD patients. *J Pharm Pharmacol* 2010b; 62:1676–1684.

Karrout Y, Neut C, Siepmann F et al. Peas starch-based film coatings for site specific drug delivery to the colon. *J Appl Polym Sci* 2011; 119:1176–1184.

Karrout Y, Dubuquoy L, Piveteau C et al. In vivo efficacy of microbiota-sensitive coatings for colon targeting: A promising tool for IBD therapy. *J Control Release* 2015; 197:121–130.

Kositprapa U, Bodmeier R. Ion-pair formation between cationic drugs and anionic surfactants. *Pharm Res* 1994; 11(10):S-235.

Kositprapa U, Herrmann J, Bodmeier R. Interactions between cationic drugs and anionic surfactant. *Pharm Res* 1993; 10(10):S-153.

Kucera SA, Felton LA, McGinity JW. Physical aging in pharmaceutical polymers and the effect on solid oral dosage form stability. *Int J Pharm* 2013; 457:428–436.

Laicher A, Lorck CA, Grunenberg PC et al. Aqueous coating of pellets to sustained-release dosage forms in a fluid-bed coater. Influence of product temperature and polymer concentration on in vitro release. *Pharm Ind* 1993; 55:1113–1116.

Lecomte F, Siepmann J, Walther M et al. Blends of enteric and GIT-insoluble polymers used for film coating: Physicochemical characterization and drug release patterns. *J Control Release* 2003; 89(3):457–471.

Lecomte F, Siepmann J, Walther M et al. Polymer blends used for the aqueous coating of solid dosage forms: Importance of the type of plasticizer. *J Control Release* 2004a; 99:1–13.

Lecomte F, Siepmann J, Walther M et al. Polymer blends used for the coating of multiparticulates: Comparison of aqueous and organic coating techniques. *Pharm Res* 2004b; 21(5):882–890.

Lecomte F, Siepmann J, Walther M et al. pH-sensitive polymer blends used as coating materials to control drug release from spherical beads: Elucidation of the underlying mass transport mechanisms. *Pharm Res* 2005a; 22(7):1129–1141.

Lecomte F, Siepmann J, Walther M et al. pH-sensitive polymer blends used as coating materials to control drug release from spherical beads: Importance of the type of core. *Biomacromolecules* 2005b; 6(4):2074–2083.

Li SP, Mehta GN, Buehler JD et al. The effect of film-coating additives on the in vitro dissolution release rate of ethyl cellulose-coated theophylline granules. *Pharm Technol* 1990; 14(3):20–24.

Lindstedt B, Ragnarsson G, Hjartstam J. Osmotic pumping as a release mechanism for membrane-coated drug formulations. *Int J Pharm* 1989; 56:261–268.

Lindstedt B, Sjoberg M, Hjartstam J. Osmotic pumping release from KCl tablets coated with porous and non-porous ethylcellulose. *Int J Pharm* 1991; 67:21–27.

Lippold BC, Lippold BH, Sutter BK et al. Properties of aqueous, plasticizer-containing ethyl cellulose dispersions and prepared films in respect to the production of oral extended release formulations. *Drug Dev Ind Pharm* 1990; 16:1725–1747.

Lippold BH, Sutter BK, Lippold BC. Parameters controlling drug release from pellets coated with aqueous ethyl cellulose dispersion. *Int J Pharm* 1989; 54(1):15–25.

Maroni A, Dorly Del Curto M, Zema L et al. Film coatings for oral colon delivery. *Int J Pharm* 2013; 457:372–394.

McAinsh J, Rowe RC. Sustained release pharmaceutical composition. U.S. Patent 4,138,475, February 6, 1979.

Muschert S, Siepmann F, Cuppok Y et al. Improved long term stability of aqueous ethylcellulose film coatings: Importance of the type of drug and starter core. *Int J Pharm* 2009a; 368:138–145.

Muschert S, Siepmann F, Leclercq B et al. Prediction of drug release from ethylcellulose coated pellets. *J Control Release* 2009b; 135:71–79.

Muschert S, Siepmann F, Leclercq B et al. Drug release mechanisms from ethylcellulose:PVA-PEG graft copolymer coated pellets. *Eur J Pharm Biopharm* 2009c; 72:130–137.

Muschert S, Siepmann F, Leclercq B et al. Dynamic and static curing of ethylcellulose:PVA-PEG graft copolymer film coatings. *Eur J Pharm Biopharm* 2011; 78:455–461.

Nakagami H, Keshikawa T, Matsumura M et al. Application of aqueous suspensions and latex dispersions of water-insoluble polymers for tablet and granule coating. *Chem Pharm Bull* 1991; 39(7):1837–1842.

Nesbitt RU. Effect of formulation components on drug release from multiparticulates. *Drug Dev Ind Pharm* 1994; 20(20):3207–3236.

Ottewill RH. Colloidal properties of latex particles. In: Candau F, Ottewill RH, eds. An Introduction to Polymer Colloids. Boston: Academic Publishers, 1990:129–157.

Paeratakul O. Pharmaceutical Applications of Aqueous Colloidal Polymer Dispersions. PhD dissertation. University of Texas at Austin, 1993.

Parikh NH, Porter SC, Rohera BD. Tensile properties of free films cast from aqueous ethylcellulose dispersions. *Pharm Res* 1993; 10(6):810–815.

Patton TC, ed. Paint Flow and Pigment Dispersion—A Rheological Approach to Coating and Ink Technology. New York: John Wiley & Sons, 1979.

Porter SC. Controlled Release Symposium, Colorcon Inc., PA, 1990:I-1–54.

Ramig A. Latex paints—CPVC, formulation, and optimization. *J Paint Technol* 1970; 47(602):60–67.

Rekhi GS, Jambhekar SS. Ethylcellulose—A polymer review. *Drug Dev Ind Pharm* 1995; 21:61–77.

Rekhi GS, Mendes RW, Porter SC et al. Aqueous polymeric dispersions for controlled drug delivery-Wurster process. *Pharm Technol* 1989; 13:112–125.

Rodriguez MT, Gracenea JJ, Saura JJ et al. The influence of the critical pigment volume concentration (CPVC) on the properties of an epoxy coating: Part II. Anti-corrosion and economic properties. *Prog Org Coat* 2004; 50(1):68–74.

Rosiaux Y, Muschert S, Chokshi R et al. Ethanol-resistant polymeric film coatings for controlled drug delivery. *J Control Release* 2013a; 169:1–9.

Rosiaux Y, Velghe C, Muschert S et al. Ethanol-resistant ethylcellulose:guar gum coatings—Importance of formulation parameters. *Eur J Pharm Biopharm* 2013b; 85:1250–1258.

Rosiaux Y, Velghe C, Muschert S et al. Mechanisms controlling theophylline release from ethanol-resistant coated pellets. *Pharm Res* 2014; 31:731–741.

Selinger E, Brine CJ. Use of thermal analysis in the optimization of polymeric diffusion barriers in controlled release delivery systems. *Thermochim Acta* 1988; 134:275–282.

Shah NH, Zhang L, Railkar A et al. Factors affecting the kinetics and mechanism of release of cilazapril from beadlets coated with aqueous and nonaqueous ethyl cellulose-based coatings. *Pharm Technol* 1994; 18(10):140–149.

Shin-Etsu Technical Information, Shin-Etsu Chem. Inc., Japan, 1991.

Siepmann F, Hoffmann A, Leclercq B et al. How to adjust desired drug release patterns from ethylcellulose-coated dosage forms. *J Control Release* 2007a; 119(2):182–189.

Siepmann F, Muschert S, Zach S et al. Carrageenan as an efficient drug release modifier for ethylcellulose-coated pharmaceutical dosage forms. *Biomacromolecules* 2007b; 8:3984–3991.

Siepmann F, Siepmann J, Walther M et al. Polymer blends for controlled release coatings. *J Control Release* 2008a; 125(1):1–15.

Siepmann F, Wahle C, Leclercq B et al. pH-sensitive film coatings: Towards a better understanding and facilitated optimization. *Eur J Pharm Biopharm* 2008b; 68:2–10.

Siepmann F, Muschert S, Leclercq B et al. How to improve the storage stability of aqueous polymeric film coatings. *J Control Release* 2008c; 126:26–33.

Siepmann J, Paeratakul O, Bodmeier R. Modeling plasticizer uptake in aqueous polymer dispersions. *Int J Pharm* 1998; 165:191–200.

Siepmann J, Siepmann F. Modeling of diffusion controlled drug delivery. *J Control Release* 2012; 161:351–362.

Siepmann J, Siepmann F. Stability of aqueous polymeric controlled release film coatings. *Int J Pharm* 2013; 457:437–445.

Sperry PR, Hopfenberg HB, Thomas NL. Flocculation of latex by water-soluble polymers: Experimental confirmation of a nonbridging, nonadsorptive, volume-restriction mechanism. *J Colloid Interf Sci* 1981; 82:62–76.

Sutter B. Aqueous ethylcellulose dispersions for preparation of microcapsules with controlled drug release. PhD Thesis, University of Düsseldorf, 1987.

Wesseling M, Bodmeier R. Drug release from beads coated with an aqueous colloidal ethylcellulose dispersion, Aquacoat, or an organic ethylcellulose solution. *Eur J Pharm Biopharm* 1999; 47(1):33–38.

Wong D, Bodmeier R. Flocculation of an aqueous colloidal ethyl cellulose dispersion (Aquacoat) with a water-soluble polymer, hydroxypropyl methylcellulose. *Eur J Pharm Biopharm* 1996; 42:12–15.

Wong D, Paeratakul O, Bodmeier R. Combination of hydroxypropyl methyl cellulose (HPMC) and aqueous latexes for coating purposes. *Pharm Res* 1991; 8(10):S-116.

Wong D. Water-Soluble Polymers in Pharmaceutical Aqueous Colloidal Polymer Dispersions. PhD dissertation, University of Texas at Austin, 1994.

Yang QW, Flament MP, Siepmann F et al. Curing of aqueous polymeric film coatings: Importance of the coating level and type of plasticizer. *Eur J Pharm Biopharm* 2010; 74:362–370.

Yang ST, Van Savage G, Weiss J et al. The effect of spray mode and chamber geometry of fluid-bed coating equipment and other parameters on an aqueous-based ethylcellulose coating. *Int J Pharm* 1990; 86:247–257.

Zentner GM, Rork GS, Himmelstein KJ. Osmotic flow through controlled porosity films: An approach to delivery of water-soluble compounds. *J Control Release* 1985; 2:217–229.

Zhang GH, Schwartz JB, Schnaare RL et al. Abstracts of papers, 17th International Symposium on Controlled Release of Bioactive Materials, Reno, NV, July 22–25, 1990:194–195.

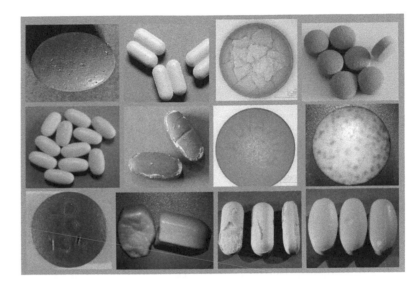

FIGURE 5.1   Common film coating defects.

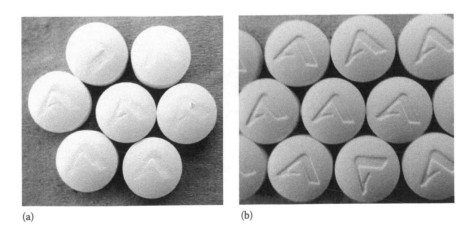

(a)   (b)

FIGURE 5.6   (a) Standard HPMC. (b) HPMC/copovidone composite. Illustration of how choice of an appropriate coating formulation can eliminate logo bridging problems.

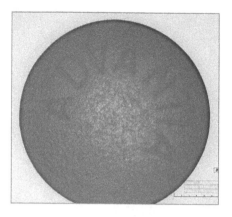

FIGURE 5.7   Example of infilling of tablet logos/break lines.

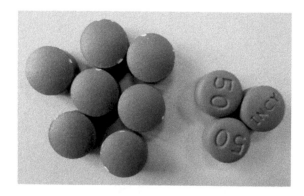

FIGURE 5.13   Example of how changing from standard concave punches to compound radius punches can eliminate tablet edge chipping.

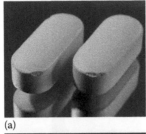

Property	Value
Film strength (MPa)	30.69
Elongation (%)	21.96
Film adhesion (N m^{-1})	22.51

(a)

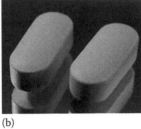

Property	Value
Film strength (MPa)	53.86
Elongation (%)	16.77
Film adhesion (N m^{-1})	17.16

(b)

FIGURE 5.14 Example of how improving coating mechanical properties can eliminate edge chipping problems. (a) The problem and (b) the solution.

FIGURE 5.18 Example of tablet pitting.

FIGURE 5.20 Example of film peeling and flaking.

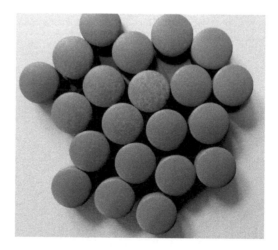

FIGURE 5.21 Example of tablet-to-tablet color variation. (© 2015 AstraZeneca, all rights reserved. Image reproduced with permission of AstraZeneca; image is copyright of AstraZeneca and may not be reproduced without written permission.)

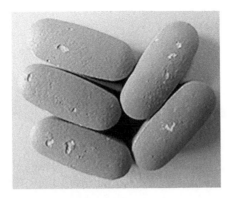

FIGURE 5.25   Example of tablets exhibiting picking.

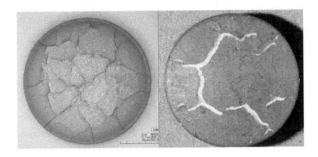

FIGURE 5.26   Examples of cracking of film coatings.

FIGURE 5.28   Example of unacceptably rough film-coated tablets

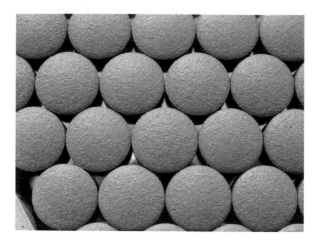

FIGURE 5.29 Example of tablets where roughness of the coating is caused by overwetting. (© 2015 AstraZeneca, all rights reserved. Image reproduced with permission of AstraZeneca; image is copyright of AstraZeneca and may not be reproduced without written permission.)

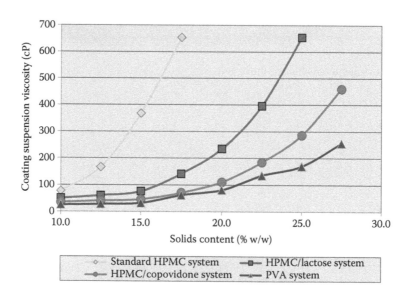

FIGURE 5.30 Illustration of how adjustment of the coating formulation can be used to reduce the viscosity of the coating liquid.

FIGURE 5.31 Example of how changing from a PVA-based coating system to an HPMC/copovidone hybrid system can reduce surface roughness.

FIGURE 5.32 Tablet discoloration caused by interaction of ingredients in the tablet core.

FIGURE 5.36 Examples of surface scuffing of film-coated tablets.

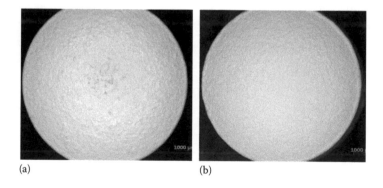

(a)   (b)

FIGURE 5.37 (a) PVA coating and (b) HPMC/copovidone coating. Example of how changing the polymers used in the coating formulation can eliminate scuffing problems.

9 Chemistry and Application Properties of Poly(meth)acrylate Systems

Brigitte Skalsky and Hans-Ulrich Petereit

CONTENTS

INTRODUCTION

HISTORY OF POLY(METH)ACRYLATES AND THEIR APPLICATIONS

More than 70 years ago, poly(methyl methacrylate) (PMMA) was invented as a crystal-clear, unbreakable organic glass of outstanding quality. With the trademark Plexiglas®, it achieved world-wide recognition as a unique synthetic material and a symbol of technical progress. The excellent biocompatibility of PMMA was detected early, and it was used for artificial limbs and implants. PMMA is also well tolerated by the skin and the mucosa so that dental prostheses and contact lenses made of PMMA are used until today. Many medical devices that come into direct contact with blood are made from PMMA or similar copolymers. Macro-porous oxirane acrylic beads, commercialized under the trade name EUPERGIT®, gained significant importance in chemical and medical applications, such as blood purification and extracorporal therapies. Enzymes covalently immobilized on EUPERGIT® serve as highly stable, recyclable catalysts in industrial biotransformation [1–3].

Functional pharmaceutical methacrylate coatings, commercialized since 1954 under the trade name EUDRAGIT®, are derived from the PMMA chemistry, enriched by functional groups providing desired in vivo behavior and produced under GMP standards. The polymerization originally followed the PMMA technology by substance polymerization and down-streaming. Recently, the substance polymerization was replaced by a continuous solution polymerization process in a fully contained production system, allowing more precise process control and traceability and thus a further reduction of residual monomers and residual solvents [4,5].

In addition to the traditional application field of oral pharmaceutical dosage forms, alternative polymers and formulations were developed meeting the requirements of nutraceuticals and enabling functional coatings in this field. In order to ease polymer applications, "ready-to-use" compositions were designed, including the polymer and functional excipients, such as glidants, surfactants, fatty acids, and optional plasticizers. These products allow fast dispersion in water to generate the functional coating suspension [6].

DEFINITIONS OF LATICES AND THEIR PHYSICOCHEMICAL SPECIFICATIONS

Historically, EUDRAGIT® polymers were applied to oral dosage forms as polymer solutions in organic solvents. The preferred options of aqueous pharmaceutical manufacturing processes are based on enhancing polymer preparations toward polymer dispersions, such as latices, pseudolatices, or colloidal solutions. EUDRAGIT® polymers, manufactured by these enhanced techniques were introduced into pharmaceutical manufacturing first in 1972 [7]. The aqueous latex dispersions are able to form functional films at mild temperatures of 25°C–35°C. Depending on the polymer, other functional excipients, such as plasticizers and glidants, may need to be added for proper processing.

The term *latex* is often used for aqueous polymer dispersions. Originally, it described rubber latex, which is called *natural latex*, in contrast to *synthetic latices*, which are preferably prepared by emulsion polymerization. The term *pseudolatex* is used for dispersions that are prepared by emulsification of organic polymer solutions in water followed by the elimination of the organic solvents [8] or direct dispersion of polymers in water [9].

Polymer dispersions or latices are characterized by particle sizes between 10 and 1000 nm. The upper limit is imposed by thermal convection and the Brownian particle movement. Both together compensate the sedimentation velocity of the particles. The lower limit is defined by the light-scattering effect of the dispersed solids, resulting in a milky appearance. Latices are characterized by low viscosity even when they have a high solids content. In the technical field of polymer applications, the terms *dispersions* and *latices* are often used synonymously. Systems of smaller particle sizes are called *micro-dispersions* or *colloidal solutions*. They are nearly transparent and exhibit the Tyndall effect in a light beam.

The film-formation mechanism of aqueous polymer dispersions is based on coalescence of (pseudo) latex particles enhanced by capillary forces created through evaporation of water [10,11]. For the formation of continuous and functional films, coating and drying conditions have significant impact and need to be well designed.

The *minimum film-forming temperature* (MFT) is the temperature in degrees Celsius above which a continuous film is formed from aqueous polymer dispersions under distinct drying conditions [12,13]. Film formation is correlated to the glass transition temperature (T_g) of the polymer itself. The T_g is defined as the temperature at which the viscosity of a melted thermoplastic polymer increases considerably while the temperature is continuously decreasing. In molecular terms, this is the temperature at which the flexibility of polymer chains and thus their material properties change. One widely used method for the determination of T_g is the differential scanning calorimetric method (DSC) described by Turi [14].

SYSTEMATICS OF NOMENCLATURE AND COMMERCIAL PRODUCTS

The trade name EUDRAGIT® is a composite of the Greek εὐ, meaning *good* or *functional* and the French *dragée*, meaning *(sugar) coated bead*; thus, the meaning of the trademark is *excellent functional coating*. The product line includes pharmaceutical copolymers from esters of acrylic or methacrylic acid whose properties are determined by specific functional groups. The individual grades differ in their proportion of neutral, alkaline, or acid groups and thus in terms of their physicochemical properties. Among soluble polymers, a distinction is made between cationic EUDRAGIT® E types, soluble in acidic fluids, and anionic EUDRAGIT® L, S, and FS types, which dissolve in neutral or alkaline fluids, respectively. Insoluble EUDRAGIT® RL/RS types carry hydrophilic quaternary ammonium groups as hydrochlorides, providing different permeability whereas the insoluble EUDRAGIT® NE/NM types include no reactive functional groups.

TABLE 9.1
Explanation of the EUDRAGIT® Letter Codes

EUDRAGIT® Type	What Does the Letter Code Stand for?
L	Lightly soluble in the intestine
S	Slightly soluble in the intestine
FS	Flexible EUDRAGIT® S
RL	Retarding Lightly
RS	Retarding Strongly
NE	Neutral Ester
NM	Neutral, for Matrix applications

These insoluble polymers absorb water from physiological fluids and swell pH-independently, controlling drug release by diffusion.

While the letters in the trade names include chemical information and functionality (Table 9.1), the numbers that follow indicate the polymer concentration (%-w/w) in the commercial product. Granules produced by continuous solution polymerization followed by extrusion may be milled or micronized to powders and are designated by the letters PO. The different physical grades are based on identical polymers and enable harmonized analytical and regulatory procedures, no matter which physical form was applied.

OTHER PHARMACEUTICAL APPLICATIONS BESIDES COATING

Besides functional coatings of oral solid dosage forms, poly(meth)acrylates are used in pharmaceutical production also for wet granulation processes, providing controlled drug release matrices [15]. Furthermore, they are applied as control layers in transdermal therapeutical systems (TTS), combining skin adhesion as well as controlled drug diffusion from matrix and membrane systems [16–19].

Co-spray drying of EUDRAGIT® dispersions or solutions enable polymer carrier systems, which improve drug efficiency by solubility enhancement and stabilization [20].

The porous structure of spray-dried EUDRAGIT® powders enable plastic deformation during compression. In matrix tablets, the polymers provide two functionalities: binder function and release control by diffusion [21–23].

Beneficial melt flow properties of solid EUDRAGIT® polymers allow economical melt processing, such as hot melt extrusion [24,25]. The main application is solubility enhancement of poorly soluble actives [26,27]. Furthermore, also sustained release applications are reported [28]. To serve the needs of the global pharmaceutical industry for more efficient and more convenient processing, "easy-to-use" or "ready-to-use" coating products were recently introduced to the market, combining the functional polymers with other crucial ingredients, such as plasticizers, glidants, and color pigments [29].

CHEMISTRY, PRODUCTION, AND QUALITY

CHEMICAL STRUCTURE

Starting monomers are synthesized following technical realization of chemical methodology [30]. The (meth)acrylic chemistry provides unique polymer properties due to the number of different esters, which can be included in the covalently linked C–C backbone by copolymerization. Thus, physicochemical properties are advantageously influenced to meet physiological needs. Chemical structure, names, and functions are summarized in Tables 9.2 and 9.3.

TABLE 9.2

Chemical Structure and Characteristics of Soluble Methacrylate Copolymers

H	CH_3	CH_3	CH_3	CH_3
|	|	|	|	|
$C=CH_2$	$C=CH_2$	$C=CH_2$	$C=CH_2$	$C=CH_2$
|	|	|	|	|
$C=O$	$C=O$	$C=O$	$C=O$	$C=O$
|	|	|	|	|
$O-CH_3$	$O-CH_3$	$O-C_4H_9$	$O-C_2H_5-N(CH_3)_2$	OH
Monomers Methylacrylate (MA)	Methylmethacrylate (MMA)	Butylmethacrylate (BMA)	Dimethyaminoethyl methacrylate (DMAEMA)	Methacrylic acid (MAA)

Scientific Name (IUPAC)	Structure	Solubility	EUDRAGIT® Types	Commercial Forms
Poly(butylmethylmethacrylate-co-(2-dimethylaminoethyl) methacrylate-co-methyl methacrylate) 1:2:1, 150,000	BMA-DMAEMA-MMA = 25:50:25	Below pH 5.0	E 100 E PO E 12,5	Granules powder Organic solution
Poly(methacrylic acid-co-ethyl acrylate) 1:1, 250,000	MAA-EA = 50:50	Above pH 5.5	L 30 D-55 L 100-55	Aqueous dispersion Powder
Poly(methacrylic acid-co-methyl methacrylate) 1:1, 135,000	MAA-MMA = 50:50	Above pH 6.0	L 100 L 12,5	Powder Organic solution
Poly(methacrylic acid-co-methyl methacrylate) 1:2, 135,000	MAA-MMA = 30:70	Above pH 7.0	S 100 S 12,5	Powder Organic solution
Poly(methylacrylate-co-methyl methacrylate-co-methacrylic acid) 280,000	MA-MMA-MAA = 65:25:10	Above pH 7.0	FS 30 D	Aqueous dispersion

Soluble Polymers

By introducing cationic or anionic functionalities as free amino or carboxylic groups, respectively, into the polymer side chain, pH-dependent solubility is achieved, which enables pH-controlled solubility of the polymer coatings and thus defined drug release. In response to the physiological pH profiles in man, these polymers enable gastrointestinal targeting to the stomach, the small intestine, or the colon.

Acid-Soluble Polymers

The cationic monomer dimethylaminoethyl methacrylate (DMAEMA), copolymerized with methyl and butyl methacrylate, results in the EUDRAGIT® E polymer and ensures its acid solubility beneath pH 5 by salt formation with anions being present in the gastric fluid (Figure 9.1).

These salts remain soluble over the entire physiological pH range and are neither precipitated by pH nor enzymes of gastric and intestinal fluids. Hence, the polymer is preferably used for protective coatings, including taste or odor masking and moisture protection. The polymer EUDRAGIT® E 100 is manufactured by solution polymerization and extrusion. It was commercialized as granules for solvent-coating processes. Its micronized grade, EUDRAGIT® E PO, with a maximum particle size of approximately 10 µm, was developed to enable aqueous coating processes from a colloidal solution [31]. Thin coatings of 10 to 20 µm thickness provide efficient taste-masking properties because the polymer is insoluble in the saliva. For moisture-protective coatings, film thicknesses

TABLE 9.3

Chemical Structure and Characteristics of Insoluble Methacrylate Copolymers

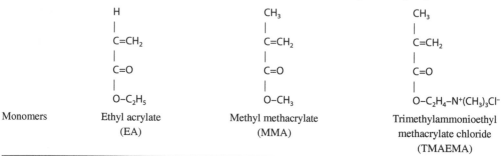

Monomers	Ethyl acrylate (EA)	Methyl methacrylate (MMA)	Trimethylammonioethyl methacrylate chloride (TMAEMA)

Scientific Name	Structure	Permeability	EUDRAGIT®	Commercial Forms
Poly(ethyl acrylate-co-methyl methacrylate) 2:1, 800,000	EA-MMA = 70:30	Medium	NE 30 D	Aqueous dispersion
			NE 40 D	Aqueous dispersion
Poly(ethyl acrylate-co-methyl methacrylate) 2:1, 600,000	EA-MMA = 70:30	Medium	NM 30 D	Aqueous dispersion
Poly(ethyl acrylate-co-methyl methacrylate-co-trimethylammonioethyl methacrylate chloride) 1:2:0.2, 150.000	EA MMA-TMAEMA = 30:60:10	High	RL 100	Granules
			RL PO	Powder
			RL 30 D	Aqueous dispersion
			RL 12,5	Organic solution
Poly(ethyl acrylate-co-methyl methacrylate-co-trimethylammonioethyl methacrylate chloride) 1:2:0.1, 150.000	EA-MMA-TMAEMA = 30:65:5	Low	RS 100	Granules
			RS PO	Powder
			RS 30 D	Aqueous dispersion
			RS 12,5	Organic solution

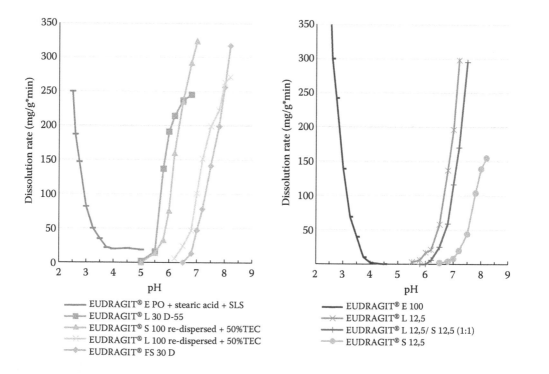

FIGURE 9.1 pH-controlled solubility of methacrylate films from aqueous (left) and solvent processing (right).

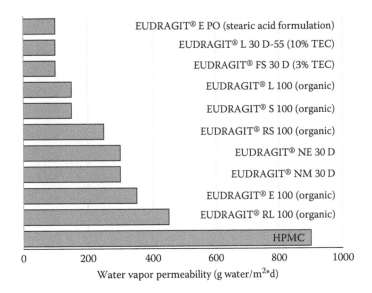

FIGURE 9.2 Water vapor permeability of pharmaceutical coating materials, measured according to DIN 53122 with isolated films of 25 μm thickness.

may be increased up to 100 μm, depending on the moisture sensitivity of the drug. The very low water vapor permeability of the polymer ensures high moisture protection effectiveness (Figure 9.2).

Alkali-Soluble Polymers

The monomer methacrylic acid (MAA) determines the solubility properties of anionic poly(meth) acrylates, enabling pH-dependent targeting along the gastrointestinal tract. Primarily, the quantities of methacrylic acid in the polymerization process determine the pH of polymer dissolution. However, the ester co-monomers significantly contribute to the dissolution and thermal properties as well.

EUDRAGIT® L 100 and EUDRAGIT® S 100 copolymers, containing 50% (w/w) and 30% (w/w) methacrylic acid, respectively, were historically developed for solvent-based coating processes providing drug release above pH 6 and pH 7, respectively. Aqueous EUDRAGIT® L/S coatings based on re-dispersion of the partly neutralized polymers shift the drug release toward lower dissolution pH values compared to corresponding organic coatings. Changing the ester monomer from methyl methacrylate (MMA) to ethyl acrylate (EA) reduced the T_g and MFT significantly (Table 9.4) and resulted in EUDRAGIT® L 30 D-55, the most commonly used polymer for enteric coatings, and its corresponding powder grade, EUDRAGIT® L 100-55. Due to the impact of the neutral ester monomer, the pH of dissolution was reduced to 5.5, providing both reliable gastric resistance and a faster drug release in the upper small intestine (Figure 9.1). Polymer chain flexibility could be increased even further by introducing methyl acrylate as an ester monomer. The aqueous dispersion of a copolymer composed of MAA:MA:MMA = 10:65:25 is commercialized as EUDRAGIT® FS 30 D. With the methylacrylate polymer, the presence of 10% methacrylic acid only provides even faster dissolution above pH 7 than EUDRAGIT® S containing 30% methacrylic acid. Both polymers allow colon targeting of drugs. Thermal properties such as T_g and MFT are significantly reduced compared to EUDRAGIT® S 100 (Table 9.4), and elongation at break is increased up to 300% [32].

In order to modulate and optimize drug release in vivo, anionic polymers can be neutralized fully or partially, which decreases the dissolution pH. If organic amines of low molecular weight are used for neutralization, the early solubility is maintained during the intestinal transit [33].

TABLE 9.4

Physicochemical Properties of Pharmaceutical Methacrylate Copolymers

Type	Mw [g/mole]	$T_{g,m}$ [°C]	MFT [°C]	Thermal Stability of Functional Group [°C][a]	Elongation at Break ε_R [%]
EUDRAGIT® E PO	~47,000	~45	NA	~220	~70[b]
EUDRAGIT® L 100-55	~320,000	~96	~25[c]	~157	~14[c]
EUDRAGIT® L 100	~125,000	>130	>100	~190	NA
EUDRAGIT® S 100	~125,000	>130	>100	~186	NA
EUDRAGIT® FS 30 D	~280,000	~43	~14[e]	~207	~300[c]
EUDRAGIT® NE 30 D	~750,000	~6	~5	NA	~600
EUDRAGIT® NM 30 D	~600,000	~9	~5	NA	~700
EUDRAGIT® RL 100	~32,000	~63	~40	~140	~300[d]
EUDRAGIT® RS 100	~32,000	~58	~45	~140	~250[d]

Source: M. Adler et al., *e-Polymers*, 055, 2004; M. Adler et al., *e-Polymers*, 057, 2005; Deutsches Institut für Normung E.V. 53455; EUDRAGIT® Application Guidelines, 12th Edition 2012.

[a] Thermogravimetry: 1% decomposition within 5 minutes.

[b] 10% (w/w) sodium lauryl sulfate, 15% (w/w) stearic acid.

[c] 10% (w/w) Triethylcitrate.

[d] 20% (w/w) Triethylcitrate.

[e] 3% (w/w) Triethylcitrate.

Insoluble Polymers

Polymerizing neutral esters of acrylic acid or methacrylic acid, including derivatives with quaternary ammonium salts, creates water-insoluble polymers. They absorb water and swell in physiological media. Coatings or matrix structures form reproducible diffusion barriers. Their physiological performance may be influenced by specific ions or osmotic pressure, but solubility remains pH independent in biological systems. Thus, these polymers gained significant importance in the formulation of time-controlled release oral single and multiunit dosage forms, such as pulsed release delivery systems.

Neutral Polymers

Neutral latices are formed by emulsion polymerization of MMA and EA. The polymer is commercialized as EUDRAGIT® NE 30 D/NE 40 D and EUDRAGIT® NM 30 D. The latter provides a broader application range due to the use of polyethylene glycol 600 stearyl ether as an emulsifier in the commercial product. Drug release is controlled by coating thickness. The superior flexibility of these polymers enables the compression of coated particles into rapidly disintegrating controlled-release tablets (Table 9.4, elongation at break).

In addition to the coating of multiparticulates, the polymer gained significant importance in the formulation of controlled-release matrix tablets. EUDRAGIT® NM 30 D is the preferred grade used in wet granulation processes as binder and matrix former.

Ionic Polymers

In order to expand the variability in release kinetics from controlled release dosage forms, water insoluble but hydrophilic polymers containing trimethylammonio ethyl methacrylate chloride (TMAEMACl) as hydrophilic moieties have been developed. EUDRAGIT® RL, carrying 10% (w/w) TMAEMACl, provides relatively high permeability while EUDRAGIT® RS

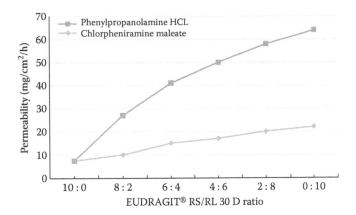

FIGURE 9.3 Diffusion rate of drugs through isolated films prepared from mixtures of EUDRAGIT® RS 30 D und EUDRAGIT® RL 30 D, tested in diffusion cells.

includes 5% (w/w) only and generates coatings and matrices with low permeability and hence strong retardation. Both polymers swell in water depending on their content of hydrophilic quaternary ammonium groups. Because quaternary ammonium groups dissociate completely in physiological media of pH 1 to 8, the permeability of coatings is pH independent. The polymers are synthesized by solution polymerization. Their particular value in controlled-release dosage forms is based on their miscibility in any ratio [9]. Thus, release kinetics can be adjusted to accommodate pharmacokinetic needs and the solubility of the active compound. Figure 9.3 shows how the ratio of EUDRAGIT® RS and RL polymers influence the diffusion rate of two model drugs.

MANUFACTURE

Bulk and Solution Polymerization

Bulk polymerization is a technically well-developed process that was initially used for the production of the hydrophilic poly(meth)acrylic esters EUDRAGIT® RL 100 and EUDRAGIT® RS 100, as well as EUDRAGIT® E 100. In this process, a mixture of liquid monomers is polymerized under controlled temperature conditions started by initiators. To obtain free-flowing granules, bulk polymers are extruded and pelletized as a final step.

Recently, a continuous polymerization process in solution was developed. Solution polymerization is a process that starts with the addition of monomers as initiator, preferably peroxides, which are added into a solvent and start the polymerization process by forming radicals and are included as end groups in the polymer chains. Solvents act as diluent and improve heat transfer of the exothermic polymerization process. Thus, process control is improved and product variability reduced. Chain transfer agents, which moderate the molecular weight, react with polymer radicals and are also found as terminal groups in the polymer chains.

To obtain free-flowing granules, the solid polymer is isolated by solvent evaporation and subsequent extrusion. During the extrusion process, volatile components of the polymer raw material, such as residual monomers, solvents, and polymerization modifiers, can be evaporated by vacuum.

Emulsion Polymerization

The anionic EUDRAGIT® L- and EUDRAGIT®S-types as well as the copolymers of neutral (meth) acrylic esters EUDRAGIT® NE 30 D/NE 40 D and EUDRAGIT® NM 30 D are produced by radical

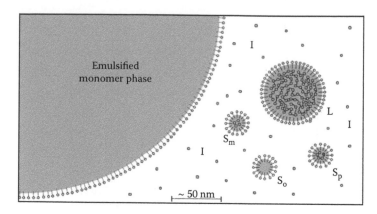

FIGURE 9.4 Mechanism of particle formation by emulsion polymerization. (°) monomer molecule in the water phase, (—•) water soluble emulsifier molecule, (I) initiator molecule, (S_o) empty micelles of emulsifier and monomer molecules, (S_m) micelles of emulsifier and monomer molecules, (S_p) micelles oligomer radicals, (L) dispersed, final, end polymerized latex particles. (From N. Suetterlin, *Macromol Chem Suppl*, 10/11, 403–418, 1985.)

emulsion polymerization [34,35] which is illustrated in Figure 9.4 [36]. The monomers are dispersed in water by stirring and emulsifier addition, which stabilizes the monomer droplets. Polymerization is started by adding a hydrophilic initiator into the water phase containing monomer molecules dissolved or solubilized in micelles if the emulsifier concentration is above the critical micelle concentration (CMC). During the initial phase, particle formation is completed rapidly. The polymer chain growth reaction is maintained by monomer diffusion from the surface of the monomer droplets via the water phase to the latex particles.

In the latex, the molecular weight of the polymer can be controlled by the concentration and the decomposition rate of the initiator and chain transfer agent. Modern test methods of molecular weight distributions are based on gel permeation chromatography using polyester gels [37,38]. The data are summarized in Table 9.4.

Typical initiators are peroxides, which start polymer chain growth by radical reactions with monomers. Finally, they are chemically bound and built into the polymers as terminal alcoholic or ester groups. The amount of residual monomers can be reduced by optimizing the polymerization conditions at the end of the process, and they can also be eliminated by evaporation.

Further Processing

Drying: Aqueous latex dispersions of methacrylic acid copolymers can be processed to free-flowing powders by spray-drying. The resulting products consist of loose agglomerates of latex particles if the temperature during the drying process is held below the MFT. Polymers with low Tg may be dried by freeze-drying in order to maintain latex particles. Such solid materials can be re-dispersed in water, forming a stable latex dispersion that corresponds to the original latex in its relevant technical characteristics, such as particle size, film formation, and coating functionality [39]. EUDRAGIT® L 100, EUDRAGIT® L 100-55, and EUDRAGIT® S 100 are produced in this manner.

Another process that can be used even with softer polymer latices is freeze-drying. The process avoids coalescence of soft dispersions, e.g., EUDRAGIT® NE 30 D and EUDRAGIT® FS 30 D. The dried agglomerates show poor flow properties due to their irregular shape and stickiness caused by low glass transition temperatures. Hence, these products are preferably used in thermal processes, such as melt extrusion [40,41] or injection molding.

Milling: Polymers from solution polymerization are available in the form of cylindrical granules because the final process step is extrusion. Hence the materials need to be milled in order to enable further processing. As the glass transition temperatures of these polymers are low, the milling process needs to be controlled exactly, particularly in terms of temperature. Pin mills or jet mills are used due to efficient cooling of the high airflow. They create suitable narrow particle size distributions with maxima between 10 and 100 μm. Finally, sieving is performed in conventional equipment.

Direct dispersion of ionic insoluble polymers: A production process was developed to directly disperse extruded granules and milled powders of EUDRAGIT® RL 100 and EUDRAGIT® RS 100 to pseudolatex dispersions and thus to allow aqueous processing [9]. The statistical distribution of the polar groups enables the direct emulsification in water without surfactants at elevated temperatures above the polymer T_g (Table 9.4). At the dispersing temperature of 80°C, the flexibility of the polymer chains is increased and, supported by the ionic effects of the quaternary groups, a self-dispersing process starts forming latices with mean particle sizes of approximately 100 nm. Although particle size distributions of thermally dispersed cationic polymers are broader than those from emulsion polymerization, reliable film formation is ensured by the addition of hydrophilic plasticizers in coating suspensions. The dispersion process is supported and accelerated by a mechanical disperser. The final products are the 30% aqueous dispersions EUDRAGIT® RL/ RS 30 D.

PHARMACEUTICAL QUALITY

Polymer Characterization and Quality Control

Pharmaceutical quality is based on the high purity of raw materials, specific manufacturing equipment, and controlled polymerization processes as described above. The procedures exceed the directions of ISO9001 and ISO14001. Quality control involves identity, functional, and purity testing. Procedures to perform the tests are described in product-specific pharmacopeial monographs (see Table 9.5).

The proof of identity is usually given by using infrared–spectroscopic methods. As in most polymers, the functional groups determine the polymeric properties in dosage forms; their assay, preferably determined by titration, serves as proof of identity and functionality. Further meaningful tests include particle size, loss on drying, viscosity, refractive index, and relative density. Purity testing includes traditional pharmaceutical tests, such as sulfated ash on ignition and heavy metals. Polymer-specific purity tests determine residual monomer content.

Residual monomer content in commercial EUDRAGIT® products for pharmaceutical purposes can be determined by liquid chromatography. The total content of residual monomers is below 0.3% in commercial products and even lower in sprayed films due to the evaporation of the volatile substances during the coating process. Product-specific limits are given in the pharmacopeial monographs or in the Evonik product specifications.

Particular physical and microbiological aspects have to be considered in the context of aqueous polymer dispersions. Physical stability, indicated by the particle size of the dispersed phase, is achieved over 18 to 24 months under ambient conditions for all polymers except for EUDRAGIT® FS 30 D, which requires cool storage. Latices are sensitive to freezing, thermal stress, the addition of electrolytes, and pH changes that affect the physical stabilization of latex particles by ionic charges or reaction by salt formation or ion exchange with the particles or their stabilizing components.

Neutral poly(meth)acrylate latexes EUDRAGIT® NE 30 D/NM 30 D are sensitive to microbial contamination of the dispersing medium water; 5–10 ppm active chlorine can be added for disinfection in the form of sodium hypochlorite solution. The weakly cationic, hydrophilic dispersions EUDRAGIT® RS 30 D/RL 30 D are preserved with 0.25% sorbic acid. Additionally, 0.1% hydrogen peroxide may be added if required. Microbial growth in anionic polymethacrylic acid copolymer latexes is avoided in acidic environments below pH 3 as a result of the free

TABLE 9.5

Registration and Toxicological Information on Pharmaceutical Methacrylate Copolymers

Type	Year of Introduction	Pharmacopoeia Monographs and DMFs
EUDRAGIT® E	1959	Basic Butylated Methacrylate Copolymer—Ph. Eur. Amino Methacrylate Copolymer—NF Aminoalkyl Methacrylate Copolymer E—JPE US DMFs 27877 (Quality) and 27770 (Safety)
EUDRAGIT® L	1954	Methacrylic Acid–Methyl Methacrylate Copolymer (1:1)—Ph. Eur. Methacrylic Acid and Methyl Methacrylate Copolymer (1:1)—NF Methacrylic Acid Copolymer L—JPE US DMFs 27796 (Quality) and 26602 (Safety)
EUDRAGIT® S	1954	Methacrylic Acid–Methyl Methacrylate Copolymer (1:2)—Ph. Eur. Methacrylic Acid and Methyl Methacrylate Copolymer (1:2)—NF Methacrylic Acid Copolymer S—JPE US DMFs 27796 (Quality) and 26602 (Safety)
EUDRAGIT® L 100-55	1972	Methacrylic Acid–Ethyl Acrylate Copolymer (1:1)—Ph. Eur. Methacrylic Acid and Ethyl Acrylate Copolymer (1:1)—NF Dried Methacrylic Acid Copolymer LD—JPE DMF 2584 (USA)
EUDRAGIT® FS	1999	DMF 13941 (USA) Canadian DMF 2006-176
EUDRAGIT® NE/NM	1972/2006	Polyacrylate Dispersion 30%—Ph. Eur. Ethyl Acrylate and Methyl Methacrylate Copolymer Dispersion—NF Ethyl Acrylate and Methyl Methacrylate Copolymer Dispersion—JPE DMF 2822 (USA)
EUDRAGIT® RL	1968	Ammonio Methacrylate Copolymer, Type A—Ph. Eur. Ammonio Methacrylate Copolymer, Type A—NF Aminoalkyl Methacrylate Copolymer, Type A—JPE US DMFs 27404 (Quality) and 26773 (Safety)
EUDRAGIT® RS	1968	Ammonio Methacrylate Copolymer, Type B—Ph. Eur. Ammonio Methacrylate Copolymer, Type B—NF Aminoalkyl Methacrylate Copolymer, Type B—JPE US DMFs 27404 (Quality) and 26773 (Safety)

Note: DMF, drug master file.

carboxylic groups in the polymer chain [42]. Thus EUDRAGIT® L 30 D-55 and EUDRAGIT® FS 30 D are self-preserved by the pH of the dispersion, provided the pH remains in the range of 2 to 3.

Toxicology and Registration

The EUDRAGIT® polymers were first introduced to the pharmaceutical industry more than 60 years ago (Table 9.5).

Owing to the carbon backbone, degradation of the polymer chains in the human body, for example, by hydrolysis, can be excluded and has never been observed. The polymeric structure is not affected by gastric acid or by digestive enzymes, such as pepsin, trypsin, chymotrypsin, amylase, and lipase. Absorption, distribution, metabolism, and excretion (ADME) studies with radiolabeled polymers confirm that the polymers are not absorbed in the intestine but are rapidly excreted in the feces and remain chemically unchanged.

Studies in different animals, including non-rodents, confirm very low to negligible acute oral and dermal toxicity (>2000 mg/kg). Polymer specific no-effect levels were investigated in chronic application studies for up to 52 weeks in rodents, dogs, and pigs. The polymers do not elicit teratogenic effects. In vitro data (Ames test, Mouse Lymphoma Assay, UDS) do not suggest any mutagenic potential.

PROPERTIES, HANDLING, AND FUNCTIONS

FILM-FORMING MECHANISMS

Aqueous latex dispersions exhibit a special film-forming mechanism. With drying progress, the dispersed latex particles move closer to each other until they form a dense sphere package. During further evaporation, the remaining water is squeezed out, and the latex particles flow together and form a homogeneous film by coalescence. One driving force of the film-forming process is the generation of surface tension energy [10]. However, also the capillary forces in the channels between the latex particles in the dense sphere package play a more important role. It can be calculated using the Laplace equation for the capillary pressure, $P = 2\,\gamma/r$, where γ is the interfacial tension between water and air, and r is the radius of the latex particles [43] or the curvature of the aqueous meniscus [11]. Both parameters—surface tension and particle radius—were postulated by Bindschaedler et al. [44] to be valid also in film-coating processes. Precondition for complete coalescence is a sufficient softness of the latex given by the polymer itself or adjusted by plasticizer addition.

The final film structure, which is reached when the interdiffusion process of the polymer chains is completed [45], can be reached within a few minutes for polymers with a high Tg or, in cases of soft polymers with glass transition temperatures below room temperature, may require several hours or even days, depending on the polymer, the coating formulation, and the processing parameters. Changes in the coating permeability, and thus in the dissolution profile, may occur during this period. Hence, final testing of a batch should be conducted when the equilibrium has been reached. Film formation can be monitored by scanning electron microscopy (SEM). Incomplete film formation may display individual latex particles in a densely packed arrangement, as shown in Figure 9.5. Under optimal film-forming conditions, films formed from latex dispersions are dense, homogeneous and free of pores. For the neutral methacrylic copolymers EUDRAGIT® NE 30 D and NM 30 D the addition of hydrophilic polymers, for example, HPMC or HPC or, alternatively, of a nonionic emulsifier (polysorbate 80) was shown [46] to reduce the curing time significantly. Traditionally, curing has been done on trays in an oven. However, a more economic method is to conduct this step in the coating equipment with defined relative humidity and product temperature [47].

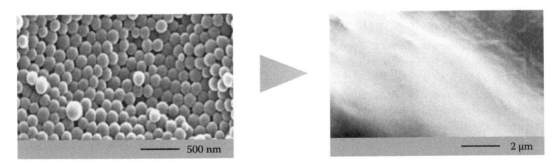

FIGURE 9.5 Dense layer of latex particles during the film-forming process (left) and homogeneous film structure after coalescence (right).

COMPATIBILITY AND FORMULATIONS

Common additives used in poly(meth)acrylate formulations and their impact on film properties and processing are described below. Examples of formulations containing various additives are shown in Table 9.6.

Plasticizers

The addition of a plasticizer lowers the MFT to a certain extent, depending on the quantity added and the plasticizer's compatibility with the specific polymer. Flexible polymers, such as EUDRAGIT® E and EUDRAGIT® NE 30 D/NM 30 D, usually do not require the addition of plasticizers due to their low glass transition temperatures. However, in special cases, plasticizers can be added, but sticking tendencies may arise. More brittle polymers, such as EUDRAGIT® L/S types, EUDRAGIT® RL/RS, and EUDRAGIT® FS 30 D, require plasticizers, usually in the range of 5% to 30%, calculated on dry polymer mass. Higher quantities can be added to achieve specific physical properties that are not required in standard film coatings. For re-dispersion of EUDRAGIT® L 100 and EUDRAGIT® S 100, the addition of 50% to 70% triethyl citrate calculated on dry polymer quantity is required to ensure a good re-dispersion process and a maximum MFT of 10°C. Not every plasticizer is effective for each polymer (Figure 9.6). Unsuitable plasticizers or insufficient amounts of plasticizer may lead to a loss in functionality due to crack formation caused by mechanical stress during the coating process or expansion effects of cores after coating. Insufficient MFT reduction is observed when combining EUDRAGIT® RL/RS 30 D with polyethylene glycols (PEGs), EUDRAGIT® FS 30 D with PEGs or polysorbate 80, and EUDRAGIT® L 30 D-55 with butyl citrates or other lipophilic esters. Besides triethyl citrate, the standard plasticizer in poly(meth)acrylate formulations, the following ones have successfully been employed: PEGs (preferably PEG 6000), acetyl triethyl citrate, to some extent butyl citrates, polysorbates (preferably polysorbate 80), dibutyl sebacate as a lipophilic plasticizer (pre-emulsification in water for aqueous formulations with 1% w/w polysorbate 80 and post-stirring over one hour recommended), and triacetin (dis-advantage: hydrolysis during storage). As a general rule, it can be stated that at equivalent concentrations, the use of hydrophilic plasticizers leads to coatings with higher permeability and faster dissolution whereas lipophilic ones tend to reduce permeability and dissolution rate.

Water-soluble plasticizers, such as triacetin or triethyl citrate, can be added directly to the latices; freely water-soluble or even hygroscopic substances, such as PEG and sorbitan esters, may be added as 20% to 35% aqueous solutions for improved physical stability. Water-insoluble plasticizers are usually emulsified in water using some latex-compatible emulsifier, such as 1% w/w polysobate 80, and mixed with the latex until equilibrium distribution is reached. Lipophilic plasticizers require longer stirring times of several hours when combined with aqueous polymer dispersions in order to ensure proper distribution and equilibrium formation between the phases.

Glidants or Anti-Tacking Agents

To avoid sticking or agglomeration of the products during coating, drying, and storage, glidants are added to the spray suspensions. These materials are suspended separately and then added to the polymer mixtures. These materials can also be added in powder form by sprinkling. For coated particles, the addition of 0.5% to 2% talc; fumed silica, for example, Aerosil® 200; or precipitated silica, for example, Sipernat® PQ can prevent tacking problems during storage. The powders can be added to the fluidizing particles after the coating process or by spray application of an aqueous suspension. Alternatively, thin aqueous coatings based on glycerol monostearate (mono- and diglycerides NF) or hydroxypropyl methylcellulose (HPMC) will produce similar effects. The most commonly used glidants for poly(meth)acrylate formulations are described below.

Talc is often used in combination with pigments. Because it is a product derived from mineral sources, there is the risk of microbial contamination. Also, it can contain free ions that may react with other components in the coating formulation and thus cause instabilities. Typical quantities are

TABLE 9.6
Example Formulations

	Immediate Release	GI Targeting			Time-Controlled Release		
	Taste-Masking and Moisture Protection	Small Intestine Release	Colonic Delivery	Flexible Coating	Low Permeability		Medium Permeability
EUDRAGIT®	E PO	L 30 D-55	FS 30 D	L 30 D-55/NE 30 D	NE/NM 30 D	RS 30 D	RL/RS 30 D
	[g]	[g]	[g]	[g]	[g]	[g]	[g]
Polymer dispersion	100.5	730.0	603.0	L 30 D-55 248.5; NE 30 D 248.5	300.0	392.0	RS 470.5; RL 52.5
Quantity dry polymer	100.5	219.0	181.0	L 30 D-55 74.5; NE 30 D 74.5	100.0	117.6	RS 141.2; RL 15.8
pH adjustment							
1N NaOH				0.03			
Citric acid				1.0			
Salt former							
Stearic acid	15.0						
Anti-tack agent							
GMS		6.5	7.0	4.5	10.0		8.0
Polysorbate 80		2.5	3.0	0.4			3.5
Talc					100.0	59.0	
Mg-stearate	35.0						
Emulsifier							
SLS	10.0						
HPMC					10.0		
Plasticizer							
TEC		22.0	9.0	15.0		23.5	31.5
Water	840.0	239.0	378.0	482.0	559.7	525.5	434.0
Total	1000.0	1000.0	1000.0	1000.0	1000.0	1000.0	1000.0
Polymer content	10.0%	21.9%	18.1%	15.0%	30.0%	11.8%	15.7%
Solid content	16%	25%	20%	17%	20%	20%	20%

Source: EUDRAGIT® Application Guidelines, 12th Edition 2012.
Note: GMS, glyceryl monostearate; HPMC, hydroxypropylmethylcellulose; SLS, sodium lauryl sulfate; TEC, triethyl citrate.

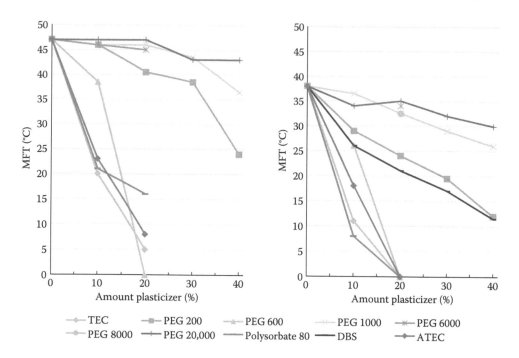

FIGURE 9.6 Plasticization effects of various plasticizers as MFT reduction on EUDRAGIT® RS 30 D (left) and EUDRAGIT® RL 30 D (right). Percentages referring to polymer quantity. MFT, minimum film-forming temperature.

25% to 100% based on dry polymer mass. With equal amounts of talc and polymer, opaque coatings are obtained. With more than 200% talc based on polymer weight, the permeability of the films decreases, and dissolution of enteric coatings in intestinal fluid will be delayed [48]. Because larger particles may reduce permeability and the dissolution rate, the desired particle size and distribution should be specified [49,50].

Glycerol monostearate (mono- and diglycerides NF) is an excellent alternative to talc or magnesium stearate as glidant in all aqueous formulations [51]. Due to its high efficacy, typically 5% to 20% based on polymer mass is sufficient to achieve comparable effects. Glycerol monostearate is water insoluble. Fine particle dispersions are prepared by emulsification in hot water (70°C–80°C) in the presence of polysorbate 80. The main advantages compared to talc are the low risk of microbiological contamination, improved compatibility, and lower effective concentrations, which reduce total coating thickness, processing time, and, hence, save production costs.

Precipitated silica (silicon dioxide NF) can be used in quantities up to 40% based on polymer mass. It has a matting effect and increases the permeability of film coatings. It is typically used for extended-release particle coatings and is not recommended for enteric coatings.

Magnesium stearate is somewhat more effective than talc and often provides low permeability and good sealing of the film coatings. However, it can only be used in organic polymer solutions or in aqueous formulations based on EUDRAGIT® NE/NM or EUDRAGIT® E because coagulation or thickening may occur with other aqueous poly(meth)acrylate dispersions. Because of the possible reaction between magnesium ions and the carboxylic groups of the polymers, magnesium stearate is incompatible with anionic poly(meth)acrylate latices.

Pigments and Dyes

The pigment-binding capacity of poly(meth)acrylates is excellent compared to polysaccharides. Up to three parts by weight of solid additives and, in specific instances, even up to 10 parts of binding dry polymer can be incorporated into one part of dry polymer without affecting the film's properties. The pigment quantity necessary to cover an underlying surface of unpleasant or irregular color is about 2 to 3 mg/cm^2. This can be incorporated into a film of 10 to 15 μm thickness, which is about 1 to 1.5 mg dry polymer/cm^2 surface area. Titanium dioxide is a commonly used white pigment that has extremely high coverage power. It is combined with color pigments to obtain the desired shades. Its hard, abrasive properties are disadvantageous as they can lead to black spots on the coatings caused by grinded metal from the pan wall. For colored film coatings, the color pigments can be added to the functional coating, which may require higher polymer application or, alternatively, be applied as a colored topcoat. When white tablets are going to be coated, smaller quantities of pigments are sufficient. Both aluminium lakes and iron oxide pigments are suitable. Colored lakes of poorer quality often contain significant amounts of water-soluble dyes, which may cause coagulation of the polymer as these dyes often are strong electrolytes. Water-soluble dyes often lead to inhomogeneous, marbled colors, which rub off during handling. Dyes show lower stability against chemical influences and light than lakes and thus tend to fade at a faster rate.

Pigments and dyes are usually pre-dispersed in water using high-shear equipment and subsequently added to the latex as a suspension under gently stirring in order to avoid coagulation.

Emulsifiers and Stabilizers

Stabilization is recommended for latex dispersions that contain pigments or electrolytes that may change the zeta potential and hence cause coagulation. This occurs immediately after mixing or when the mixture is sheared. Emulsifiers and stabilizers, such as polysorbates, low viscous carboxymethylcellulose sodium (CMC-Na), polyvinylpyrrolidone, or sodium dodecyl sulfate, are often added to pigment suspensions. Depending on the quantity and quality of destabilizing excipients in the formulation, 2.5% to 10% (calculated on dry polymer mass) stabilizers are required. Freshly dispersed pigments are most reactive when primary particles are ruptured, and new surfaces are produced by intensive milling processes. Such pigment dispersions should be mixed with some of the same emulsifier used in the latex and stored overnight for aging. Stabilizing agents used in formulations of anionic latices are preferably nonionic emulsifiers, such as sorbitan esters, which are also used in emulsion polymerization processes. They also act as plasticizers and are normally incorporated in the pigment suspension to improve their compatibility. The best stabilizing principle for anionic dispersions is partial neutralization using alkaline substances, for EUDRAGIT® L 30 D-55 1N NaOH is preferred at a level of less than 10% referring to the quantity of carboxylic groups. Because the stability of anionic methacrylate latices and their compatibility with additives are to some extent higher at pH 5 than at pH 2 to 3, which is the specified pH of EUDRAGIT® L 30 D-55, the pH may even be raised to about 5.0 to 5.4. When different anionic poly(meth)acrylates are going to be mixed, pH modification should be done before mixing the polymer dispersions and by using the same agent.

Flavors and Sweeteners

The addition of flavors to film coatings is mostly limited due to their complex chemistry and limited stability. Sweet-tasting film coatings can be prepared by the addition of saccharin sodium, usually in amounts of 2% to 10%, based on dry polymer mass, depending on the desired sweet taste intensity.

Easy-to-Use Additives

Designing the coating formulation by using pure substances provides the highest flexibility to formulators and allows tailored formulation compositions. However, as for many coating formulations, standard compositions with defined quantities of all excipients related to the polymer quantity provide the desired performance, and the use of pre-formulated excipient mixtures generates two benefits: the increase of preparation efficiency and related cost reduction as well as the reduction of risks for failures.

PlasACRYL® stands for a portfolio of formulation ingredients needed for aqueous poly(meth)-acrylate coatings. Basis for all products is the state-of-the-art glidant glycerol monostearate, finely dispersed in water and combined with specific quantities of triethylcitrate (plasticizer) and a stabilizer. The PlasACRYL® compositions are tailored to the corresponding EUDRAGIT® standard coating formulations. Hence, the spray suspension preparation requires only the addition of PlasACRYL® to the commercial polymer dispersion.

POLYMER COMBINATIONS AND FUNCTIONALITY MODIFICATIONS

Poly(meth)acrylate Mixtures

Lehmann and Dreher [51] provided a detailed summary of possible poly(meth)acrylate combinations. In the following paragraphs, the most common combinations are described. Beyond these, other combinations are possible as well. In general, however, when combining aqueous latices, the ionic character of the polymers and the pH values of the dispersions have to be considered in terms of possible instabilities.

Combinations of anionic (meth)acrylate polymers: Mixing organic solutions of different anionic poly(meth)acrylates allows the adjustment of the dissolution pH of the film coating for the purpose of gastrointestinal targeting (Figure 9.1). The combination of organic solutions of EUDRAGIT® L and EUDRAGIT® S is of particular importance. However, when mixing aqueous re-dispersed latices thereof, the situation is different. The resulting film coatings contain domains of both polymers that dissolve at their specific pH values. Between the lower and the higher solution pH, one polymer can act as pore former by dissolving and releasing the drug slowly. Beyond the higher dissolution pH, both polymers dissolve, resulting in enhanced drug release.

Combinations of anionic and neutral poly(meth)acrylates: When highly flexible enteric coatings are required, especially for the preparation of multiparticulate tablets, higher plasticization of the brittle EUDRAGIT® L 30 D-55 becomes necessary. Because it is not useful to increase the plasticizer level beyond 20%, the combination with EUDRAGIT® NE 30 D or EUDRAGIT® NM 30 D as the highest flexible poly(meth)acrylate polymers is the solution of choice [52]. With up to 50% of the neutral polymer in the film, the coating maintains its enteric behavior and shows significant increase in film flexibility. At higher amounts, the properties of this polymer become predominant, and dissolution of the coating will be delayed. In these instances the MFT of the combination decreases below 25°C, which enables coatings without the addition of plasticizer.

Mixtures of anionic and neutral poly(meth)acrylates can also be useful when preparing controlled-release formulations of weakly basic actives showing high solubility in acidic media and reduced solubility at higher pH values. In the stomach, the polymer combination ensures low permeability for a highly soluble active whereas in neutral to weakly alkaline intestinal fluid, the entero-soluble component dissolves and compensates for the reduced solubility of the drug, resulting in constant release rates over the entire pH range. With minor amounts of entero-soluble polymer in the film, the permeability only increases when the pH enters the range above the dissolution pH of this component, but the film remains stable. These effects can be used in many variations for controlled-release preparations, especially for the coating of small particles when disintegration effects play a minor role and the release is mainly diffusion controlled by the permeability of the encapsulating membrane [53].

When preparing such mixtures, it has to be considered that the polymer dispersions have different pH values (EUDRAGIT® L 30 D-55: pH 2–3/EUDRAGIT® NE 30 D: pH 7–8). In order to avoid coagulation, it is necessary to adjust the pH of both dispersions to the same level before mixing. If there is no specific requirement, the pH is adjusted to 5. For this purpose, 1N NaOH is added to EUDRAGIT® L 30 D-55. The pH of EUDRAGIT® NE 30 D can be reduced by the addition of 20% (w/w) citric acid solution or 1N HCl. To ensure a stable pH, the dispersions should be post-stirred for 20 minutes with pH control. Afterwards, the neutralized EUDRAGIT® NE 30 D is poured slowly into the EUDRAGIT® L 30 D-55 while stirring gently. Additionally, dilution of the dispersions before mixing can enhance stability. The remaining preparation is done as usually for aqueous spray suspension preparation.

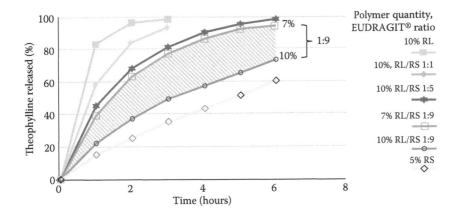

FIGURE 9.7 Drug-release variations by permeability adjustment in combining EUDRAGIT® RL/RS 30 D in various ratios and applying different coating amounts.

Combinations of insoluble ionic poly(meth)acrylates: EUDRAGIT® RL and EUDRAGIT® RS can be mixed with each other in any ratio, either in organic solutions or as aqueous dispersions to adjust the intermediate permeability and to obtain a specific release pattern [54]. The quality and quantity of excipients to be used in formulating mixed films are the same for both. Because the EUDRAGIT® RL features are predominant in these combinations, the quantity of EUDRAGIT® RS in the combination usually is much higher for extended-release effects. Typical ratios are RS:RL = 95:5 or 90:10 or 80:20. An example using coated theophylline granules is shown in Figure 9.7. By varying polymer ratio and coating quantity, two effective controls are provided that allow maximum flexibility for tailor-made formulation design.

Poly(meth)acrylates in Combination with Soluble Substances

To increase the permeability of film layers or to modify the tortuosity of matrix structures made of poly(meth)acrylates, several water-soluble or water-swellable substances can be added, such as sucrose, lactose, or other saccharides; starch; micronized cellulose; soluble cellulose ethers; poly(vinylpyrrolidone); poly(vinyl alcohol); or PEGs and PEG derivatives. However, water-soluble cellulose ethers have limited compatibility. They stimulate slow agglomeration and coagulation within several hours or days. Alternatively, fumed or precipitated silica as a water-insoluble but hydrophilic agent can be added to coating formulations in order to increase film permeability.

Poly(meth)acrylates in Combination with Other Polymers

Hydrocolloids, such as hypromellose (HPMC), carboxymethylcellulose (CMC-Na), or hydroxypropyl cellulose (HPC), can be used in mixtures with methacrylic copolymers. Typical is the addition of HPMC or HPC in EUDRAGIT® NE 30 D or EUDRAGIT® RL/RS 30 D coating formulations. Depending on the quantities of the hydrocolloid, the effects will be the enhancement of curing processes and related stabilization of the coating (A6) or, in higher quantities, modification of the release profile through pore formation. Lecomte and Siepmann investigated the physicochemical characteristics and dissolution modifications of coatings based on combinations of EUDRAGIT® L with Ethylcellulose in controlled release formulations and compared this to EUDRAGIT® L-NE combinations [55,56]. In colored enteric coatings CMC-Na is commonly used as stabilizer of the aqueous coating suspension to prevent coagulation of the dispersion caused by insoluble pigments. The physical stability of such formulations has to be investigated individually. To increase flexibility and to optimize film formation of cellulose-based coatings, the soft types EUDRAGIT® NE 30 D or EUDRAGIT® NM 30 D can be added [54].

SPRAY SUSPENSION PREPARATION

General Considerations

When processing aqueous polymeric latices, high-shear forces must be prevented. In general, solid excipients should be suspended and homogenized separately from the aqueous polymer dispersion before being combined. Useful tools are rotor/stator systems, for example, Ultra Turrax®, Silverson®, or toothed colloid mills. Simple propeller stirrers cannot deagglomerate solid particles, which may lead to uneven or rough coating surfaces or marbled coloration in the case of coarse pigments. However, dispersing solid excipients in organic polymer solutions can be done directly. Preparation steps are the same for solvent- and aqueous-based spray suspensions: The diluent is placed into a container, and all other excipients except the polymers are added while homogenizing with a high-speed stirrer. The latex dispersion is weighed into another vessel, then diluents and, if required, a stabilizer are added. Finally, the excipient suspension is poured into the latex dispersion while stirring gently. To avoid sedimentation, suspensions should be stirred continuously during the coating process with conventional propeller stirrers.

Commonly, aqueous-based formulations contain 20%–25% solids whereas EUDRAGIT® E PO colloidal solutions are adjusted to 15%, the ready-to-use version EUDRAGIT® E PO ReadyMix to 20% solids. Higher solids content requires extremely well-controlled coating processes in order to guarantee good film formation.

To some extent, aqueous dispersions are sensitive to microbial contamination and hence should be used within 24 hours after preparation. However, longer storage times may be evaluated individually. Another possible risk for aqueous latex systems at longer storage times is physical instability, resulting in changed properties of the coatings, and/or processing issues such as nozzle blockage during spraying.

Redispersion of Anionic Polymer Powders

EUDRAGIT® L 100-55, EUDRAGIT® L 100, and EUDRAGIT® S 100 are spray-dried powders that consist of spray agglomerates. In order to process them as aqueous dispersions, they need to be re-dispersed in water to form nano-sized latex particles [58]. The solid agglomerates can be re-dispersed in water and processed from aqueous formulations in analogy to original dispersions [59]. Re-dispersion is effected by adding small quantities of alkali or ammonia, respectively, to the aqueous suspension of the polymer powder. They partially neutralize the carboxylic groups in the polymer and thus enhance de-agglomeration of the latex particles while forming a regular latex dispersion. The polymer types require different alkali types and processing times (Figure 9.8). Owing

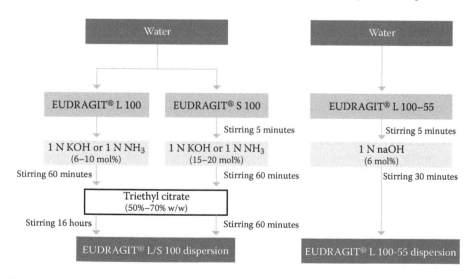

FIGURE 9.8 Re-dispersion procedure of anionic poly(meth)acrylates: EUDRAGIT® L/S 100 (left) and EUDRAGIT® L 100-55 (right).

to the influences of coating parameters and other excipients in re-dispersed EUDRAGIT® L/S 100 coating formulations, the degree of neutralization and plasticizer quantities can be increased to the upper levels given in Figure 9.8 in order to enhance film formation. However, plasticizer levels of 70% calculated on dry polymer mass may increase sticking tendencies and possibly film permeability, too. Hence, increasing the neutralization level is the preferred approach. The particle size in re-dispersed latices is around 100 nm as in the commercial dispersions. The gastro-resistance and entero-solubility of the coatings, resulting from re-dispersed EUDRAGIT® L 100-55, are equivalent to those from the commercial aqueous dispersion EUDRAGIT® L 30 D-55 [59]. Re-dispersed latices can be used for gastro-resistant coatings in the same way as the commercial polymer dispersions. The most simple and highly efficient re-dispersion process is provided when using Acryl-EZE®, a fully formulated coating system containing EUDRAGIT® L 100-55 as a functional polymer. To get the spray suspension the powder product is simply stirred in water within 30 minutes.

For spray suspension preparation of EUDRAGIT® L/S 100, the plasticizer is added to the latex dispersion. As given in Figure 9.8, the plasticizer amounts are higher than for other formulations. The plasticizer has to be stirred with the polymer dispersion for at least 60 minutes before adding the anti-tacking agent and other excipients. In case of EUDRAGIT® L 100-55, the resulting dispersion can be handled like the commercially available EUDRAGIT® L 30 D-55. The plasticizer and other excipients are homogenized separately and then added to the dispersion.

For larger-scale re-dispersion, effective but slow-moving stirring equipment should be used. The stirrer should be equipped with speed control to adapt the stirring speed to the viscosity of the system, which normally means a higher speed for the higher viscosity in the beginning and a lower speed when the viscosity decreases toward the end of the latex-forming process. Incorporation of air bubbles should be avoided during all stages of the re-dispersion process.

Mixtures of re-dispersed EUDRAGIT® L 100 and EUDRAGIT® S 100: Both re-dispersions are conducted separately as described above using 1N NH$_3$. With moderate stirring, the EUDRAGIT® S 100 re-dispersion is poured into the re-dispersed EUDRAGIT® L 100. Stirring is continued for another 15 minutes before further excipients are added to prepare the spray suspension.

Colloidal Solution of EUDRAGIT® E PO

Standard EUDRAGIT® E PO coating suspensions contain 10% sodium lauryl sulfate (SLS) as a wetting and dispersing agent and 15% stearic acid, which forms a soluble salt with the polymer. It is highly recommended to use stearic acid of powder-grade quality for optimal processing. With stearic acid and SLS, EUDRAGIT® E PO forms a colloidal solution in water that appears almost clear or pale yellow and shows the Tyndall effect. Having a low viscosity, the polymer solution is processable like the commercial aqueous EUDRAGIT® dispersions. Furthermore, the colloidal solution is not sensitive to shear forces, which allows the use of high-shear homogenizers for preparation. The preparation of EUDRAGIT® E PO spraying suspensions follows the scheme in Figure 9.9.

First, the water is put into a vessel. SLS, stearic acid and EUDRAGIT® E PO are added. Adding stearic acid prior to EUDRAGIT® E PO helps the polymer dissolve faster. Conventional propeller stirrers, dissolver plates, and high shear mixers (e.g., Ultra Turrax or Silverson) can be used as stirring devices. However, due to the low efficiency of conventional stirrers, the preparation procedure can take a couple of hours whereas when using dissolver plates or high-shear mixers, time can be reduced significantly. In any case, foam formation should be avoided. Otherwise, antifoaming agents can be added.

After achieving a yellowish, slightly turbid colloidal solution, an anti-tacking agent (preferably talc) and optional pigments are added. By using a conventional propeller stirrer for preparing the colloidal solution, it is necessary to suspend both separately in water to achieve ideal homogeneity and then to add into the colloidal solution while homogenizing. In the other cases, anti-tacking agent and pigments are added right after the colloidal solution has been obtained. Alternatively, magnesium stearate can be used as anti-tacking agent, mainly if additional protection against water vapor permeability is required.

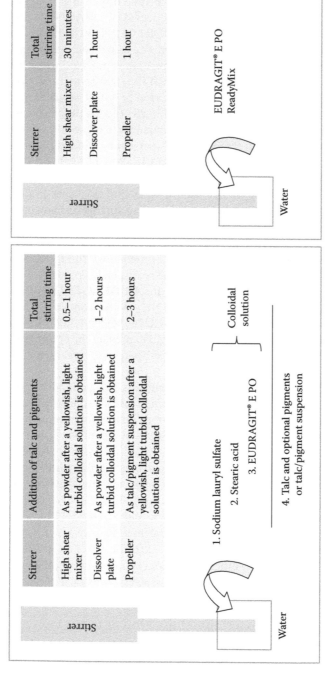

FIGURE 9.9 Preparation of EUDRAGIT® E PO (left) or EUDRAGIT® E PO ReadyMix (right) spray suspensions.

EUDRAGIT® E PO is also available as a customized ready-to-use powder blend (EUDRAGIT® E PO ReadyMix) with retained protective functionality and immediate release characteristics. The coating suspension can easily be prepared in one step by simply dispersing the powder in water using a conventional stirrer or a high-shear mixer (Figure 9.9).

CORE MATERIALS

Low viscosity of organic and aqueous EUDRAGIT® formulations enable efficient and reproducible coating processes on different core materials, which have sufficient hardness and low friability. Core size may vary from a few millimeters to more than 1500 mm. Smaller particles are preferably processed in fluidized bed systems [60], and larger tablets or capsules are coated in perforated drum coaters. Tablets may have different shapes and sizes but need a certain hardness in order to avoid liberation of films [61] or abrasion, which may affect functionality of film coatings. Also capsules can nicely be coated with EUDRAGIT® polymers; suitable are hard and soft gelatin capsules and also cellulosic shells [62].

PROCESS PARAMETERS

Except for temperatures, aqueous- and organic-based coating suspensions have the same processing conditions as long as the formulations are within the recommended solids content range in order to guarantee a low viscous coating liquid [29].

Product Bed Temperature as a Main Control Parameter

To ensure appropriate film formation, the product temperature during spraying should be at least 10–20 K above the MFT of the dispersion. Recommended product bed temperatures in fluid bed processes are 20°C–30°C for aqueous coating processes and organic-based processes. Higher temperatures, especially in combination with turbulences, can lead to spray-drying effects. If particles from either spray-drying, abrasion, or broken substrates are incorporated into the coating, they may act as channeling agents and thus lead to increased permeability. Especially organic formulations tend to form porous film structures at increased bed temperatures.

Spray Rate

In aqueous processes, the main effort is to prevent the inclusion of water into the cores and any subsequent interaction with moisture-sensitive actives or reduced storage stability. To achieve good film formation under mild processing conditions, the cores are heated to about 30°C to 40°C prior to coating. It is recommended to spray at lower rates initially. For most substrates, it is useful to start with approximately 75% of the usual spray rate. After 30 to 60 minutes, the first polymer layer forms a thin film, which isolates the core against water penetration. From this point on, the spray rate can be increased to the usual equipment related optimized spray rate. Spraying too fast will cause overwetting combined with sticking tendencies and may also cause stability issues. If processing water is not completely evaporated from the core, the included humidity will plasticize the film [63] and hence cause sticking issues and changes in the permeability of the coating during storage. Thicker coatings in particular will trap the solvent or water and severely hinder evaporation. Loss on drying as in-process control is highly recommended.

Atomization and Pattern Air Pressure

Aqueous latex formulations have very low viscosities comparable to pure water and hence do not require high atomizing air pressures like cellulosic coatings. The optimum level for atomizing air pressure is 1 to 2 bar (14–28 PSI). Beyond that, spray-drying of the coating suspension may occur, causing loss of coating material and functionality due to incomplete film formation or incorporating of dried dust particles into the functional coating. In tablet coating, spray nozzles are usually designed with a second air channel to form the spray pattern. This pressure should usually not exceed the value of the atomizing air pressure in order to generate an oval-shaped beam and to prevent the division of the spray zone into two separate beam sections ("lying eight"). Depending on the spray nozzle model and the air cap design, the ideal pressure for the pattern air can be between 50% and 100% of the atomizing air. It is useful to install flow meters for the atomizing air and pattern air on every spray gun in order to indicate blocking tendencies during the coating process.

Inlet Air Humidity

High inlet air humidity will slow down water evaporation. In this case, it is recommended to preferably reduce the spray rate or to increase the inlet air temperature moderately, in order to reduce the relative humidity of the incoming drying air. Installation of a dehumidification system guarantees reproducible conditions independent from the environmental conditions.

Drying Air Volume

Sufficient air volume effectively evaporates the water or solvent. The pan pressure should not exceed 150 Pa (~1.5 mbar, ~1.1 mm Hg, ~0.02 PSI). Higher pressures lead to poor cascading of tablets or capsules in the coating pan. A recommended drying air capacity in pan coating processes is 0.3–0.5 m³/min/kg product. In fluid-bed coaters the air volume must be adjusted to get proper fluidization of the material. Depending upon the fluid-bed technique and flow properties of the substrate, the values may vary in a wider range and need to be adjusted accordingly.

Pan Coater Setup

It is recommended to have a distance of 10 cm (lab scale) to a maximum of 25 cm (production scale) between spray nozzle and product bed at an angle of 90° between bed surface and spray stream in the upper third of the tablet bed in the rotating drum. Shorter distances may lead to over-wetting and inhomogeneous coatings while longer distances lead to spray-drying effects, especially with turbulent air flow inside the pan. Uncoated tablets have a relatively rough surface and thus higher friction. The pan rotation speed must be optimized to ensure gentle movement of the tablets. Higher pan rotation speeds will increase mechanical stress and lead to chipping and cracking of the tablet edges and the film surface. Lowering the pan speed can lead to over-wetting because the tablets move more slowly in the pan and thus are exposed to the spray zone for longer periods of time. For comparable process conditions, the pan rotation is set slower in production scale equipment than in lab scale depending on the drum diameter. The factor for calculating rotation speed predictively is based on the physical speed of the individual outer pan surface.

Pump System

The spray suspension based on aqueous polymer dispersions are sensitive to shear forces and should be delivered by peristaltic pumps to the spray nozzles using tubes with internal diameters as small as possible (down to 2 mm) in order to achieve a high flow speed, which prevents sedimentation. For the same reason, the tubing should be as short as possible, especially if tubes with wider diameters are used. Alternatively, piston pumps or pressure vessels can be used instead. Organic polymer solutions can be processed in gear pumps, airless systems, or vacuum pumps, too.

POSTCOATING TREATMENT AND CURING

Residual traces of water can act as a plasticizer [63] and may have an influence on the coating permeability. Furthermore, unsuitable processing, that is, excessive atomizing air pressure or high temperatures, can cause incomplete film formation during the coating process. Therefore, a validated drying process either in the coating equipment or in external drying facilities is recommended. Removal of water from the coated dosage forms may be delayed significantly if it has penetrated into the cores during coating because of high spray rates.

Curing enhances film formation from aqueous dispersions after coating by facilitating coalescence of the latex particles. Duration and processing requirements depend upon the polymer characteristics, plasticizer content, temperature, and environmental relative humidity [64]. Furthermore, Zheng and McGinity [53] as well as Guiterrez-Rocca and McGinity [65] reported that curing duration varies by polymer combination. A standard post-drying of one hour at 40°C is recommended. Specific post-coating treatment is required for EUDRAGIT® RL/RS 30 D and EUDRAGIT® NE/ NM formulations. Conventionally, curing has been performed on trays at 40°C over 24 hours. However, curing can be done more efficiently in the coating equipment [66]. Relative humidity, temperature, and process time have to be evaluated and optimized during product development with consideration given to the specific product, equipment, and environmental conditions. The curing progress can be monitored by dissolution tests. The end point and thus storage stability is reached when dissolution profiles become static with storage time. Mechanical stress should be kept at minimum levels in order not to damage the coatings. For EUDRAGIT® NE and NM coatings the addition of low quantities of HPMC and polysorbate 80 enhances the curing process [29].

MULTILAYER COATINGS

It is feasible to apply several coating layers onto a substrate in succession. Multilayer coatings become relevant when the serial addition of different functionalities is required. Doing this smartly allows the generation of specific targeted release profiles. A selection of relevant applications is provided in the subchapter on drug delivery systems. A simple and common application for two subsequent coating layers is the application of a separating subcoat in order to prevent possible interactions between substrate and the functional coating. To reduce interactions between the different layers or to enhance the storage stability, it may be necessary to apply individual coating layers with an intermediate drying step after each coating. The drying process should be evaluated for each coating layer applied.

SWITCH FROM ORGANIC TO AQUEOUS FORMULATIONS

Except for EUDRAGIT® FS 30 D, EUDRAGIT® NE 30/40 D, and EUFDRAGIT® NM 30 D, which are only available as aqueous dispersions, all other poly(meth)acrylates can be formulated as both organic and aqueous systems. When an organic formulation is to be replaced by a bioequivalent aqueous one, specific distinguishing in vitro dissolution test methods need to be developed. Differences in film density and especially composition modifications may result in changed drug release profiles and hence require adaptation of the formulation. Usually, the switch can be done easily by formulating the same polymer from a different physical form for EUDRAGIT® RL/RS to aqueous dispersion EUDRAGIT® RL 30 D/RS 30 D and adapting the coating formulation. For aqueous coating processes with EUDRAGIT® E the micronized powder grade EUDRAGIT® E PO is used. Processibility, film formation, and functionality are provided by including a surfactant, fumed silica and partial neutralization with a fatty acid, such as stearic acid. When changing an organic enteric coating based on EUDRAGIT® L, EUDRAGIT® L 100-55, or EUDRAGIT® S to a re-dispersed aqueous system, an increased dissolution speed in buffer media may be observed, mainly caused by the higher plasticizer content (50%–70% instead of 10% in the organic formulation) as well as by partial neutralization. Adaptation of the formulation needs to be done accordingly.

FUNCTIONAL DOSAGE FORMS

IMMEDIATE RELEASE: TASTE MASKING AND MOISTURE PROTECTION

The highly efficient and most simple approach to ensure proper taste masking or protection against moisture uptake for tablets, capsules, and particles is the application of polymeric film coatings.

TABLE 9.7

Protection Potential of Different EUDRAGIT® Films Given as Weight Gain Dry Polymer per Unit Surface Area Substrate

	EUDRAGIT® E PO and E PO ReadyMix (mg/cm²)	EUDRAGIT® L 30 D-55 and L 100-55 (mg/cm²)	EUDRAGIT® RL 30 D (mg/cm²)
Sealing	~1	~1[a]	~1[a]
Taste masking	1–2	~1[a]	~1[a]
Moisture protection	4–10	~1[a]	~1[a]

[a] Beyond this coating quantity, the main functionality of the polymer may show up.

Among the different poly(meth)acrylates, the acid-soluble EUDRAGIT® E PO and its ready-to-use product EUDRAGIT® E PO ReadyMix are most suitable [67]. EUDRAGIT® L 30 D-55 and EUDRAGIT® RL 30 D can be used in thinner layers beneath the functional threshold for this purpose as well, the latter possibly in combination with soluble cellulose ethers for disintegrating coatings [68]. Table 9.7 shows the usual application quantities for the various protection targets.

Because EUDRAGIT® E PO dissolves in the acidic conditions of the stomach, thicker coatings up to 10 mg/cm² can be applied without delaying drug release. For both EUDRAGIT® L 30 D-55 and EUDRAGIT® RL 30 D, not more than approximately 1 mg/cm² should be applied in order to avoid modified release effects. Despite the superiority of EUDRAGIT® E PO for protective coatings, the use of EUDRAGIT® L 30 D-55 may become necessary for cationic drugs in order to avoid ionic interactions.

Aqueous EUDRAGIT® E PO coatings are highly flexible and can be applied to all kinds of substrates, including small particles, such as granules, pellets, or active pharmaceutical ingredient (API) crystals, etc. The coated particles can be compressed into rapidly disintegrating tablets without damaging the film coating. Most pediatric formulations, such as dispersible or chewable tablets or single-use dry syrup formulations can be designed using this polymer. Usually, 1–2 mg/cm² of polymer application provides excellent taste-masking properties. With EUDRAGIT® E PO ReadyMix a ready-to-use powder is available that allows a fast and easy preparation of the spray suspension ensuring the same taste-masking level as the formulated polymer.

For taste-masking applications, an alternative to film coating is the neutralization of the bitter taste of basic drugs or salts thereof by targeted ionic interaction with acidic polymers. Powerful alkaline salts of basic drugs react with anionic poly(meth)acrylates, for example, EUDRAGIT® L 100, and bind to the copolymer by ion exchange principles. The manufacturing process is a regular aqueous high-shear mixer granulation operation. The resulting polymer-active granules are often insoluble in water and have an almost neutral taste and odor. This allows the formulation of dry syrups. Furthermore, after binding to EUDRAGIT® L 100, chemically unstable actives often show improved stability with no further additives needed for stabilization even in liquid formulations. Suitable drugs are those that have at least one basic functional group for reaction with the anionic groups of the polymer. Because the functional principle is based on molecular interactions, limitations for this process are high dose, high molecular weight of the API, and steric hindrance of the functional group in the active. The optimal drug/EUDRAGIT® L 100 mixing ratio must be determined experimentally for every product. Bitter drugs with a low-to-medium dose are preferred for this technique.

Gastro-Resistance and Gastrointestinal Targeting

For simple enteric coatings that quickly dissolve in the small intestine, EUDRAGIT® L 30 D-55, EUDRAGIT® L 100-55, or the ready-to-use product AcrylEZE® containing EUDRAGIT® L 100-55 as functional polymer are typically used. If the drug is supposed to be released in the lower sections of the small intestine, EUDRAGIT® L and EUDRAGIT® S can be used in mixtures to create specific dissolution pH values (Figure 9.1). Furthermore, the aqueous dispersions EUDRAGIT® L 30 D-55 and EUDRAGIT® FS 30 D can be combined with each other for that purpose. However, drug-release profiles will differ from the organic profiles. For pharmaceutical forms that target drug release in the colon, EUDRAGIT® grades that dissolve above pH 7 (EUDRAGIT® S or the highly flexible EUDRAGIT® FS 30 D) are used. For the safe application of gastro-resistant formulations, it is important that the films remain largely impermeable in the acidic environment of the stomach. For particles, stomach transit times are typically in the range of 30 to 120 minutes, and for tablets, transit times can be up to several hours, depending on the core size and on the type and quantity of food in the stomach [69]. Anionic poly(meth)acrylate films meet these requirements with minimum layer thicknesses of 40–50 μm (i.e., 4–5 mg dry polymer/cm^2). It is crucial that coatings of critical areas, such as corners or edges, conform to the required minimum layer thickness because these areas as sites of the lowest wall thickness would otherwise contribute to a premature dissolution of the film. If thicker layers or polymers with higher dissolution pH are used, a delayed release in the small intestine can be achieved.

EXTENDED RELEASE BY TIME CONTROL

The poly(meth)acrylates that are used for sustained-release film coatings and matrix tablets are EUDRAGIT® RL (highly permeable), EUDRAGIT® RS (low permeable), EUDRAGIT® NE, and EUDRAGIT® NM (both low permeable). After contact with gastrointestinal fluids, the film coatings swell, independent of pH, and release the active ingredient by a diffusion control. EUDRAGIT® RL and EUDRAGIT® RS can be mixed with each other in any ratio in either organic or aqueous form to adjust permeability and obtain specific release patterns. Because the EUDRAGIT® RL features are dominant in these combinations, the share of EUDRAGIT® RS polymer is usually much higher, in order to achieve extended-release effects. For typical ratios, see the section entitled Poly(meth) acrylate Mixtures and Figure 9.7. EUDRAGIT® NE and NM have no reactive functional groups because all carboxylic groups are esterified with neutral short chain alcohols. Here, drug release is mainly controlled by the coating thickness. Two-phase drug release can be designed by applying a drug-containing immediate-release topcoat onto the controlled-release coating.

For controlled-release matrix tablets, both the pH-independent polymers EUDRAGIT® NE 30 D, EUDRAGIT® NM 30 D, EUDRAGIT® RL, and EUDRAGIT® RS as well as the anionic types, EUDRAGIT® L 30 D-55, EUDRAGIT® L, EUDRAGIT® S, and EUDRAGIT® FS 30 D are used. Under physiological conditions, the EUDRAGIT® L polymers provide matrix tablets with higher pH effects than EUDRAGIT® S and EUDRAGIT® FS. Poly(meth)acrylates can be processed via all common granulation techniques. Also, direct compression can be used to manufacture poly(meth) acrylate matrix tablets from solid polymer powders [70,71]. With higher degrees of distribution, increasing retardation effects are achieved (Figure 9.10). Depending on drug solubility, usually 5% to 20% of dry polymer substance based on tablet weight is sufficient to control drug release over a period of six to eight hours. In contrast to film coating, wet granulation with aqueous latex dispersions can be done without the addition of plasticizers. However, the addition of a plasticizer enhances the coalescence of the latex particles and hence increases the retardation effect.

In order to ensure extended drug release, matrix formulations should not contain strong disintegrants. The quantities of polymers necessary to achieve the desired effects of the polymer matrix on drug release characteristics are significantly smaller in wet granulation than in direct compression of powders. During compression, they are embedded in a sponge-like network of thin polymer layers, which first control the penetration of digestive fluid into the matrix and later the diffusion of the dissolved drug through pores, channels, and capillaries in the matrix. Insoluble polymers, such as EUDRAGIT® NM 30 D, form inert matrices. Their release mechanism is pH-independently controlled by diffusion and results in straight lines in the plot of dissolved drug versus square root

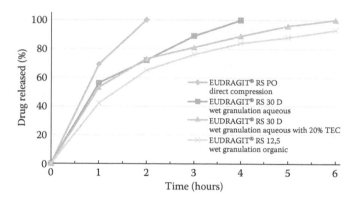

FIGURE 9.10 Influence of processing technique on drug release from EUDRAGIT® RS matrix tablets shown with diprophylline as model drug.

of time. However, when matrices from anionic poly(meth)acrylates start to dissolve at higher pH via salt formation, erosion effects increase drug release whereas release is only based on diffusion at lower pH. Finally, complete dissolution or disintegration of the tablet is achieved. The low permeability of poly(meth)acrylate latices allow the production of sustained-release matrix tablets containing more than 80% active drug. This is of particular importance for highly dosed actives. In such matrix formulations, the compression force normally has little influence on the release rate.

The release pattern of poly(meth)acrylate matrixes can additionally be modified by a thin functional poly(meth)acrylic top coating, which will preferably reduce the release rate in the first phase and thus provide more linear release profiles [72].

MULTIPARTICULATE TABLETS

Fast-disintegrating multiunit tablets generally show superior biopharmaceutical behavior with less variation in gastrointestinal transit time and less food effect compared to monolithic dosage forms. Besides sufficient substrate hardness, the main precondition is high flexibility of the controlled-release coating in order to prevent cracking of films during compression. Among the different poly(meth)acrylates, EUDRAGIT® E, EUDRAGIT® FS 30 D, EUDRAGIT® NE 30 D, and EUDRAGIT® NM 30 D stand out for their excellent flexibility whereas the anionic poly(meth)-acrylates show more brittle characteristics (Table 9.4). As a point of reference, coating formulations with 100% elongation at break provide sufficient flexibility for later compression [73]. The highly flexible polymers can be formulated without the addition of plasticizers whereas the more brittle anionic types require the addition of plasticizers or the combination with soft polymers EUDRAGIT® NE 30 D, NM 30 D, or also FS 30 D to achieve the required elasticity. However, the quantities of the flexible polymer should be kept on levels that do not significantly change the release behavior.

Outer-phase excipients with plastic properties, such as microcrystalline cellulose or lactose, provide additional protective effects on the coated multi-particulates. In addition, the functions of the outer-phase excipients are to prevent direct contact of coating layers, reduce friction during compression, improve compressibility, and ensure rapid disintegration after administration. The amount necessary to fill the intermediate spaces and to protect the coated particles during compression can be estimated by testing the tapped density of mixtures from particles and tableting excipients. Most useful mixtures should have the maximum tapped density. If the amount of the outer phase is less than 30%, excessive amounts of coated particles are likely to break during compression. In development, possible changes in dissolution profiles caused by mechanical stress in the tableting process should be controlled thoroughly. Differences should be less than 10% in order to ensure reproducible drug release. Beckert [74] investigated extensively the different aspects of preparing multi-particulate tablets based on particles coated with poly(meth)acrylates.

DRUG DELIVERY SYSTEMS

Particular value to approved therapies is added by the development of drug delivery systems that provide optimized drug release tailored to the specific treatment. The benefits are increasing therapeutic success and improved patient compliance. These advantages support the development of new chemical entities, product life cycle extension, and conceptions of added-value generics.

Accelerated Enteric Release

Regular enteric coatings ensure that there is less than 10% drug released in the acidic environment of the stomach over two hours followed by at least 80% release after 30 minutes in buffer pH 6.8. Usually, EUDRAGIT® L 30 D-55 is used as functional polymer for these coatings. However, for selected applications a faster dissolution of the API right after stomach transit may be required. To speed up drug release in the upper small intestine a bilayer coating technology has been developed. Beneath the regular EUDRAGIT® L 30 D-55 coating an anionic partially neutralized poly(meth)acrylic sub-coat is applied that additionally contains a buffer agent and enhances rapid dissolution of the enteric coating after the drug product has left the stomach [75–77]. The technology whose effectiveness was proven in man [78] is marketed under the brand name EUDRATEC® ACR.

Colonic Delivery Systems

Colonic delivery has gained importance for the treatment of local diseases but also for the oral delivery of proteins and peptides. Long transit times of dosage forms through the colon call for a targeted, time-controlled drug release in order to optimize therapeutic effects. Thus a novel multiunit delivery system was developed by combining the pH characteristics of anionic poly(meth) acrylates and diffusion-controlled kinetics, marketed as EUDRATEC® COL. The multiparticulate dosage forms provide a relatively constant transit through the intestine and consist of a drug-layered pellet core coated with an inner layer of a pH-independent diffusion barrier from an aqueous coating of EUDRAGIT® NE 30 D or EUDRAGIT® RL/RS 30 D. This layer enables controlled drug release throughout the colon up to a 20-hour period. The outer layer of the pH-dependent EUDRAGIT® FS 30 D triggers the start of release at the ileo cecal junction. The particle core can be prepared by powder layering or extrusion/and spheronization. In vitro proof of concept studies using 5-aminosalicylic acid (5-ASA) as the active compound and mixtures of EUDRAGIT® RL/RS 30 D as inner coatings confirmed the variability of the delivery system by statistical modulation using the central composite design [79,80]. Outer coating thickness (EUDRAGIT® FS 30 D dry polymer), inner coating composition, and inner coating thickness were found to control drug release reproducibly within a 95% confidence interval and function as a basis for optimization [81].

A clinical study in healthy volunteers with dosage forms containing 200 mg of caffeine as a pharmacokinetic marker, which is well absorbed from both small and large intestine, and ^{13}C-lactose-ureide for determining the oro cecal transit time demonstrated excellent in vitro in vivo correlations. Plasma profiles of caffeine were significantly prolonged for the pH- and timed-controlled delivery system in comparison with the pH-only based system (Figure 9.11). Compared to traditional drug products approved for the treatment of ulcerative colitis (UC), drug release from this new multiunit dosage form with a double layer poly(meth)acrylate coating offers a new dimension for the oral treatment of mid-to-distal UC [82,83].

When formulating the EUDRATEC® ACR double coating technology with EUDRAGIT® S that dissolves at pH 7, accelerated drug release in the ileo cecal region can be generated [78,84].

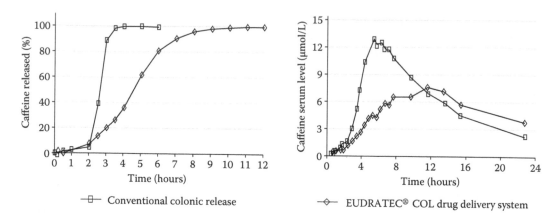

—☐— Conventional colonic release   —◇— EUDRATEC® COL drug delivery system

FIGURE 9.11 EUDRATEC® COL, a pH-triggered diffusion-controlled colonic delivery technology. In vitro dissolution (left) and in vivo plasma profiles (right) in comparison to conventional colonic release.

Modulated Controlled-Release Systems

Certain diseases have predictable cyclic, circadian rhythms, and timing of dose regiments can improve therapies in selected chronic conditions of diseases, such as bronchial asthma, arthritis, duodenal ulcers, cancer, diabetes, and neurological disorders. Thus, needs were identified that were aimed at improved, time-programmed oral therapeutic systems [85]. Conventional sustained-release systems provide release profiles following first-order or square root of time kinetics due to a constant diffusion barrier. The permeability modulation of coatings over time enables therapeutically optimized release profiles in vitro and in vivo.

The permeability of hydrophilic EUDRAGIT® RL and EUDRAGIT® RS coatings can be influenced by the interaction of anions with the polymeric quaternary ammonium groups. Basic investigations confirmed that the mechanism of drug release involves an immediate penetration of water into the hydrophilic polymer layer followed by an instant exchange of chloride ions against anions present in the dissolution medium. Dependent on the attraction of the anions to quaternary ammonium groups, permeability was influenced by exchanging anions. Strong attraction, reported for nitrate, sulfate, and citrate, resulted in a low water flux and thus reduced coating permeability for drugs [86]. Weak attractions typical for acetate and succinate ions induced a high water flux and thus accelerated drug diffusion [87]. These effects were used for the development of sigmoidal or pulsed oral drug delivery systems [88].

Figure 9.12 shows the formulation variation options. Using theophylline as model drug, cores were prepared containing different amounts and types of organic acid salts and were finally coated with conventional EUDRAGIT® RS 30 D spray suspensions at 10%–40% polymer weight gain. With succinate and acetate as modulators, sigmoidal or pulsed release profiles were achieved. The slope can be controlled by the type and amount of modulator ions in the core. Coating thickness defines the lag time of pulses. Thus ionic interactions of active ingredients or excipients can be controlled, and the release kinetic can be optimized.

Water-soluble salts of organic acids as modulating agents provide increased efficiency and flexibility when formulating pulsatile drug delivery systems. The option to not only use organic acids but also salts thereof broadens the applicability of the technology to active compounds with lower solubility in physiological media [89]. In vitro–in vivo correlation (IVIVC) was confirmed by several in vivo studies [90].

This intrinsic effect confirms that the anions in the formulation interact with the quaternary ionic groups of the polymer and thus will react very much independent of the different physiological media initiating the burst drug release.

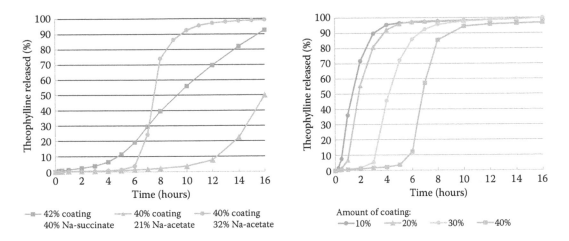

FIGURE 9.12 EUDRATEC® MOD with EUDRAGIT® RS 30 D: Adjustment of the slope of the release profile through different organic salts (left) and effect of coating thickness on the lag time of steep pulses (right).

Further acceleration of pulsatile release kinetics can be achieved by including hydrophilic or soluble compounds into the EUDRAGIT® RL/RS coating. This works in both aqueous and organic solvent coating processes.

Precipitated silica particles smaller than 50 μm, added at a 5%–50% level referred to polymer enhance the ionic interaction of modulator and polymer and thus enhance the pulsed drug release [91]. A similar effect is observed when combining anionic polymers, such as EUDRAGIT® L or L 100-55, into the release-controlling EUDRAGIT® RL/RS coating layer [92].

Initiated by the increasing physiological pH level in the intestine, the anionic polymers get solved out of the release-controlling polymer coating and thus accelerate drug release. The combination of pH-independent polymers with anionic pH-dependent polymers impacts the release kinetics, which may provide advantageous therapeutic effects to oral pharmaceutical formulations.

Based on the anion exchange at the quaternary ammonium groups in EUDRAGIT® RL 30 D and EUDRAGIT® RS 30 D, multilayered particles were developed that allow the modulation of drug release from first-order kinetics to linear, zero-order, or even accelerated profiles. The mechanism is based on kinetical control of the ion exchange–induced permeability changes by separating the drug and the modulating ions or salts by an additional polymer layer acting as diffusion barrier for the modulating anions [93]. The formulations, manufactured via conventional pharmaceutical processes and equipment, consist of EUDRAGIT® NE 30 D–coated salt cores that are further layered with the drug and finally coated with EUDRAGIT® RL/RS 30 D as shown in Figure 9.13. During

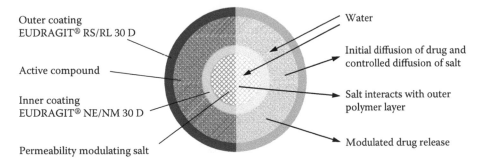

FIGURE 9.13 Structure (left) and function (right) of a modulated release particle, providing permeability modulation based on the EUDRATEC® MOD technology.

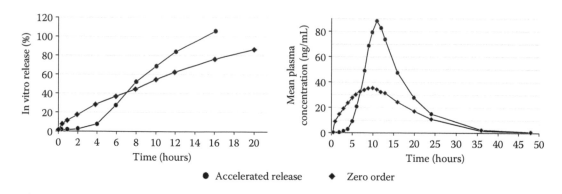

FIGURE 9.14 EUDRATEC® MOD technology: In vitro dissolution (left) and in vivo plasma profiles (right) with linear and accelerated drug release of metoprolole succinate. (From H. Ravishankar et al., *J Control Release*, 111, 65–72, 2006.)

release, the controlled flux of modulating salts allows for time-controlled permeability modulation of the outer EUDRAGIT® RL/RS layer.

A multi-particulate dosage form was developed that provided a linear release of terbutaline sulfate over a period of eight hours using trisodium citric acid crystals as cores. Citrate ions inhibit the hydration of the outer EUDRAGIT® RS film, controlling drug release to the desired diffusion pattern. In vivo studies in healthy, adult volunteers, including statistical analysis, confirmed biorelevant IVIVC on level A. The simulation of steady-state plasma concentrations did not show a significant difference compared to a commercial product [94].

Another multiunit dosage form was formulated that provided accelerated drug release of metoprolol succinate in vitro over a period of 16 hours as shown in Figure 9.14. In vivo plasma profiles confirmed significantly higher bioavailability than a commercial zero-order release product. Data processing by the numerical deconvolution method confirmed reliable IVIVC of level A, high predictability, and the value of statistics as a development tool for these delivery systems [95,96].

Alcohol-Resistant Formulations

Although poly(meth)acrylates are soluble in alcohol, they may be used to form CR matrix tablets being robust in hydro-alcoholic media. This is true for both the time-controlled release polymers such as

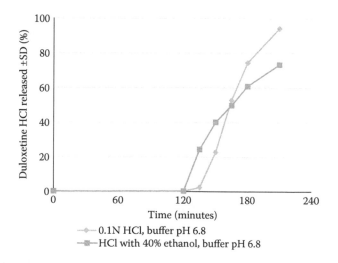

FIGURE 9.15 EUDRATEC® ADD dissolution profiles of delayed release duloxetine pellets with and without 40% alcohol in 0.1N HCl over two hours, followed by buffer pH 6.8.

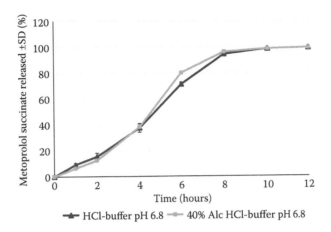

FIGURE 9.16 EUDRATEC® ADD dissolution profiles of extended release metoprolol succinate pellets with and without 40% alcohol in 0.1N HCl over two hours, followed by buffer pH 6.8,

EUDRAGIT® RL/RS and EUDRAGIT® NE/NM as well as for the anionic, pH-dependent types EUDRAGI® L/S.

Because of the higher surface area, multiparticulates with a controlled release coating are more sensitive to hydro-alcoholic media. However, combining the main functional poly(meth)acrylate with a second polymer enables the preparation of coated multiparticulates that resist 40% hydro-alcoholic acid over two hours [97,98]. The second polymer can have poly(meth)acrylate chemistry or can be a suitable hydrocolloid (Figures 9.15 and 9.16) [99,100].

FUNCTIONAL FORMULATIONS BEYOND COATINGS

GRANULATES

Because poly(meth)acrylates provide excellent mechanical stability with remarkable flexibility, they can be used as binders in immediate-release formulations and in matrix tablets. Both organic polymer solutions and aqueous latex dispersions can be applied alone or in combination with the polymer powders in order to increase process efficiency by reducing the solvent volume. In contrast to coating processes, the addition of plasticizers is not required. However, plasticizers will increase the softness of the system and result in different matrix structures that usually are characterized by higher distribution and hence stronger retardation effects (Figure 9.10). Due to the low viscosity of the latex dispersions, granulation can be performed in any common granulation equipment. Fluid bed processes show advantages, such as simultaneous drying and homogeneous distribution. Even dry granulation (e.g., roller compaction) can be used with the EUDRAGIT® powder grades.

DERMAL AND TRANSDERMAL THERAPY SYSTEMS

In order to provide uniform blood levels over a period of up to several days, transdermal therapeutic systems have been developed, preferably based on matrix structures [101]. Neutral and hydrophilic poly(meth)acrylate dispersions EUDRAGIT® NE/NM 30 D and EUDRAGIT® RL/RS 30 D can be applied in combination with auxiliaries, including plasticizers, by continuous blade or roller coating processes on foils [17,18]. Thus, aqueous-based manufacturing processes are possible. Drugs incorporated into the aqueous polymer dispersion as solutions or suspensions get embedded in the polymer matrix upon drying. The final transdermal systems may include the active compound in dissolved or dispersed form.

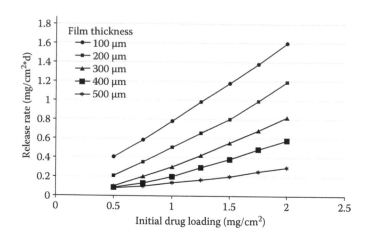

FIGURE 9.17   Principle of release control from cast-free (meth)acrylate films, included in dermal or trans-dermal systems. Drug release per surface area is a linear function of drug loading and layer thickness.

The release kinetics of drugs embedded in insoluble poly(meth)acrylate films follow Fick's second law of diffusion. Thus, release can be controlled by drug loading and layer thickness. Figure 9.17 reports the calculated diffusion data from model experiments in vitro, demonstrating controlled drug release of propranolol from different EUDRAGIT® RS layers by initial loading and matrix thickness [19].

EUDRAGIT® polymers can be used as release rate controlling agents in both matrix and membrane transdermal systems. For the latter ones, application processes can be adapted to any patch layering technique and polymer type. Layer thickness provides an additional parameter for controlling the diffusion rate of the active compound to the skin from a reservoir [45].

Anionic dispersions, such as EUDRAGIT® L 30 D-55, can be used for molecular entrapment of drugs, particularly in combination with the neutral EUDRAGIT® NE/NM 30 D, to regulate the diffusion rate of drugs by their ionic influence on drug diffusion through the matrix [87]. Further options to control diffusion of embedded drugs from patches to the skin are the addition of plasticizers and penetration enhancers.

Poly(meth)acrylate matrices proved high binding capacity for incorporated materials, including drugs, pigments, fillers, or other functional excipients. By absorbing water up to 60% of their own weight, they avoid influencing natural skin transpiration. Particularly, matrices of EUDRAGIT® NE 30 D have shown good clinical skin tolerance and were selected as a carrier in transdermal formulations [102].

SOLUBILITY ENHANCEMENT

Typically, solid solutions show a high potential to increase the solubility of poorly soluble drugs. Utilizing EUDRAGIT® polymers as the carrier, the drug can be dispersed in the polymer as crystals, amorphous particles or molecularly dispersed. As the particle size is decreased to the molecular level, solid solutions provide the highest effectiveness for solubility enhancement.

The cationic EUDRAGIT® E polymer is able to stabilize the amorphous state of the drug and thus maintain improved solubility. It is used either alone or for additional effects in combination with a second polymer, for example, EUDRAGIT® NE/NM 30 D. According to the polymer solubility, the EUDRAGIT® E matrix gets dissolved in the acidic environment of the stomach and drug molecules are released, providing increased solubility and hence enhanced bioavailability. The formulations may be processed by melt extrusion or spray drying [20,27] (Figures 9.18 and 9.19).

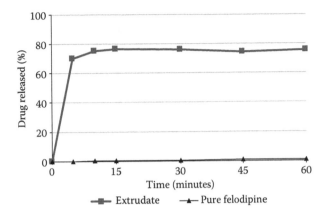

FIGURE 9.18   Dissolution profiles of felodipine as pure crystalline substance and as extruded compound with EUDRAGIT® E 100 and 10% felodipine content in 0.1N HCl.

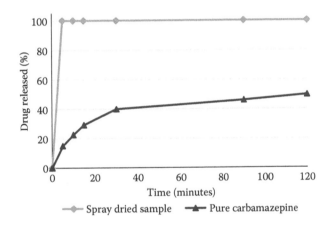

FIGURE 9.19   Dissolution profiles of carbamazepine as pure crystalline substance and as spray dried compound with EUDRAGIT® E 100 and 10% carbamazepine content in 0.1N HCl.

SUMMARY

Over more than 60 years, EUDRAGIT® polymers have become a standard for modified release formulations. In thin coatings, they provide reliable functionalities for delayed and extended release as well as for protective coatings. By combining the polymers in one or also in subsequent multiple layers, practically any kind of drug release profile can be tailored. Coating of oral solid dosage forms is still the standard application for these polymers. However, additional innovative applications have been developed and have achieved remarkable importance in drug delivery. Recent innovations focused on easy- or ready-to-use polymer products and associated coating additives as well as on drug delivery technologies based on EUDRAGIT®.

REFERENCES

1. Katchalski-Katzir E, Kramer DM. EUPERGIT® C, a carrier for immobilization of enzymes of industrial potential. *J Mol Catal B* 2000; 10:157–176.
2. Bosch T, Wendler T. Efficacy and safety of DALI-LDL-apheresis in two patients treated with the angiotensin II-receptor 1 antagonist losartan. *Ther Apher Dial* 2004; 8:269–274.

3. Solomon B, Raviv O, Leibman E, Fleminger G. Affinity purification of antibodies using immobilized FB domain of protein A. *J Chromatogr* 1992; 597:257–262.

4. EP 2627677B1 Process for preparing a (meth)acrylate copolymer containing tertiary amino groups by free-radical polymerization in solution.

5. EP2627684B1 Process for preparing a (meth)acrylate copolymer containing quaternary ammonium groups by free-radical polymerization in solution).

6. WO2011012161A1, Powdery or granulated composition comprising a copolymer, a dicarboxylic acid and a fatty monocarboxylic acid.

7. Lehmann K. Magensaftresistente und retardierende Arzneimittelüberzüge aus wässrigen Acryl-harzüberzügen, *Acta Pharm Technol* 1975; 21(4):255–260.

8. Banker GS. The new water based colloidal dispersions. *Pharm Technol* 1981; 5(4):12–19.

9. Lehmann K. In Wasser dispergierbare, hydrophile Acrylharze mit abgestufter Permeabilität für diffusionsgesteuerte Wirkstoffabgabe aus Arzneiformen. *Acta Pharm Technol* 1986; 32(3):146–152.

10. Frenkel J. Viscous flow of crystalline bodies under the action of surface tension. *J Phys (UDSSR)* 1943; 9:385.

11. Brown GL. Formation of films from polymer dispersions. *J Polymer Sci* 1956; 12:423–434.

12. ISO 2115:1996 Plastics—Polymer dispersions—Determination of white point temperature and minimum film-forming temperature.

13. DIN 53, 787. *Prüfung von wässrigen Kunststoff-Dispersionen—Bestimmung der Mindest-Filmbildetemperatur und des Weißpunktes* E.V.

14. Turi EA, ed. Thermal Characterization of Polymeric Materials. New York: Academic Press, 1981.

15. Lehman K, Petereit HU. Application of aqueous poly(meth)acrylate dispersions for matrix tablet granulates, *Acta Pharm Technol* 1988; 34(4):189–195.

16. Petereit HU. Application of aqueous latex dispersions of methacrylic ester copolymers for the preparation of dermal and transdermal drug delivery systems, Third European Congress of Biopharmaceutics and Pharmcokinetics—Proceedings, Vol. 1 *Biopharm* 1987.

17. Rafiee-Tehrani M, Safaii N, Toliat T, Petereit HU, Beckert T. Effect of plasticizers and enhancers on release behaviour of estradiol from unilaminate TDD patch. 26th International Symposium on Controlled Release of Bioactive Materials, Boston, 1999.

18. Beckert TE, Kahler S, Bergmann G., Fillinger M, Petereit HU. A new system for the development of hydrophilic transdermal therapy systems based on EUDRAGIT®, *Proc Control Release Soc* 1998; 25:517–572.

19. Rafiee-Tehrani M, Safaii N, Toliat T, Petereit HU, Beckert T. Acrylic resins as rate-controlling membranes in estradiol transdermal unilaminate patch, *Proc Control Release Soc* 1998; 25:28–29.

20. Leuner C, Dressman J. Improving drug solubility using solid dispersions, *Eur J Pharm Biopharm* 2000; 50(1):47–60.

21. Rahman AAA. Evaluation of ibuprofen controlled release tablets, *Eur J Pharm Biopharm* 1988; 38(2):71–77.

22. Cameron CG, McGinity JW. Controlled release tablet formulations containing acrylic resins, II. Combination of resin formulations, *Drug Dev Ind Pharm* 1987; 13(8):1409–1427.

23. Cameron CG, McGinity JW. Controlled release tablet formulations containing acrylic resins III. Influence of filler excipients, *Drug Dev Ind Pharm* 1987; 13(2):303–318.

24. Aitken-Nichol C, Zhang F, McGinity JW. Hot melt extrusion of acrylic films, *Pharm Res* 1996; 13(5):804–808.

25. Young CR, Koleng JJ, McGinity JW. Production of spherical pellets by a hot-melt extrusion and spheronization process, *Int J Pharm* 2002; 242:87–92.

26. Nollenberger K, Gryczke A, Morita T, Ishii T. Using polymers to enhance solubility of poorly soluble drugs, *Pharm Technol* 2009; 33:20–25.

27. Miller DA, McConville JT, Yang W, Williams RO, McGinity JW. Hot-melt extrusion for enhanced delivery of drug particles, *J Pharm Sci* 2007; 96(2):S. 361–376.

28. Zhu Y, Mehta KA, McGinity JW. Influence of plasticizer level on the drug release from sustained release film coated and hot-melt extruded dosage forms, *Pharm Dev Technol* 2006; 11:285–294.

29. EUDRAGIT® Application Guidelines, 12th Edition 2012.

30. Houben-Weyl. Methoden der Organ. Chemie [Methods of Organ. Chemistry], Vol. 14, Part 1, 4th Ed., pp. 133ff, Georg Thieme Verlag, Stuttgart 1987.

31. EP1368007 B1, Coating and binding agent for pharmaceutical formulations having improved storage stability

32. Lehmann K, Süfke T. New methacrylic acid copolymers for improved coating technology. *Pharm Res* 1995; 12(9) (Suppl):137.

33. EP 1848751B1 Partly Neutralized Anionic (Meth)acrylate Copolymer.
34. Odian G. Principles of Polymerization, 4th edn. Hoboken: John Wiley & Sons, 2004.
35. Elias HG, Stafford JW, eds. Macromolecules, Vol. 2. New York and London: Plenum Press, 1977, pp. 761–798.
36. Suetterlin N. Structure and properties of emulsion copolymers, *Macromol Chem Suppl* 10/11, 403–418 (1985).
37. Adler M, Pasch H, Meier C et al. Molar mass characterization of hydrophilic copolymers, 1: Size exclusion chromatography of neutral and anionic (meth)acrylate copolymers. *e-Polymers* 2004; 055.
38. Adler M, Pasch H, Meier C et al. Molar mass characterization of hydrophilic copolymers, 2: Size exclusion chromatography of cationic (meth)acrylate copolymers. *e-Polymers* 2005; 057.
39. Bauer, Lehmann, Osterwald, Rothgang. Coated Pharmaceutical Dosage Forms. CRC Press, 1999.
40. McGinity JW, Zhang F. Meltextruded controlled-release dosage forms. *Drug Pharm Sci* 2003; 133: 183–208.
41. McGinity JW, Zhang F, Repka MA, Koleng JJ. *Am Pharm Rev* 2001; 4(2):25–36.
42. Lehmann K. Formulation of controlled release tablets. *Acta Pharm Fenn* 1984; 93:55–74.
43. Dillon RE, Matheson LA, Bradford EB. Sintering of synthetic latex particles. *J Colloid Sci* 1951; 6:108–117.
44. Bindschaedler C, Gurny R, Doelker E. Theoretical concepts regarding the formation of films from aqueous microdispersions and application to coatings. *Lab Pharm Probl Tech* 1983; 31(331):389–394.
45. Lin AY, Muhammed NA, Pope D, Augsburger LL. A Study of the effects of curing and storage conditions on controlled release Diphenhydramine HCl pellets coated with EUDRAGIT® NE 30 D, *Pharm Dev Tech*, US, 2003; 8(3):277–287.
46. Baer H. Untersuchung des Curings- und Alterungsverhaltens von EUDRAGIT® NM 30 D Filmüberzügen auf Metoprololtartrat Pellets, FH Bingen (2009).
47. Dassinger Th, Albers J, Jautze S. A quality-by-design approach to optimize in-process curing of EUDRAGIT® NM 30 D, *Pharm Technol Solid Dosage Excipients* 2014; 48–52.
48. Felton LA, McGinity JW. Influence of insoluble excipients on film coating systems, *Drug Dev Ind Pharm* 2002; 28(3):225–243.
49. Maul KA, Schmidt PC. Influence of different-shaped pigments on bisacodyl release from EUDRAGIT® L 30 D-55. *Int J Pharm* 1995; 118:103–112.
50. Maul KA, Schmidt PC. Influence of different-shaped pigments and plasticizers on theophylline release from EUDRAGIT® RS 30 D and Aquacoat ECD30 coated pellets. *STP Pharma Sci* 1997; 7:498–506.
51. Petereit HU, Assmus M, Lehmann K. Glycerol monostearate as a glidant in aqueous film-coating formulations. *Eur J Pharm Biopharm* 1995; 41(4):219–228.
52. Lehmann K, Dreher D. Mixtures of aqueous polymethacrylate dispersion for drug coating. *Drugs Made Ger* 1988; 31:101–102.
53. Zheng W, McGinity JW. Influence of EUDRAGIT® NE30 D blended with EUDRAGIT® L 30 D-55 on the release of phenylpropanolamine hydrochloride from coated pellets. *Drug Dev Ind Pharm* 2003; 29(3):357–366.
54. Amighi K, Moes AJ. Evaluation of thermal and film forming properties of acrylic aqueous polymer dispersion blends; application to the formulation of sustained-release film coated theophylline pellets. *Drug Dev Indust Pharm* 1995; 21(20):2355–2369.
55. Lecomte F, Siepmann J et al. Polymer blends used for the coating of multiparticulates: Comparison of aqueous and organic coating techniques, *Pharm Res* 2004; 21(5):882–890.
56. Lecomte F, Siepmann J et al. pH-sensitive polymer blends used as coating materials to control drug release from spherical beads: Elucidation of the underlying mass transport mechanisms, *Pharm Res* 2005; 22(7):1129–1141.
57. Goodhart FW, Harries MR, Murthy KS, Nesbitt RU. An evaluation of aqueous film-forming dispersions for controlled release. *Pharm Technol* 1984; 8(4):64–70.
58. Lehmann K, Petereit HU. Film coatings based on aqueous polymethacrylate dispersions for sustained release in the intestinal tract. *Drugs Made Ger* 1994; 37(1):19–21.
59. Lehmann K. Acrylic latices from redispersable powders for peroral and transdermal drug formulations. *Drug Dev Ind Pharm* 1986; 12(3):265–287.
60. Rubino O. Fluid-bed technology, *Pharm Tech* 1999; 6:104–113.
61. Felton LA and McGinity JW. Adhesion of polymeric films to pharmaceutical solids, *Eur J Pharm Biopharm* 1999; 47:3–14.
62. Felton LA, Haase MM, Shah NH, Zhang G, Infeld MH, Malick AW, McGinity JW. Physical and enteric properties of soft gelatin capsules coated with Eudragit L30 D-55, *Int J Pharm* 1995; 113:17–25.

63. Bodmeier R, Paeratakul O. Mechanical properties of dry and wet cellulosic and acrylic films prepared from aqueous colloidal polymer dispersions used in the coating of solid dosage forms. *Pharm Res* 1994; 11(6):882–888.

64. Amighi K, Moes A. Influence of plasticizer concentration and storage conditions on the drug release from EUDRAGIT® RS 30 D film-coated sustained-release theophylline pellets. *Eur J Pharm Biopharm* 1996; 42(1):29–35.

65. Guiterrez-Rocca JC, McGinity JW. Influence of aging on the physical-mechanical properties of acrylic resin films cast from aqueous dispersion and organic solutions. *Drug Dev Ind Pharm* 1993; 19:315–332.

66. EP1781252 B1. Method for producing coated drugs having a stable profile for the release of active ingredients.

67. Joshi S, Petereit H-U, Film coatings for taste masking and moisture protection. *Int J Pharm* 2013; 457:395–406.

68. EP 0,955,041 B1 and US 6,656,507 B2. Aqueous dispersion suitable for the production of coatings and binders for solid oral drugs.

69. Coupe AJ, Davis SS, Wilding IR. Variation in gastrointestinal transit of pharmaceutical dosage forms in healthy subjects. *Pharm Res* 1991; 8(3):360–364.

70. McGinity JW, Cameron CG, Cuff GW. Controlled-release theophylline tablet formulations containing acrylic resins I. Dissolution properties of tablets. *Drug Dev Ind Pharm* 1983; 9(1/2):57–68.

71. Gohel MC, Patel TP, Bariya SH. Studies in preparation and evaluation of pH-independent sustained-release matrix tablets of verapamil HCl using directly compressible EUDRAGIT®s. *Pharm Dev Tech* 2003; 8(4):323–333.

72. Lehmann K, Dreher D. Permeable Acrylharzlacke zur Herstellung von Depot-Arzneiformen. *Pharm Ind* 1969; 31:319–322.

73. DIN 53455. *Prüfung von Kunststoff* E.V.

74. Beckert TE. Verpressen von magensaftresistent überzogenen Pellets zu zerfallenden Tabletten Dissertation. University of Tuebingen, 1995.

75. Liu F, Lizio R, Schneider UJ, Petereit HU, Blakey P, Basit AW. SEM/EDX and confocal microscopy analysis of novel and conventional enteric-coated systems, *Int J Pharm* 2009; 369:72–78.

76. Liu F, Lizio R, Meier Ch, Petereit UH, Blakey P, Basit AW. A novel concept in enteric coating: A double-coating system providing rapid drug release in the proximal small intestine, *J Control Release* 2009; 133:119–124.

77. Liu F, Basit AW. A paradigm shift in enteric coating: Achieving rapid release in the proximal small intestine of man, *J Control Release* 2010; 147:242–245.

78. Varum FJO, Hatton GB, Freire AC, Basit AW. A novel coating concept for ileo-colonic drug targeting: Proof of concept in humans using scintigraphy, *Eur J Pharm Biopharm* 2013; 84:573–577.

79. Gupta VK, Beckert TE, Price JC. A novel pH- and time based multiunit potential colonic drug delivery system. I. *Dev Int J Pharm* 2001; 213:83–91.

80. Rudolph MW, Klein S, Beckert TE, Petereit HU, Dressman JB. A new 5-aminosalicylic acid multi-unit dosage form for the therapy for ulcerative colitis. *Eur J Pharm Biopharm* 2001; 51:183–190.

81. Gupta VK, Assmus M, Beckert TE, Price JC. A novel pH- and time-based multiunit potential colonic drug delivery system. I Development. *Int J Pharm* 2001; 213:93–102.

82. Bott C, Rudolph MW, Schirrmacher S et al. In vivo evaluation of a novel pH- and time-based multi-unit colonic drug delivery system. *Aliment Pharmacol Ther* 2004; 20:347–353.

83. Klein S, Rudolph MW, Skalsky B, Petereit HU, Dressman JB. Use of BioDis to generate a physiologically relevant IVIVC, *J Control Release* 2008; 130(3):216–219.

84. Liu F, Moreno P, Basit AW, A novel double-coating approach for improved pH-triggered delivery to the ileo-colonic region of the gastrointestinal tract, *Eur J Pharm Biopharm* 2010; 74:311–315.

85. Lemmer B. Chronopharmacokinetics: Implication for drug treatment. *J Pharm Pharmacol* 1999; 51: 887–890.

86. Wagner KG, McGinity JW. Influence of chloride ion exchange on the permeability and drug release of EUDRAGIT® RS 30 D films. *J Control Release* 2002; 82(2/3):385–397.

87. Wagner KG, Gruetzmann R. Anion-induced water flux as drug release mechanism through cationic EUDRAGIT® RS 30 D film coatings. *AAPS J* 2005; 7(3):Article 67.

88. Narisawa S, Nagata M, Danyoshi C et al. An organic acid-induced sigmoidal release system for oral controlled release preparations. *Pharm Res* 1994; 11:111–116.

89. US 6878397 B1 Coated medicament forms with controlled active substance release.

90. Narisawa S, Nagata C, Hirakawa Y, Kobayshi M, Yoshino H. An organic acid-induce sigmoidal release system for oral controlled release preparations, permeability enhancement of EUDRAGIT® RS coating led by the physico chemical interaction with organic acid. *J Pharm Sci* 1996; 85:184–188.

91. EP 2051704 B1 Pharmaceutical composition with controlled active ingredient delivery for active ingredients with good solubility in water.

92. US 9011907B2 Coated pharmaceutical or nutraceutical preparation with enhanced pulsed active substance release.

93. Ravishankar H, Patil P, Petereit HU, Lizio R. Controlled release by permeability alteration of cationic ammonio methacrylate copolymers using ionic interactions, *Drug Dev Ind Pharm* 2006; 32:709–718.

94. Ravishankar H, Iyer-Chavan J. Clinical studies of terbutaline controlled release formulation prepared using EUDRAMODETM. *Drug Del Tech* 2006; 6(6):50–56.

95. Ravishankar H, Patil P, Samel A, Petereit HU, Lizio R, Iyer-Chavan J. Modulated release metoprolol succinate formulation, based on ionic interactions: In vivo proof of concept. *J Control Release* 2006; 111:65–72.

96. Ravishankar H, Patil P, Petereit HU, Renner G. Modulated release system EUDRAMODETM: A novel approach to sustained release oral drug delivery systems. *Drug Del Tech* 2005; 5(9):48–55.

97. EP2187875 B1, pH-dependent controlled release pharmaceutical composition for non-opioids with resistance against the influence of ethanol.

98. EP2326313 B1, pH-dependent controlled release pharmaceutical composition for non-opioids with resistance against the influence of ethanol.

99. WO2014032741 A1, Gastric resistant pharmaceutical or nutraceutical composition with resistance against the influence of ethanol.

100. WO2014032742 Pharmaceutical or nutraceutical composition with sustained release characteristic and with resistance against the influence of ethanol.

101. Bindschaedler C, Gurny R, Doelker E. Theoretical concepts regarding the function of films from micro-dispersions and application to coatings. *Lab Pharm Probl Tech* 1983.

102. Chien YW. Logics of Transdermal Controlled Drug Administration. *Drug Dev In Pharm* 1983; 9:497.

10 Application of HPMC and HPMCAS to Aqueous Film Coating of Pharmaceutical Dosage Forms

Sakae Obara, Anisul Quadir, and Hiroyasu Kokubo

CONTENTS

INTRODUCTION

The first application of hypromellose, also known as hydroxypropyl methylcellulose (HPMC), for film coating appeared in a patent by Singiser [1] of Abbott Laboratories in 1962. Film coatings using HPMC have become popular, taking the place of the conventional sugar coating of tablets because they give a superior appearance, act as protection for fragile tablets, and mask the unpleasant taste of drug substances. The main reason for the extensive use of HPMC as a film-coating polymer is that it is soluble in some organic solvents and also in water over the entire biological pH range. Film coating can therefore be done using an organic solvent or an aqueous-based system, and the film formed will dissolve in the digestive juices, leading to complete release of the active ingredients.

However, the lowest viscosity of HPMC available in the early 1960s was 50 mPa sec (viscosity of a 2% solution at 20°C). It was too viscous to prepare a coating solution having a high concentration of the polymer. Thus, the coating cost was relatively high. In 1965, low-viscosity types of HPMC (3, 6, and 15 mPa sec) were developed by Shin-Etsu Chemical Co., Ltd. (Tokyo, Japan). This contributed significantly to the worldwide growth of film coating using HPMC in subsequent years. The use of an organic solvent system in film coating was long considered to be inevitable. The solvent systems most commonly used in film coating with HPMC were mixtures of a chlorinated hydrocarbon and an alcohol. A typical solvent blend consisted of a mixture of methylene chloride and ethanol. However, the use of such organic solvents has been considered undesirable for the following reasons:

1. Solvents are difficult to remove completely from the coated preparations and may present a health hazard.
2. Regulations on the discharge of organic solvents into the atmosphere have become more severe as environmental concerns have increased in recent years.
3. Regulations on the exposure of factory workers to organic solvent vapors have become more stringent.
4. Economic considerations, such as organic solvent cost and the provision of facilities to avoid the risk of explosion during film coating, are also important.

The main reason for using organic solvents originally in film coating was to avoid possible decomposition of the active ingredients and problems such as "picking" or degradation of dosage forms during the coating operation, which might occur if water was used. Research in the mid-1970s demonstrated that the decomposition of active ingredients and possible coating difficulties were not a serious concern in the actual application of aqueous film coating using HPMC. The latent heat of evaporation of water (539 kcal/kg) is about three times higher than that of ethanol (204 kcal/kg), and this value raised concerns that a much longer coating time would be required in aqueous coating. This problem was largely overcome by equipment modifications including the side-vented coating apparatuses, which have a higher drying efficiency.

A point to which special attention should be paid in aqueous coating using HPMC is that the ideal ranges of coating conditions are somewhat narrow compared with those used in organic solvent coating, and improper coating conditions sometimes result in damage to coating batches, which makes them unsuitable for reprocessing. In the following sections, the properties of HPMC and fundamental aspects of the application of HPMC in aqueous film coating are discussed.

Enteric coating from aqueous systems has also been attractive to pharmaceutical manufacturers for the same reasons mentioned above. Because enteric materials are essentially insoluble in water, the use of an aqueous emulsion or suspension system seemed to be the best approach for aqueous coating. Hypromellose phthalate (HPMCP), an enteric polymer derived from HPMC, has long been used for solvent-based enteric coating. An approach to use this material for aqueous coating by suspending the micronized particles in water was studied. However, this material was not optimal for an aqueous system because it is not well dispersible in water. An alternative material, hypromellose acetate succinate (HPMCAS), was subsequently developed. HPMCAS is also derived from HPMC and characteristically has good compatibility with plasticizers. It can dissolve over a wide range of pH values higher than 5.5 by controlling substitution in the polymer structure. This means that films having good gastric resistance can readily be produced in usual coating operations, and HPMCAS can be used not only for enteric coating but for the preparation of prolonged-release preparations. Some results of basic and applied research on HPMCAS are presented in the second half of this chapter.

PROPERTIES OF HPMC

TYPES OF HPMC FOR FILM COATING

The chemical structure of HPMC is shown in Figure 10.1. HPMC is classified according to the content of substituents and its viscosity. Commercially available HPMC includes several substitution types such as those shown in Table 10.1, that is, HPMC 1828, 2208, 2906, and 2910. Of the four digits in each number, the first two represent the median percentage content of methoxy groups and the last two represent that of the hydroxypropoxy groups. The selection of proper substitution type is important for some pharmaceutical applications. The substitution affects the

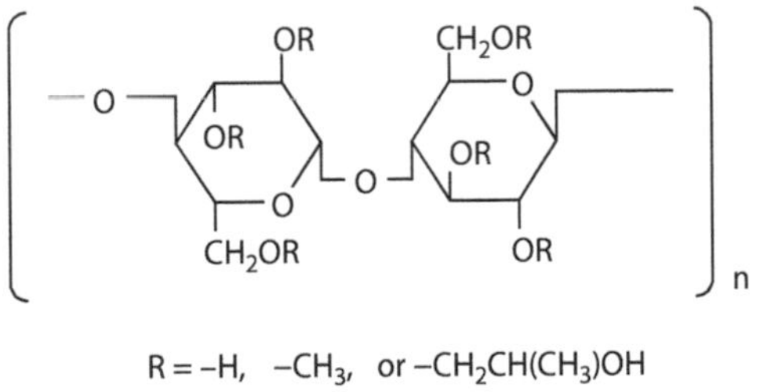

$R = -H, \ -CH_3, \ or \ -CH_2CH(CH_3)OH$

FIGURE 10.1 Chemical structure of hydroxypropyl methylcellulose.

TABLE 10.1
Standards on the Contents of Substituents of HPMC

Substitution Type	Methoxy (%)		Hydroxypropoxy (%)	
	Minimum	Maximum	Minimum	Maximum
1828	16.5	20.0	23.0	32.0
2208	19.0	24.0	4.0	12.0
2906	27.0	30.0	4.0	7.5
2910	28.0	30.0	7.0	12.0

solubility–temperature relationship. Among the three grades, 2208, 2906, and 2910, which have long been commercially available worldwide, 2910 has the best solubility in organic solvents, and so it has often been used for organic solvent–based coating. Even though aqueous coating has been replacing solvent-based coating and the solubility in organic solvents is of less importance, the 2910 grade is still widely used. Substitution grades other than 2910 are also applicable for aqueous coating, but there are few suitable commercial products of those substitution grades having low viscosity.

Another important parameter of HPMC is its molecular weight. Size exclusion chromatography (SEC) is commonly used to determine the molecular weight of water-soluble polymers. However, measuring the molecular weight by SEC is not a routine quality control for HPMC manufacturers due to difficulties in obtaining reproducible results and the fact that the SEC requires expensive apparatus. Because viscosity of the HPMC solution is directly correlated with its molecular weight, viscosity measurements are used for quality control. Labeled viscosity (nominal viscosity) is usually utilized as a parameter that represents viscosity grades. It is based on apparent viscosity of a 2% aqueous solution at 20°C. In the previous United States Pharmacopeia (USP) and Japanese Pharmacopeia (JP), the apparent viscosity was specified to be measured using a capillary viscometer whereas the use of a rotational viscometer was the method stated in the European Pharmacopeia (Ph.Eur). The harmonization of the method for viscosity measurement for HPMC was discussed, and the three pharmacopeias have reached an agreement to use the Ubbelohde viscometer to measure viscosity less than 600 mPa sec and the Brookfield-type viscometer for 600 mPa sec and higher. This is based on a collaborated study, which found that one viscometer cannot cover the whole viscosity range of the current commercially available HPMC products with sufficient reproducibility. In this chapter, all labeled viscosities are based on the Ubbelohde viscosity, as only low-viscosity grades are discussed. A "6-mPa sec grade" means HPMC having a labeled viscosity of 6 mPa sec. Labeled viscosity does not mean the exact viscosity value of a product lot. In the compendial monograph, the apparent viscosity of a low-viscosity HPMC product is specified to be from 80% to 120% of the labeled viscosity. HPMC 2910 of low labeled viscosity (3–15 mPa sec) is commonly used in film coating. The low-viscosity grades of HPMC are typically produced by depolymerization of high-viscosity grades. Examples of commercially available products of HPMC for film coating widely used throughout the world are Pharmacoat 603, 645, 606, and 615 (Shin-Etsu Chemical Co. Ltd., Tokyo, Japan) and Methocel E3, E5, E6, and E15 (Dow Chemical Company, Midland, MI). As an example, the specifications of Pharmacoat are summarized in Table 10.2.

TABLE 10.2
Specifications of HPMC

	603	645	606	615
Substitution type	2910	2910	2910	2910
Labeled viscosity	3 mPa sec	4.5 mPa sec	6 mPa sec	15 mPa sec
Specification range of viscosity (mPa sec) (2% solution at 20°C)	2.4–3.6	3.6–5.1	4.8–7.2	12.0–18.0
Appearance	Fibrous or granular powder			
Color	White to slightly off-white			
pH	5.5–8.0			
Loss on drying	Not more than 5.0%			
Residue on ignition	Not more than 1.5%			
Methoxy content	28.0%–30.0%			
Hydroxypropoxy content	7.0%–12.0%			

Characteristics of **HPMC** Aqueous Solutions

Figure 10.2 illustrates the relationships between the concentration of various viscosity grades of HPMC and their solution viscosity. The required viscosity of a solution for aqueous film coating is commonly less than 100 mPa sec. The maximum concentrations of 3, 6, and 15 mPa sec grades, which can be used in film coating, are therefore approximately 14%, 7.5%, and 4.5%, respectively. Thus, the maximum concentrations used in coating operations depend on the viscosity grade of HPMC used although there are other factors that should be taken into consideration in practical applications, such as aqueous solutions of HPMC gel upon heating. The thermal gelling temperature, which is close to the clouding point, depends on the level of substitution, and it is also affected by such factors as viscosity, concentration, heating rate, and the addition of salts. In Figure 10.3, the

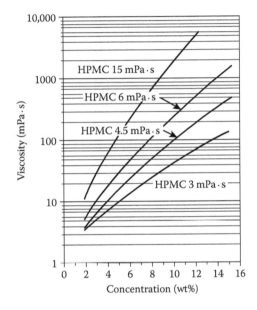

FIGURE 10.2   Viscosity–concentration curve of HPMC.

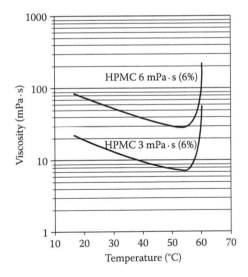

FIGURE 10.3   Effect of temperature on the viscosity of aqueous solutions of HPMC.

temperature–viscosity relationships of two 6% solutions of HPMC are shown. Dramatic increases in viscosity are observed near 60°C, which indicates the occurrence of gelation. Problems might be encountered if the solutions were at around this temperature. Preparation temperature of the coating solution should be less than 40°C for complete dissolution of HPMC particles.

MOLECULAR WEIGHT AND ITS DISTRIBUTION

Rowe [2] determined the molecular weight distribution of HPMC using SEC (also known as gel permeation chromatography or GPC). The data on molecular weight were represented based on polystyrene as a reference standard, and there is a possibility that molecular association occurred in dimethyl sulfoxide, which was used as the mobile phase, resulting in a very wide distribution; the ratio M_w/M_n (weight–average molecular weight/number–average molecular weight) was greater than 10. Kato et al. [3] determined the molecular weight distribution by aqueous SEC based on the use of a series of polyethylene oxide standards. The weight–average molecular weights of HPMC of 3, 6, 15, and 50 mPa sec grades were 12,600, 29,400, 64,800, and 104,000, respectively. The ratio of M_w and M_n ranged from 4 to 5. Figure 10.4 and Table 10.3 show molecular weight data of HPMC using SEC with the multi-angle laser light scattering (MALLS) technique. These results and previous reports indicate that molecular weight distribution is dependent on the measuring method and conditions.

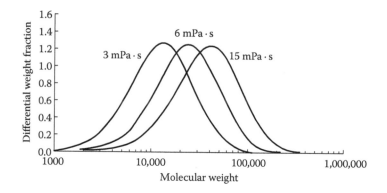

FIGURE 10.4 Molecular weight distribution of HPMC using a SEC-MALLS technique. MALLS, multi-angle laser light scattering; SEC, size exclusion chromatography.

TABLE 10.3
Molecular Weight of HPMC

Sample	Mw[a]	Mw/Mn	Viscosity[b] (mPa sec)
Pharmacoat 603	16,000	2.0	3.0
Pharmacoat 645	22,600	1.8	4.6
Pharmacoat 606	35,600	1.6	6.0
Pharmacoat 615	60,000	1.9	15.0
Metolose 60SH-50[c]	76,800	2.6	53.9

Note: MALLS, multi-angle laser light scattering; SEC, size exclusion chromatography.
[a] Weight-average molecular weight measured by the SEC-MALLS method.
[b] Ubbelohde viscosity of 2% aqueous solution at 20°C.
[c] 50 mPa sec grade of HPMC.

PHYSICAL PROPERTIES OF HPMC POWDER AND FILMS

Callahan et al. [4] classified various pharmaceutical excipients according to their hygroscopicity by measuring equilibrium moisture content at 25°C. The results on HPMC and related cellulose derivatives are illustrated in Figure 10.5. According to their classification, HPMC is considered "very hygroscopic," which is defined as follows [4]: "Moisture increase may occur at relative humidities as low as 50%. The increase in moisture content after storage for one week above 90% relative humidity (RH) may exceed 30%." Therefore, moisture absorption of HPMC-coated pharmaceuticals may occur at very high humidity. In such cases, they should be packed in a moisture-resistant material.

Various methods have been proposed for measuring water vapor permeability (WVP) of polymer films. Figure 10.6 shows a cell developed by Hawes [5] for measuring WVP. The WVP of various viscosity grades of HPMC and hydroxypropylcellulose (HPC) was measured and the results are shown in Table 10.4. Two kinds of film specimens were tested; one was a film prepared by casting an aqueous polymeric solution (free film), and the other was a film that was applied to tablets (applied film); both were 0.1 mm thick. Vials, each sealed by a sample of the film, were stored in a desiccator at 20°C with one side of the test film exposed to 0% relative humidity and the other to 75% RH. After an equilibrium period of eight to 12 hours, the samples were weighed at intervals over a test period of 72 hours. The moisture permeability value of HPMC differed slightly depending on viscosity grade. The WVP tended to decrease as viscosity decreased. The WVP of applied films was always higher than that of free films, which might reflect higher porosity. HPC showed a tendency to have smaller WVP values than HPMC in both free and applied films.

Table 10.5 shows the mechanical properties of HPMC films. The properties vary with viscosity grade. Tensile strength and elongation of films (100 μm in thickness) prepared by casting of various viscosity grades of HPMC and HPC were measured using an Instron-type tensile tester at 20°C and 65% RH. Both tensile strength and elongation of HPMC films decreased as the viscosity decreased, and elongation of the 3 mPa sec grade showed an especially small value compared with that of 6 mPa sec grade. These observations suggest that the possibility of crack formation in coated films should be taken into consideration when an HPMC of lower viscosity grade such as 3 mPa sec is used. In contrast, HPC films exhibited very low tensile strength and comparatively higher elongation due to its plasticity.

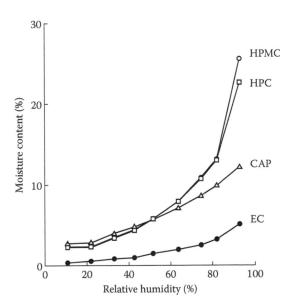

FIGURE 10.5 Equilibrium moisture curves for HPMC and related polymers at 25°C. Abbreviations: CAP, cellacefate; EC, ethylcellulose; HPC, hydroxypropylcellulose. (From J. C. Callahan et al., *Drug Dev Ind Pharm*, 8, 355–369, 1982.)

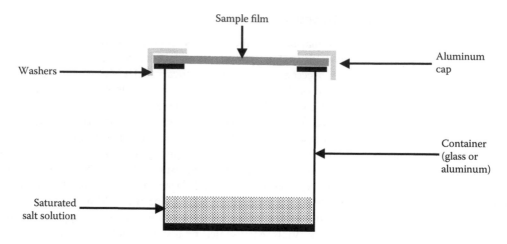

FIGURE 10.6 Water vapor permeability cell. (From M. R. Hawes, The effect of some commonly used excipients on the physical properties of film forming used in the aqueous coating of pharmaceutical tablets, Paper at the Panel of the Pharmaceutical Society of Great Britain, 1978.)

TABLE 10.4

Water Vapor Permeability of HPMC and HPC Films

Polymer, Labeled Viscosity	Sample	Water Vapor Permeability (g/m²/24 Hours)
HPMC, 50 mPa sec	Free film	219
HPMC, 15 mPa sec	Free film	207
HPMC, 6 mPa sec	Free film	194
HPMC, 6 mPa sec	Applied film	273
HPMC, 3 mPa sec	Applied film	192
HPC, 8 mPa sec	Free film	106
HPC, 8 mPa sec	Applied film	202

Note: HPC, hydroxypropylcellulose; HPMC, hydroxypropyl methylcellulose.

TABLE 10.5

Mechanical Properties of HPMC and HPC Films

Polymer, Labeled Viscosity	Tensile Strength (MPa)	Elongation (%)
HPMC, 50 mPa sec	82.3	38.8
HPMC, 15 mPa sec	66.6	27.0
HPMC, 6 mPa sec	55.9	22.6
HPMC, 3 mPa sec	48.0	3.3
HPC, 8 mPa sec	10.8	35.5

Note: HPC, hydroxypropylcellulose; HPMC, hydroxypropyl methylcellulose.

TABLE 10.6
Dissolution Time of Films from Pharmacoat 606 in Various Fluids

Test Fluid	Dissolution Time (min)[a]		
	20°C	37°C	50°C
JP 1st fluid (pH 1.2)	2.0	2.1	6.2
Water	1.9	1.8	3.3
0.1 M Phosphate buffer (pH 7.5)	2.3	3.8	>60[b]
Kolthoff's buffer (pH 10)	2.0	2.5	40–45

[a] Average of six measurements.
[b] Film remained in small fragments.

Table 10.6 shows the dissolution time of HPMC films (6 mPa sec grade, 80 μm in thickness) at various pH and temperature conditions. In film coatings soluble in gastric fluid, the dissolution properties of the films over the entire biological pH range directly influence the bioavailability of the active ingredients. There was no marked difference at 20°C. At 37°C, slight prolongation of the dissolution time was observed in 0.1 M phosphate buffer (pH 7.5) and Kolthoff buffer (pH 10). The dissolution time was dramatically delayed at 50°C. These changes are due to a salting-out effect. At 50°C, the temperature is close to the thermal gelling temperature, so the film becomes less soluble and the films disintegrated but remained in small fragments. From these data, it is expected that the films can be readily dissolved in the stomach at 37°C.

APPLICATION OF HPMC TO FILM COATING OF PHARMACEUTICALS

HPMC forms transparent, tough, and flexible films from aqueous solutions. The films dissolve completely in the gastrointestinal tract at any biological pH, and HPMC provides good bioavailability of the active ingredients. The safety of HPMC has been proven by more than 40 years of application in the food and pharmaceutical industries. Animal toxicological studies of HPMC have been published since the 1950s. The most recent study was carried out under Good Laboratory Practice (GLP) [6].

EFFECT OF MOISTURE ON THE STABILITY OF ACTIVE INGREDIENTS

When aqueous coating first appeared in the pharmaceutical field, questions arose as to whether it could be applied to water-sensitive drugs and whether moisture absorption by the product during coating might degrade the drug. The results of studies on degradation of active ingredients, effect of moisture content, and long- term stability of aqueous film–coated tablets containing aspirin and ascorbic acid, both of which degrade in the presence of water, are given in Table 10.7. In these studies, almost no degradation of the active ingredients during coating was observed. The moisture content after coating was slightly lower than that before coating in this case. Moisture present in the tablet can be partially removed by drying during the coating process. Although tablets often take up moisture to various extents during the coating operation, the moisture content can be restored to the initial level through post-drying.

A slight decrease was observed in the content of active ingredients during a storage test, as shown in Table 10.7, but there was no difference between coated tablets and uncoated tablets, so the coating operation did not affect the stability of the active ingredients.

TABLE 10.7

Stability of Aspirin/Ascorbic Acid Tablets Coated with HPMC in an Aqueous System[a]

Storage Conditions	Items Analyzed (%)	Tablet Samples	After Coating	After 30 Days	After 90 Days
37°C, 75% RH	Salicylic acid	Uncoated	0.07	0.23	0.56
		HPMC-coated	0.09	0.24	0.53
	Ascorbic acid	Uncoated	8.55	8.27	8.25
		HPMC-coated	8.46	8.39	8.28
	Moisture	Uncoated	0.49	0.92	0.90
		HPMC-coated	0.18	1.16	1.21
37°C, in closed bottle	Salicylic acid	Uncoated	0.07	0.11	0.27
		HPMC-coated	0.09	0.13	0.22
	Ascorbic acid	Uncoated	8.55	8.56	8.54
		HPMC-coated	8.46	8.46	8.56
	Moisture	Uncoated	0.49	0.45	0.41
		HPMC-coated	0.18	0.20	0.21

Note: RH, relative humidity. Tablet formulation: acetyl salicylic acid (250 mg/tablet), ascorbic acid (27.5), microcrystalline cellulose (40.5), Talc (15.0), tablet weight (333 mg/tablet), tablet size (9.5 mm in diameter), Monsanto hardness (7–8 kg), disintegration time (1 minute), coating amount (3%), apparatus (Hi-Coater HCF 100).

[a] Tablets were coated with Pharmacoat 603.

SELECTION OF VISCOSITY GRADE

Among the many viscosity types of HPMC, the 15, 6, 4.5, and 3 mPa sec grades are popular for aqueous film coating with the 6 mPa sec grade being the most popular. The 3 mPa sec grade, having a low degree of polymerization, is capable of providing high-concentration polymer solutions, but film strength is quite inferior and peeling may occur during the coating operation if fragile tablets are used or if the pigment load is high. Thus, it is necessary to confirm, when using this viscosity grade especially, that such problems do not occur. In the case of 15 mPa sec grades, a high polymer concentration is difficult to use, and it is not economical. However, the film is so strong that it is sometimes useful for coating fragile tablets. The 4.5 mPa sec grade may be used to decrease the coating time without causing a decrease in film strength. For pellet coating, a low-viscosity coating solution is more appropriate in order to prevent the pellets from sticking during the coating operation. Therefore, the 3 mPa sec grade is suitable for pellet coating.

SELECTION OF ADDITIVES

Plasticizers are not required when tablets with sufficient hardness and low friability are used and little or no pigment is contained in the coating formulation. If fragile tablets are coated or if large levels of pigment are added to the coating formulation, the film will adhere poorly to the tablet surface, and film peeling may occur or engraving on the tablets may not appear sharp. These problems may be avoided by the addition of plasticizers. Polyethylene glycol (PEG), especially a high molecular weight type such as PEG 6000, is a suitable plasticizer. Liquid-type PEG, such as PEG 400, is also applicable particularly for peeling and for avoiding logo bridging. Although a greater effect is expected as the content of plasticizer increases, it should preferably be added at the minimum effective level (usually 20%–30% with respect to the polymer). Excessive amounts of plasticizer may cause tablet tacking, plasticizer bleeding, color depletion, or interaction with the active ingredients. Propylene glycol is also effective as a plasticizer to some extent but tends to volatilize during the coating process and storage.

If titanium dioxide or a lake pigment is used, it is necessary to first disperse it in water in a ball mill or colloid mill. As inter-brand differences are observed in the dispersion properties of titanium dioxide, switching to another brand is sometimes effective in improving the properties of the dispersion.

Lake pigments, such as erythrosine aluminum lake powder, are sometimes hard to wet. The addition of a small amount of alcohol to the pigments or the addition of surfactants to the water can aid dispersion. Water-soluble dyes have deep coloring effects but may color the tongue on oral administration of the coated preparations. The use of iron oxide pigments as coloring materials has become popular, but they are apt to precipitate in the coating solution, and comparatively strong agitation is required during the coating process.

To provide tablets with suitable slipping characteristics so that blister packaging can proceed smoothly, the addition of talc is also effective; 20% to 30% with respect to the polymer is sufficient for that purpose. In pellet coating, the addition of talc is effective for avoiding pellet tacking, but in this case, more than 100% with respect to the polymer may be required to give the best performance.

PREPARATION OF THE COATING SOLUTION

A typical concentration of 6 mPa sec grade of HPMC is approximately 6% (this is not always reflected in the coating examples described below) to form a smooth surface film in tablet coating. The concentration may be increased to 8% to 10%. Higher concentrations than this are not recommended. If the active ingredient is highly water-soluble and its content is very high, the active ingredient may dissolve in the spray mists during the operation, resulting in the active ingredient being included in the film. This is often inconvenient, especially if the active ingredient has a bitter taste. Although a method to prevent this phenomenon completely has not yet been found for all cases, a fairly effective method is to keep the particle size of the spray mist small and to use a low spray rate to maintain a dry core surface.

To dissolve HPMC in water, the HPMC powder is first dispersed into a partial amount (one half to one third of the total amount used) of hot water, and then cold water is added. A clear solution is obtained after cooling. The temperature of the hot water should preferably be over 70°C to prevent lumping. On a production scale, moderate agitation is better than vigorous agitation while the powder is being added to the hot water because vigorous agitation may cause severe foaming, which may be difficult to remove. If the polymer concentration is less than 10%, even if the dispersion contains some powder aggregates, it will turn into a clear solution within a day on standing at room temperature. Therefore, if the coating solution is to be employed the next day, hot water does not always have to be used. Long-term storage of a coating solution may result in mold formation. Although no means of complete prevention of mold growth has yet been found, the addition of sorbic acid (final concentration 0.1%) is effective.

COATING EQUIPMENT

Many types of equipment can be used for aqueous film coating. As a result of the high latent heat of water evaporation, the coating time depends on drying efficiency. A side-vented pan is most suitable for coating. For the spray equipment, an air-atomizing spray is recommended. In an airless spray system, which is useful for organic solvent coating, control of the spray rate is difficult and maintenance is time-consuming. Coating equipment is described in detail in a separate chapter of this book.

COATING OPERATION

In a typical coating procedure with a commercial-scale side-vented pan, tablets are preheated, and spraying is initiated when the outlet temperature rises above 40°C. The feed rate of the coating solution is controlled so as to keep the outlet temperature over 40°C. If slight picking occurs due to over-wetting as a result of improper operating conditions, the situation can be normalized by adjusting the processing parameters. However, extreme over-wetting will damage the whole batch, so the entire operation must be carried out with great care.

When using extremely fragile tablets, the initial tablet temperature should be increased (e.g., to 50°C). Then pan rotation should be initiated at low speed and spraying started simultaneously. If the outlet temperature decreases and picking occurs, both pan rotation and spraying are stopped, and the tablets are reheated. These processes are repeated several times. After the film has developed to a reasonable strength, the operation is continued under normal conditions.

Besides the side-vented pan, a conventional pan can also be used for coating. If a continuous spray causes over-wetting of the tablets in a conventional pan, intermittent spraying can be employed. In both cases, drying aeration should be continuous. When the tablets are fragile, the pan speed should not be increased until the film is partially formed. Tablets should be preheated to about 40°C and kept at this temperature during the coating operation. Picking may occur at lower temperatures.

The following examples of tablet coating illustrate the use of both laboratory-scale and production-scale machines.

Dria Coater (Powrex Co., Ltd., Japan)

DRC-1200 (120 kg batch size) and DRC-500 (5 kg batch size) were used. The formulation and properties of the tablets used in the study are shown in Table 10.8. The operating conditions are shown in Tables 10.9 and 10.10. The evaluation was performed by determining the surface roughness of coated tablets and color variation. The results are shown in Figures 10.7 and 10.8. These data suggest that in the first 30 minutes, tablets were a mixture of coated and uncoated tablets.

Hi-Coater (Freund Industry, Japan)

Vitamin B_2 tablets were coated with Pharmacoat 645 and 606 using a Hi-Coater. The operating conditions shown in Figure 10.9, which shows the release profiles of vitamin B_2 from the coated tablets at various pH. No significant difference in drug release was observed for different pH values (Tables 10.11 and 10.12).

TABLE 10.8
Formulation and Properties of Core Tablets

Tablet Formulation

Spray-dried lactose	79.5%
Cornstarch	15.0%
L-HPC	5.0%
Magnesium stearate	0.5%
Total	100.0%

Tablet Properties

Size	7 mm in diameter, 9 mm R
Weight	137.8 mg (CV = 1.68%)
Hardness	6.9 kg
Loss on drying	2.2%
Disintegration time	3.3 min
Friability	0.03% (Roche's friabilator, 25 rpm, 10 min, 20 tablets)
Surface roughness	R = 0.73 μm

Note: L-HPC, low-substituted hydroxypropylcellulose.

TABLE 10.9
Operating Conditions (Dria Coater DRC-1200)

	Parameter
Apparatus	DRC-1200 (Powrex, Japan)
Batch size	120 kg
Spray gun	Devilbiss × 2, nozzle diameter 1.4 mm
Spray air	Atomizer 3.2 kg/cm^2 250 L/min Pattern 3.4 kg/cm^2 200 L/min
Gun distance	25 cm
Drying air flow	30 m^3/min
Spray rate	220 g/min
Inlet air temperature	75°C
Outlet air temperature	51°C
Tablet temperature	46°C
Pan speed	8 rpm
Post-drying	50°C, 30 min

TABLE 10.10
Operating Conditions (Dria Coater DRC-500)

	Parameter
Apparatus	DRC-500 (Powrex, Japan)
Batch size	5 kg
Spray gun	Devilbiss × 2, nozzle diameter 1.4 mm
Spray air	Atomizer 3.2 kg/cm^2 250 L/min Pattern 3.4 kg/cm^2 200 L/min
Gun distance	25 cm
Drying air flow	3.5 m^3/min
Spray rate	25 g/min
Inlet air temperature	70°C
Outlet air temperature	51°C
Tablet temperature	43°C
Pan speed	15 rpm
Post-drying	50°C, 30 min

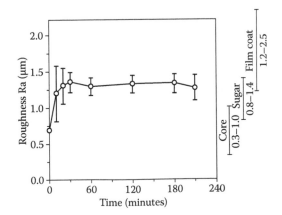

FIGURE 10.7 Changes in surface roughness of tablets coated with Pharmacoat 645. Surface roughness of 10 tablets was measured using a surface-measuring instrument (Surfcom 334A, Tokyo Seimitsu Co., Japan) and average roughness (Ra: μm) was calculated. Error bars indicate standard deviation.

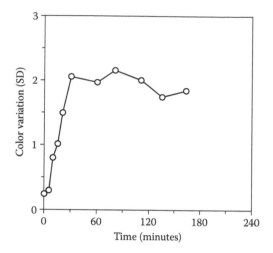

FIGURE 10.8 Changes in color variation of coated tablets. Intertablet color variation of 20 tablets was measured using a color computer (SM-4, Suga Test Instruments, Japan), and the color variation was evaluated in terms of the standard deviation of ΔE (Hunter Lab).

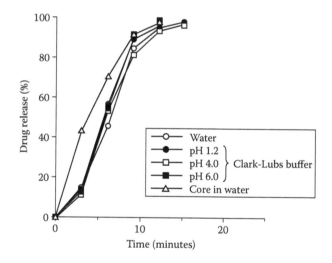

FIGURE 10.9 Dissolution characteristics of coated tablets. The coated tablet (190 mg) contains 3.2 mg of riboflavin (vitamin B_2). Paddle speed: 100 rpm.

PELLET COATING IN A FLUIDIZED BED

For pellet coating using a fluidized bed, care must be taken such that the pellets do not adhere to each other during the coating operation. For this reason, a low-viscosity HPMC such as Pharmacoat 603 or Methocel E3 is better to use than a higher viscosity grade. To avoid tacking, an inorganic compound such as talc should be added to the formulation or an organic solvent should be employed.

A recent study has shown that methylcellulose (MC), rather than HPMC, is useful for pellet coating because it has less stickiness [7]. MC is also water-soluble and has similar characteristics to HPMC. A low-viscosity grade of MC such as Metolose SM-4 (Shin-Etsu Chemical Co., Ltd., Tokyo, Japan) is commercially available for aqueous film coating. Figure 10.10 shows a comparison of agglomeration during pellet coating between MC and HPMC. The coating conditions are shown

---

TABLE 10.11
Operating Conditions (New Hi-Coater HCT-48N)

Coating Solution 1

Pharmacoat 606	6%
Water	94

Coating Solution 2

Pharmacoat 645	10%
Water	90

Operating Conditions

Apparatus	
Batch size	5 kg
Pan diameter	480 mm
Pan speed	16 rpm
Spray gun	ATF × 1, nozzle diameter, 1.2 mm
Spray air	150 L/min
Spray air pressure	2.0 kg/cm^2
Gun distance	15 cm
Spray rate	30 g/min
Airflow rate	2.5 m^3/min
Inlet air temperature	70°C
Outlet air temperature	47°C
Tablet bed temperature	40°C
Post-drying	50°C, 30 min

Results

	Pharmacoat	
	606	**645**
Coating time (3% coating-based tablet weight)	83 min	50 min
Coating solution consumption	2490 g	1500 g
Pharmacoat consumption	3.6 mg/tab	
Disintegration time		
Before coating	2.5 min	
After coating	3.9 min	

---

in Table 10.13. Using MC, the spray rate can be increased without granule agglomeration compared to HPMC. These results were from a top spray coating, but performance can be further improved by a bottom spray technique using a Wurster column. Table 10.14 shows an example of bottom spray coating.

POSSIBLE DIFFICULTIES IN AQUEOUS COATING USING HPMC

Various problems that arise during aqueous coating using HPMC can be attributed to an improper coating formulation or processing conditions. Some of the problems and suggestions to overcome them are discussed below.

Picking

Picking is the removal of film fragments from the tablet surface. It is caused by insufficient drying or excessive spraying and can be avoided by decreasing the spray rate and/or increasing the drying

TABLE 10.12

Operating Conditions (New Hi-Coater HC-130N)

	Operating Conditions
Apparatus	HC-130N (Freund Industry, Japan)
Batch size	120 kg
Pan diameter	1300 mm
Pan speed	8 rpm
Spray gun	AT × 3, nozzle diameter 1.2 mm
Spray air	Atomizer 170 L/min
Spray air pressure	2.0 kg/cm^2
Atomizer + pattern	250 L/min (at 5.3 kg/cm^2)
Gun distance	30 cm
Spray rate	70 g/min × 3
Airflow rate	15 m^3/min
Inlet air temperature	80°C
Outlet air temperature	47°C
Tablet bed temperature	46°C
Post-drying	50°C, 30 min

Results

	Pharmacoat	
	606	**645**
Coating time (3% coating-based tablet weight)	286 min	171 min
Coating solution consumption	60 kg	35.9 kg
Pharmacoat consumption	3.6 mg/tab Disintegration time	
Before coating	2.5 min	
After coating	3.9 min	

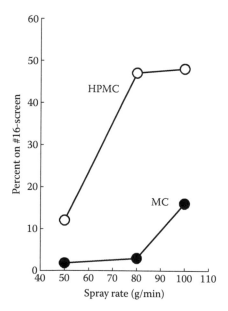

FIGURE 10.10 Effect of spray rate on agglomeration of pellets. Comparison between HPMC (Pharmacoat® 603) and MC (Metolose® SM-4). Conditions are provided in Table 10.13. HPMC, hydroxypropyl methylcellulose; MC, methylcellulose.

TABLE 10.13
Operating Conditions for Pellet Coating (Top Spray)

Formulation	
Core pellets	Spherical pellets containing theophylline (60%), 16-mesh pass
Coating solution	Pharmacoat 603 or Metolose SM-4, 7% aqueous solution
Operating Conditions	
Apparatus	FLO-5 (Freund Industry)
Batch size	5 kg
Spray nozzle	Schlick 1.2 mm in diameter
Spray air pressure	3.0 kg/cm^2
Gun distance	40 cm
Spray rate	50, 80, 100 g/min
Inlet air temperature	80°C
Outlet air temperature	47, 39, 35°C
Bed temperature	51, 44, 39°C
Coating amount	8%

TABLE 10.14
Operating Conditions for Pellet Coating (Bottom Spray)

Parameter	
Apparatus	Midi-Glatt with Wurster column
Batch size	200 g
Spray nozzle	970 with 0.5 mm port size
Spray air pressure	2.2–2.8 bar
Partition height	20 mm
Bottom plate	1-122-001-2-1
Retaining screen	100 mesh
Filter blow-out duration	Every 8 seconds
Process air flow	55.0 m^3/hours
Inlet air temperature	55°C
Product temperature	37°C–41°C
Spray rate	3.0–4.0 g/min

temperature or air flow. In some cases, a decrease in the concentration of the coating solution or the addition of sugar (over 10% with respect to HPMC) is effective. Some tablet formulations may suffer severe picking, and in these cases, the use of a high-viscosity grade may overcome the problem.

Cracking

Cracking can be observed during coating or storage of coated tablets. It occurs when the stress in the coating overcomes the tensile strength and adhesion of the coating film. The following suggestions, either alone or in combination, are effective in preventing cracking.

- Add plasticizers such as PEG 6000 (over 20% with respect to HPMC).
- Use a higher viscosity grade of HPMC.
- Use tablets with less friability. Friable loss reduces the adhesive strength between film and tablet.

Bridging

In the coating of engraved or scored tablets, the film often fails to follow the tablet contours. This occurs when the stress in the coating film overcomes the adhesive strength. Addition of PEG 6000 (20%–30% with respect to HPMC) can prevent this problem. It is also a good idea to adjust processing parameters to avoid overwetting.

Mottling

Mottling is a nonhomogeneous distribution of color on the surface of the tablet. To avoid this problem, the pigment should be dispersed completely before preparing the coating solution.

Orange Peel

Unsuitable formulation of the coating solution or coating operation frequently causes the surface of the coat to resemble the peel of an orange. Lowering the polymer concentration or decreasing the spray rate may prevent this problem. It can also be caused by an incorrectly adjusted coating apparatus, such as eccentric positioning of the needle in the gun nozzle or pulse pumping, which results in an unusual distribution of the spray mist.

Inter-Tablet Color Variation

Inter-tablet color variation corresponds to inter-tablet variation of coating. Changing the formulation of the coating solution rather than altering the operating conditions is the best way to prevent it. For example, the addition of titanium dioxide or an increase in its content or the use of a lake pigment instead of a water-soluble dye is effective although a slight change in the color tone may occur. These methods are based on reducing the dependency of color concentration on the coating amount.

HPMCAS: A POLYMER FOR AQUEOUS ENTERIC COATING

Hypromellose acetate succinate, also known as hydroxypropyl methylcellulose acetate succinate (HPMCAS), is an enteric aqueous coating polymer developed by Shin-Etsu Chemical Co., Ltd. in Japan. This enteric polymer is soluble in aqueous media at a pH higher than 5.5, owing to the presence of carboxyl groups. The chemical structure of HPMCAS is shown in Figure 10.11.

This material was first approved in 1985 in Japan and has been listed in *Japanese Pharmaceutical Excipients* (JPE) since 1988 and in *the National Formulary* (NF) since 2005.

PHYSICAL AND CHEMICAL PROPERTIES OF HPMCAS

HPMCAS for aqueous coating is a mechanically milled fine powder with an average particle size of approximately 5 μm that can be dispersed readily in water. The characteristics of HPMCAS

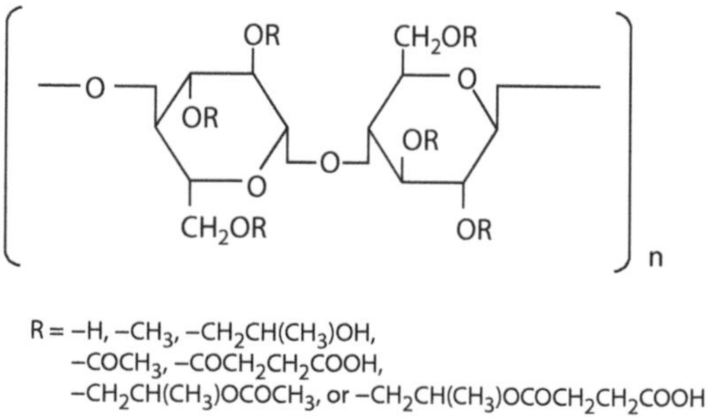

$R = -H, -CH_3, -CH_2CH(CH_3)OH,$
$-COCH_3, -COCH_2CH_2COOH,$
$-CH_2CH(CH_3)OCOCH_3, or -CH_2CH(CH_3)OCOCH_2CH_2COOH$

FIGURE 10.11   Chemical structure of hypromellose acetate succinate.

are related to the level of two substituents, that is, succinoyl and acetyl groups. Table 10.15 shows commercially available types of HPMCAS having different levels of content of substituents. There are three types—AS-LF, AS-MF, and AS-HF—depending on the ratio of succinoyl substitution to acetyl substitution (SA ratio). The SA ratio is highest in AS-LF, whereas AS-HF has the lowest SA ratio. Other specifications are also included in Table 10.15.

Figure 10.12 shows the equilibrium moisture content of HPMCAS at various humidities. Each type of HPMCAS differs in its equilibrium moisture content. These data indicate that the hydrophobicity of this polymer increases as the succinoyl content decreases or the acetyl content increases. The moisture content of AS-LF is similar to that of HPMCP under the same conditions. AS-MF and AS-HF exhibit lower equilibrium moisture contents.

Figure 10.13 shows the chemical stability of HPMCAS in comparison with other enteric polymers HPMCP and cellacefate, also known as cellulose acetate phthalate (CAP). As a measure of chemical stability, the formation of free acid from the polymers at 60°C, 100% RH was determined. The data indicate that HPMCAS is more stable than CAP and HPMCP.

Figure 10.14 shows the relationship between the dissolution time of HPMCAS films and substitution type. The polymer dissolves at the lowest pH for AS-LF, followed by AS-MF and AS-HF. The

TABLE 10.15
Specification of HPMCAS (Shin-Etsu AQOAT)[a]

Type	AS-LF	AS-MF	AS-HF
Acetyl content (%)	5.0–9.0	7.0–11.0	10.0–14.0
Succinoyl content (%)	14.0–18.0	10.0–14.0	4.0–8.0
Methoxyl content (%)	20.0–24.0	21.0–25.0	22.0–26.0
Hydroxypropoxyl content (%)	5.0–9.0	5.0–9.0	6.0–10.0
Viscosity[b]		2.4–3.6 cP	
Heavy metals	Not more than 10 ppm		
Free acid	Not more than 1.0%		
Loss on drying	Not more than 5.0%		
Residue of ignition	Not more than 0.2%		

Note: HPMCAS, hypromellose acetate succinate.
[a] Commercial name of Shin-Etsu Chemical Co., Ltd.
[b] 2% in NaOH solution, Ubbelohde viscometer, 20°C.

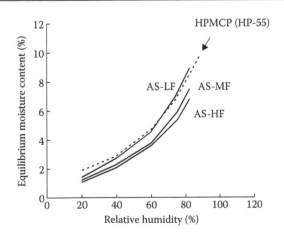

FIGURE 10.12 Equilibrium moisture content of HPMCAS and HPMCP.

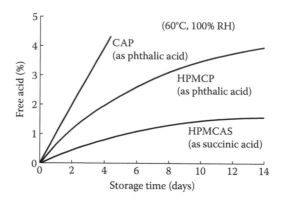

FIGURE 10.13 Formation of free acid from enteric coating polymers.

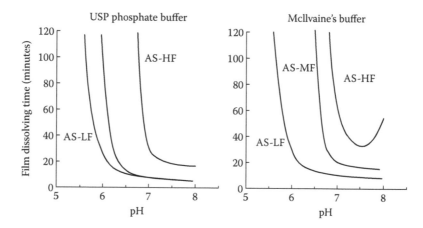

FIGURE 10.14 pH-dependent dissolution patterns of films prepared from various types of HPMCAS in USP phosphate buffer and McIlvaine's buffer. HPMCAS, hypromellose acetate succinate; USP, United States Pharmacopeia.

pH value at which the polymer dissolves depends on the buffer system, but the order is not different. The ionic strength of the buffer seems to affect the dissolving profile. There is a unique phenomenon for the dissolution of AS-HF, when the time for the polymer to dissolve increases as the pH increases above 7 in a high concentration buffer.

To further explain the mechanism of the pH-dependent solubility of HPMCAS, the electrolytic properties of carboxylic groups in the polymer were investigated. pH titration data provide some useful information on this matter. An aqueous dispersion of HPMCAS was directly subjected to pH titration. It took three to four minutes, however, to reach pH equilibrium, so continuous titration was not possible. Thus, the titration was carried out by adding a calculated amount of alkali (NaOH) to the polymeric dispersion to make a certain degree of neutralization (α) and measuring the pH of the dispersion at equilibrium. The results are shown in Figure 10.15. At $\alpha = 0.5$, the equilibrium pH was the highest for AS-HF, followed by AS-MF and AS-LF, indicating that the nature of the dissociation is different. In the case of a monobasic weak acid, pH at $\alpha = 0.5$ is equivalent to its pK_a. Assuming that this theory can be applied to HPMCAS, the results indicate that the dissociation constant is the highest for AS-HF, followed by AS-MF and AS-LF. In Figure 10.15, solid lines represent the regions in which the polymer is soluble.

Figure 10.16 shows the pH titration curves of HPMCAS (AS-LF) in the presence of NaCl. The pH at $\alpha = 0.5$ was decreased as the salt concentration increased, which indicates greater dissociation

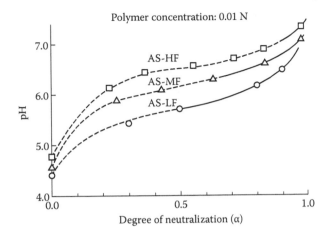

FIGURE 10.15 pH titration curves of HPMCAS. HPMCAS, hypromellose acetate succinate.

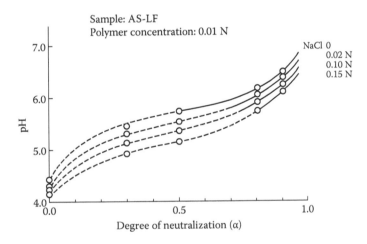

FIGURE 10.16 Effect of NaCl on pH titration of HPMCAS. HPMCAS, hypromellose acetate succinate.

at higher salt concentrations. Such a phenomenon is due to the difference in exchange of the sodium ion and the proton at the carboxyl group and is common to polyelectrolytes. Similar patterns were obtained with other types of HPMCAS.

Figure 10.17 shows the relationship between equilibrium pH at $\alpha = 0.5$ and the SA ratio in the presence or absence of NaCl. As the SA ratio was increased, the pH decreased. These results suggest that dissociation increases as the number of carboxylic groups increases and that an increase in acetyl groups inhibits the dissociation at the succinoyl group.

Partially neutralized HPMCAS films, having various α values, were then prepared and their dissolution behavior in water or 0.1N NaCl solution was investigated. Figure 10.18 shows the dissolution time of the films as a function of the degree of neutralization. There were thresholds of degree of neutralization for rapid dissolution of the films. It is suggested that as the polymer becomes more hydrophobic as the result of increasing the acetyl group content, a greater degree of neutralization is required for its dissolution. In the presence of NaCl, the threshold was shifted to high values for every polymer type, which was probably due to salting out.

Figure 10.19 shows the pH titration curves for all types of HPMCAS in the presence or absence of NaCl. At equilibrium, α defines (or reflects) the pH of the solution. Assuming that dispersed

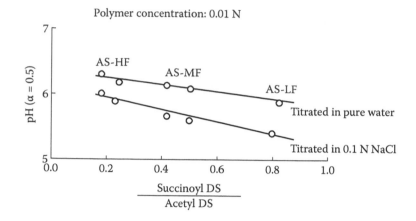

FIGURE 10.17   Correlation of equilibrium pH at $\alpha = 0.5$ and SA ratio of HPMCAS.

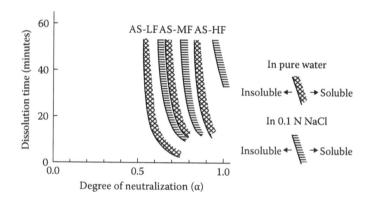

FIGURE 10.18   Effect of neutralization and the addition of NaCl on the dissolution pattern of HPMCAS.

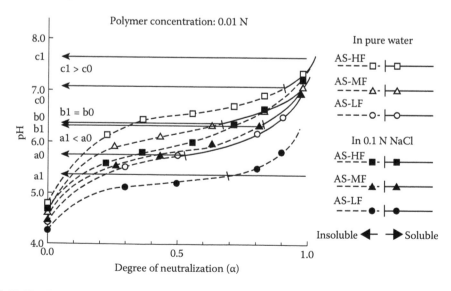

FIGURE 10.19   Relationship between pH titration curve and dissolution of HPMCAS.

particles and films dissolve in the same manner, the dissolution pH of AS-LF films is predicted to be approximately 5.8 in the absence of NaCl. In the presence of NaCl, this value is decreased. On the contrary, the AS-HF films are predicted to dissolve at pH 7.0 in the absence of NaCl, and its dissolution pH in the presence of NaCl is more than 7.0. Thus, the pH at which HPMCAS dissolves depends on the SA ratio. This is probably related to the amount of acetyl group, which gives the hydrophobic nature to the polymer.

FILM FORMATION OF HPMCAS FROM AQUEOUS MEDIA

Selection of a Plasticizer

The selection of a suitable plasticizer is important for coating with aqueous dispersions of HPMCAS because the polymer will not form a film without being plasticized. Triethyl citrate (TEC) has been found to be a suitable agent for HPMCAS based on the results of experiments described below.

Effect of Plasticizers on Film Appearance

Aqueous dispersions of HPMCAS and micronized HPMCP powder containing various plasticizers (20%–100% based on polymer) were cast on a glass plate and dried at 40°C to form films. Clear films were obtained from dispersions containing TEC, triacetin, and propylene carbonate at 30% or higher (Table 10.16).

Stability of a Plasticizer

Tablet coating was then performed by spraying the polymeric dispersions containing the plasticizers described above. The content of plasticizer was 30% with respect to the polymer. After coating, the amounts of polymer and plasticizer on the coated tablets were determined. The results are shown in Table 10.17. TEC and triacetin remained in the coated tablets in the same proportion as in the coating dispersions. Propylene carbonate remained to the extent of only 51% as it evaporated during the spraying process. Tablets coated with HPMCAS plasticized with propylene carbonate showed insufficient gastric resistance. The content of TEC in the coated tablets was not changed after storage for 15 days at 40°C and 75% RH whereas the content of triacetin decreased to 70%, probably due to degradation during the storage. Therefore, TEC was found to be the most suitable plasticizer for HPMCAS.

TABLE 10.16

Effect of Various Plasticizers on Film Formation from Aqueous Dispersions of HPMCAS (AS-MF) and HPMCP

	HPMCAS				HPMCP			
Plasticizers	20%	30%	50%	100%	20%	30%	50%	100%
Dibutyl phthalate	N	N	N	T	N	N	N	N
Triethyl citrate	T	C	C		T	C	C	
Triacetin	T	C	C		T	C	C	
Ethylene glycol monoethyl ether	N	N	N	C	N	N	N	C
Propylene carbonate	T	C	C		T	C	C	

Note: C, clear film formed; HPMCAS, hypromellose acetate succinate; HPMCP, hypromellose phthalate; N, no film formed; T, turbid film formed.

TABLE 10.17

Recovery of Plasticizers after Coating and Storage and Gastric Resistance of Coated Tablets

	Plasticizer		
	Triethyl Citrate	**Triacetin**	**Propylene Carbonate**
Recovery after coating (%)[a]	99	96	51
Recovery after storage (%)[b]	98	70	–
Gastric resistance[c]	Good	Good	Poor

Note: HPMCAS, hypromellose acetate succinate; RH, relative humidity.

[a] Tablets were coated with dispersion of the following formulation: HPMCAS (AS-MF 10%), plasticizer (3%), water (87%).

[b] 40°C, 75% RH, 15 days.

[c] Tablets were subjected to disintegration test in simulated gastric fluid (the first fluid in Japanese Pharmacopeia) to see if the tablets disintegrate (poor) or remain intact for one hour (good).

PROPERTIES OF TEC AS A PLASTICIZER FOR HPMCAS

Figure 10.20 shows the solubility limits of HPMCAS and HPMCP in TEC as functions of temperature and water content of TEC. HPMCAS dissolved in anhydrous TEC at 23°C, and as the water content was increased, this temperature limit of solubility shifted upward. HPMCP dissolved at 90°C in anhydrous TEC, but the curve showed a minimum at a water content of 3%. Because the usual temperature for tablet coating is 30°C to 40°C, the solubility range of HPMCAS is wider than that of HPMCP, so that films from HPMCAS form more readily than from HPMCP. Figure 10.21, which covers TEC concentrations in the coating dispersion, shows that the solubility of TEC in water increased at lower temperatures.

FILM FORMATION

The mechanism of film formation in aqueous latex systems has been suggested by several researchers. The particles get closer during the drying process and the capillary force makes the particles

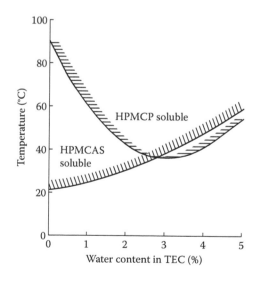

FIGURE 10.20 Solubility of HPMCAS (AS-MF) and HPMCP as a function of temperature and water content of TEC. HPMCAS, hypromellose acetate succinate; HPMCP, hypromellose phthalate; TEC, triethyl citrate.

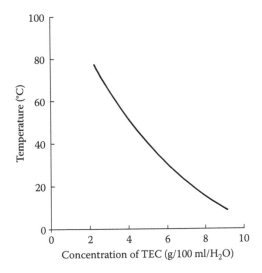

FIGURE 10.21   Solubility of TEC in water. TEC, triethyl citrate.

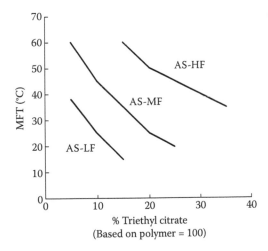

FIGURE 10.22   Minimum film formation temperature of HPMCAS at various plasticizer content.

eventually coalesce with each other. It is considered that this theory can be applied for the film formation of HPMCAS, but due to its larger particle size compared with other latex emulsions, the mechanism may be slightly different. A suggested theory of film formation from the aqueous dispersion of HPMCAS is that the plasticizer is separated from the water phase during the drying process and it dissolves or gelates the particles of HPMCAS. The particles then fuse to each other to form a film. Figure 10.22 shows the minimum film formation temperature of HPMCAS at various plasticizer contents.

MECHANICAL PROPERTIES OF HPMCAS FILMS

Table 10.18 shows the mechanical properties of free films of enteric polymers. Except for HPMCP, the films were prepared by spraying the dispersion onto a Teflon sheet, under a controlled spray rate

TABLE 10.18

Mechanical Properties of Enteric Polymeric Films

Polymer	Plasticizer	Tensile Strength (MPa)[a]	Elongation (%)[a]
HPMCAS AS-LF	Triethyl citrate 20%	15.4 ± 1.0	2.8 ± 0.2
AS-MF	Triethyl citrate 28%	16.2 ± 0.7	2.9 ± 0.2
AS-HF	Triethyl citrate 30%	13.4 ± 0.9	3.4 ± 0.2
HPMCP HP-55[b]	None	18.5 ± 1.0	5.4 ± 0.8
CAP (Aquateric)[c]	Diethyl phthalate 35%	6.7 ± 1.4	9.2 ± 2.0
Eudragit L 30 D-55[d]	Triethyl citrate 20%	25.4 ± 5.0	2.3 ± 0.6

Note: CAP, cellulose acetate phthalate; HPMCAS, hypromellose acetate succinate; HPMCP, hypromellose phthalate.

[a] Mean ± SD of at least five experiments.

[b] Cast film from organic solution.

[c] Cellacefate commercially available from FMC Corporation. Films were prepared by spraying aqueous dispersion.

[d] Acrylic resin commercially available from Evonik-Degussa GmbH. Films were prepared by spraying aqueous dispersion.

and temperature [8]. The cast method, which is commonly used for film preparation, is difficult to apply for aqueous dispersions of HPMCAS. In this method, the dispersed particles settle during the drying process, resulting in heterogeneous film formation [9].

AQUEOUS ENTERIC COATING USING HPMCAS

PREPARATION OF COATING DISPERSIONS

Table 10.19 gives a conventional enteric coating formulation using HPMCAS. The optimum content of plasticizer (TEC) depends on the type of HPMCAS; 20% based on polymer weight is suitable for AS-LF, 28% for AS-MF, and 35% for AS-HF.

Figure 10.23 shows how to prepare the coating dispersion. Prior to adding ingredients, temperature of the water should be below 25°C (a). Under stirring, dissolve TEC and sodium lauryl sulfate in the water first (b). After TEC is completely dissolved, add the powder of HPMCAS and talc gradually (c). After the powder is uniformly dispersed, the coating fluid is ready to use (d). The dispersion should be gently stirred throughout the coating process so that the dispersed particles do not settle. It is also recommended that the dispersion be kept at a temperature below 25°C to avoid coagulation of polymer and plasticizer. When the dispersion is pumped to a gun nozzle at high temperature, coagulation sometimes happens inside the nozzle, which will lead to a gun blockage. This drawback has recently been improved by using an alternative method, which will be discussed later.

TABLE 10.19

A Conventional Coating Formulation

Ingredients	
HPMCAS (Shin-Etsu AQOAT) AS-MF	7.0%
Triethyl citrate	1.96%
Talc	2.1%
Sodium lauryl sulfate	0.21%
Water	88.73%
Total	100.0%

Note: HPMCAS, hypromellose acetate succinate.

(a)   (b)

(c)   (d)

FIGURE 10.23 Preparation of coating dispersion (conventional method).

TABLET COATING

There are several key points to be noted for coating processes using HPMCAS:

1. The spray gun needs to be closer to the tablet bed surface than for an organic solvent coating.
2. The following "two-stage" coating is recommended. In the initial stage, which uses approximately 25% of the total coating dispersion, spray slowly and keep the tablet surface relatively dry. The conditions in this stage are similar to those used in coatings with HPMC. A thin layer of polymer surrounds the cores and protects the tablet surface from over-wetting in the next stage. After a weight gain of 2% has been applied, double the spray rate. The outlet product temperatures should be approximately at 40°C.
3. Once the desired amount of coating has been applied, the tablets must be dried to complete the coalescence. This typically takes about 30 minutes at an inlet temperature of 70°C until the outlet temperature reaches 50°C.

Table 10.20 shows characteristics of core tablets in this example. Core tablets should have low friability to remain intact during aqueous coating. If defective tablets are found in the coating process, a subcoating with HPMC is recommended.

TABLE 10.20
Formulation and Properties of Core Tablets (Placebo)

Ingredients	
Lactose	73%
Corn starch	18%
Povidone (K30)	3%
L-HPC (LH-11)	5%
Mg stearate	1%
Total	100%

Note: Diameter: 8 mm; thickness: 4 mm; radius: 12 mm; weight: 200 mg; hardness: 100N; disintegration time: 3 min; friability: 0.05%. L-HPC, low-substituted hydroxypropylcellulose.

The process conditions for lab-scale and production-scale machines are shown in Tables 10.21 and 10.22, respectively. Figure 10.24 shows coating amount and gastric resistance for tablets. After enteric coating, additional "over-coating" is often effective for the prevention of tacking during accelerated testing under high temperature and humidity.

FLUID-BED COATING

Table 10.23 shows the formulation of pellets used in the study. Table 10.24 shows processing conditions for granule coating using fluid-bed coating machines. In the fluid-bed coating, the spray gun

TABLE 10.21
Conditions for Tablet Coating at Laboratory Scale

Apparatus	New Hi-Coater HCT-48N (Freund Industry, Japan)	
Batch size	5 kg	
Spray gun	ATF × 1	
Nozzle diameter	1.8 mm	
Nozzle distance	16 cm from bed surface	
Spray air pressure	200 kPa	
Spray airflow	150 L/min	
	Initial Stage	**Second Stage**
Pan speed	16 rpm	20 rpm
Inlet air temperature	75°C	79°C
Inlet airflow	2.5 m³/min	3.0 m³/min
Outlet air temperature	42°C	40°C
Spray rate	25 g/min	45 g/min
Spray time	35 min	89 min
Coating amount—weight gain	0%–2%	2%–11%
Final coating amount—weight gain		11%
Total spraying time		124 min
Post-drying		Inlet 75°C, 30 min

TABLE 10.22
Conditions for Tablet Coating at Production Scale

Apparatus	New Hi-Coater HCF-100N (Freund Industry, Japan)	
Batch size	60 kg	
Spray gun	ATF × 2	
Nozzle diameter	1.8 mm	
Nozzle distance	20 cm from bed surface	
Spray air pressure	400 kPa	
Spray airflow	160 L/min	
	Initial Stage	**Second Stage**
Pan speed	16 rpm	20 rpm
Inlet air temperature	70°C	80°C
Inlet airflow	13 m³/min	15 m³/min
Outlet air temperature	41°C	40°C
Spray rate	110 g/min	220 g/min
Spray time	80 min	233 min
Coating amount—weight gain	0%–2%	2%–11%
Final coating amount—weight gain		11%
Total spraying time		313 min
Post-drying		Inlet 80°C, 30 min

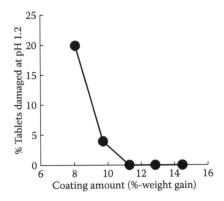

FIGURE 10.24 Gastric resistance of tablets coated with HPMCAS (AS-MF). One hundred tablets were treated with a simulated gastric fluid (pH 1.2) containing a red dye, which colored defective positions of the coated surface. Data represents the number of defective tablets.

TABLE 10.23
Formulation and Properties of Core Pellets (Digestive Enzyme)

Ingredients	
Pancreatin	60.0%
Lactose	25.6%
Corn starch	6.4%
Hydrated silicone dioxide[a]	5.0%
HPC[b]	3.0%
Total	100.0%

Note: Pellet diameter: 0.8 mm; shape: cylindrical; disintegration time: 11 min.
[a] Carplex®, Shionogi, Japan.
[b] Hydroxypropylcellulose Type L, Nisso, Japan.

TABLE 10.24
Conditions for Pellet Coating with a Fluidized Bed (Laboratory Scale)

Parameter	
Apparatus	Flow Coater FLO-1 (Freund Industry, Japan)
Batch size	1.5 kg
Spray gun	Schlick × 1
Nozzle diameter	1.8 mm
Nozzle distance	12 cm from bed surface
Spray air pressure	200 kPa
Spray airflow	120 L/min
Inlet air temperature	80°C
Inlet air flow	2.5 m³/min
Outlet air temperature	36°C
Product temperature	34°C
Spray rate—polymer dispersion	60 g/min
Spray time	71 min
Coating amount—weight gain	32%
Post-drying	Inlet 75°C, 30 min

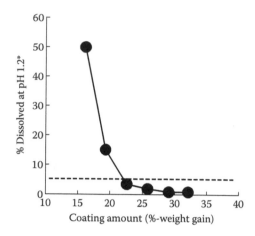

FIGURE 10.25 Gastric resistance of pellets coated with HPMCAS (AS-MF). *Percentage dissolved in a simulated gastric fluid (pH 1.2) at two hours.

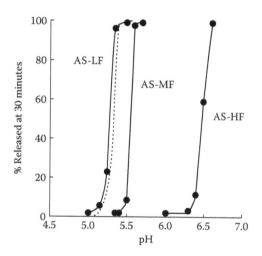

FIGURE 10.26 Dissolution of vitamin B_2 at 30 minutes from granules coated with HPMCAS at various pH. Dotted line represents granules coated with HPMCP (HP-55) for comparison. HPMCAS, hypromellose acetate succinate; HPMCP, hypromellose phthalate.

needs to be set in a closer position as mentioned in the tablet coating, but two-step coating is not necessary. Figure 10.25 shows gastric resistance at various coating amounts. Figure 10.26 shows the release of vitamin B_2 coated with HPMCAS at various pH in comparison with HPMCP-coating.

AQUEOUS COATING WITH A CONCENTRIC DUAL-FEED SPRAY NOZZLE

Since HPMCAS reached the marketplace, several pharmaceutical manufacturers have launched their products successfully using it. It should be pointed out, however, that this product has a major drawback, that is, the requirement for cooling. Increase in temperature of the coating dispersion causes coagulation of the polymer and plasticizer, which sometimes leads to gun blocking. Therefore the temperature of the dispersion during the coating process has been recommended to be below 25°C, preferably less than 20°C.

One method was developed to overcome this drawback [10]. Because the coagulation occurs by strong binding of the polymer and plasticizer at high temperature, in the new approach, the plasticizer is sprayed separately to avoid coagulation. Using this technique, nozzle clogging does not occur, and it is not necessary to chill the coating dispersion. Moreover, the polymer can be applied in greater concentrations than in the conventional method. Therefore, shorter processing times can be achieved.

Figure 10.27 shows the whole scheme of this coating method. The plasticizer (TEC) and the polymeric dispersion without plasticizer are separately sprayed using a newly developed "concentric dual-feed spray nozzle." Figure 10.28 shows the structure of the spray nozzle. Currently, this type of nozzle is commercially available from Spraying Systems Co., Japan. This nozzle has a triple layer tip, consisting of an inner nozzle tip for the polymer dispersion without plasticizer, a middle one for the plasticizer, and an outer one for the atomizing air. Two pumps are required, and each of them should be set to a proper speed to supply materials in a desired ratio. Table 10.25 shows processing parameters in tablet coating. The polymer concentration can be increased up to 15% whereas the conventional aqueous dispersion coating can only use 7% at maximum. Figure 10.29 shows

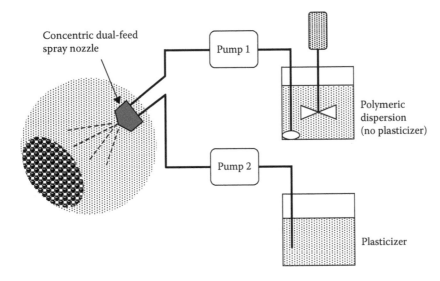

FIGURE 10.27 Aqueous coating of HPMCAS using a concentric dual-feed spray nozzle.

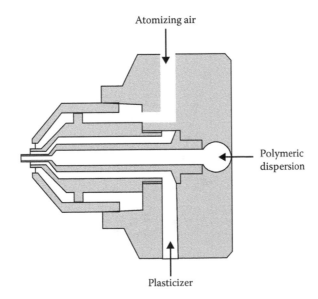

FIGURE 10.28 Concentric dual-feed spray nozzle.

TABLE 10.25

Conditions for Tablet Coating Using a Concentric Dual-Free Spray Nozzle

Coating Formulations

Polymer dispersion	
HPMCAS (AS-MF)	15.0%
Talc	4.5%
Sodium lauryl sulphate	0.15%
Water	80.35%

Plasticizer

Triethyl citrate	28% with regard to HPMCAS

Operating Conditions

Apparatus	New Hi-Coater HCT-48N (Freund Industry)
Batch size	5 kg
Spray nozzle	ATFM concentric dual spray nozzle × 1
Nozzle distance	6 cm from bed surface
Spray air pressure	300 kPa
Spray airflow	100 L/min

	Initial Stage	Second Stage
Pan speed	16 rpm	20 rpm
Inlet air temperature	70°C	75°C
Inlet airflow	2.4 m³/min	2.8 m³/min
Outlet air temperature	43°C	38°C
Spray rate—polymer dispersion	25 g/min	50 g/min
Spray rate—plasticizer	1.05 g/min	2.10 g/min
Spray time	13 min	47 min
Coating amount—weight gain	0%–1.6%	1.6%–13%
Post-drying		Inlet temperature 75°C, 30 min
Final coating amount—weight gain		13%
Total processing time		60 min

Note: HPMCAS, hypromellose acetate succinate.

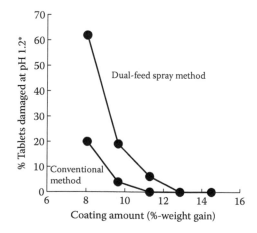

FIGURE 10.29 Gastric resistance of tablets coated with HPMCAS (AS-MF) using a concentric dual-feed spray nozzle. *One hundred tablets were treated with a simulated gastric fluid (pH 1.2) containing a red dye, which colored defective positions of the coated surface. Data represents the number of defective tablets.

gastric resistance at various coating amounts in comparison with the conventional and the dual-feed method. Because gastric resistance is better at a lower polymer concentration, the dual-feed method requires slightly more coating to obtain sufficient gastric resistance. However, because the polymer concentration is higher, the processing time is significantly shorter.

PARTIALLY NEUTRALIZED COATING FORMULATIONS

More user-friendly aqueous coating techniques using HPMCAS are continuously being developed. Partially neutralized formulation is one example of that approach. Using this method, there is no need to chill the coating fluid, and the polymer concentration can be higher than that used in the conventional formulation. Table 10.26 gives an example of a partially neutralized formulation using ammonia. The ammonia partially neutralizes the acidic groups of HPMCAS to stabilize the polymeric dispersion. Recently, it was found that basic amino acids such as L-arginine are also suitable as a neutralizing agent to provide a stable dispersion in combination with a plasticizer. This approach may be preferred as ammonia may be difficult to handle at the production scale due to its volatile nature and hazardous vapor. Table 10.27 and Figure 10.30 show an example formulation and procedure for using L-arginine as a neutralizer. The plasticizer is added at the end of the dispersion process. When the dispersion was passed through #60 mesh, no coagulation was found. Moreover, the polymer concentration was able to be increased to 12% without any nozzle clogging. Table 10.28 shows the coating condition used to coat the tablets along with the performance of the resulting products. The weight gain of the tablets was around 8% in terms of polymer content and the coated tablets went through rigorous analytical testing to evaluate the enteric effect of HPMCAS. Figure 10.31 shows the drug release when coated onto aspirin tablets.

TABLE 10.26
A Partially Neutralized Coating Formulation Using Ammonia

Ingredients	
HPMCAS (AS-MF)	10%
Triethyl citrate	1.5%
Talc	3.0%
Sodium lauryl sulfate	0.3%
Ammonia	0.04% (*)
Water	85.16%
Total	100%

*Equivalent to 20% neutralization level of succinoyl group. Added as NH_4OH.

TABLE 10.27
A Partially Neutralized Coating Formulation Using L-Arginine

Ingredients	
HPMCAS (AS-MF)	12%
Triethyl citrate	3.6%
Talc	3.6%
Sodium lauryl sulphate	0.12%
L-Arginine	0.33%
(TiO₂ or Iron oxide)	0.5%
Water	ad. 100

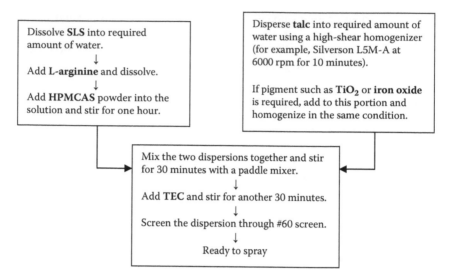

FIGURE 10.30   Preparation of partially neutralized coating fluid using L-arginine.

TABLE 10.28

Conditions for Tablet Coating Using Partially Neutralized Formulation with L-Arginine and the Performance of the Coated Products

	Parameter
Core tablets	Placebo tablets (8 mm-d, 200 mg /tab)
Coating condition	
Apparatus	GMPC-1 (Glatt)
Batch size	500 g
Pan size	0.8 L
Spray nozzle diameter	0.8 mm
Nozzle distance	10 cm
Atomizing pressure	0.6 bar
Pan speed	7–13 rpm
Inlet temperature	50°C–55°C
Outlet temperature	33°C–36°C
Spray rate	3.5–4.5 g/min
Total weight gain	13% (8% polymer basis)
Results	
Gastric resistance (pH 1.2)	Intact after 2 hours
Uptake of gastric fluid (pH 1.2)	3.5%–4.3% after 1 hour
Disintegration time at pH 6.8	7–8 min.
Stability in uptake of gastric fluid	2.6% (after 3 months, 40°C, 75% RH)

APPLICATION FOR pH-DEPENDENT SUSTAINED-RELEASE DOSAGE FORM

Enteric coatings are used for the protection against digestive enzymes and also can be used for sustained-release dosage forms by combination with uncoated components. The pH-dependent sustained-release dosage forms are less popular due to the patient-to-patient variation in gastric pH; however, these systems have been extensively used for antibiotics because these dosage forms have few bioavailability problems.

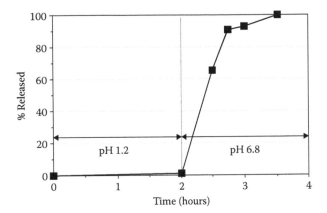

FIGURE 10.31 Drug release of enteric-coated aspirin tablets using HPMCAS partially neutralized with L-arginine. Aspirin tablets (content 260 mg, tablet weight 314 mg/tab) were coated in the same condition as above.

Figure 10.32 shows a typical blood concentration curve of a pH-dependent sustained release formulation [11]. It consists of uncoated pellets and enteric-coated pellets in a certain ratio. Although the uncoated components rapidly release the active ingredient, the coated component is emptied from the stomach over a wide time span. The coated pellets release the active ingredient at a certain pH in the small intestine. As the emptying time of each pellet is variable, the blood concentration profile exhibits a broad curve. To get a desired release pattern, it is important to design the ratio of the uncoated and coated pellets and to select a polymer type that has a suitable dissolution pH.

The following is an example of a pH-dependent sustained release dosage form of cephalexin, a widely used antibiotic, coated with HPMCAS [12]. The cephalexin pellets were prepared by an extrusion-spheronization process, and HPMCAS was coated on the pellets using a fluidized bed. The coating amount was 25% (polymer basis) with respect to the core pellets. Three types of coated pellets were prepared using AS-LF, AS-MF, and AS-HF.

An in vitro evaluation was performed using a dissolution test (Japanese Pharmacopoeia, rotary basket method). The results are shown in Figure 10.33. The graph represents the relationship between the dissolution at 30 minutes (D_{30}) and pH of the test fluid. A difference was seen in the dissolution pH of each substitution type.

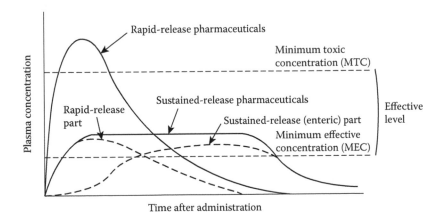

FIGURE 10.32 Blood concentration time curves after administration of sustained release pharmaceutical dosage forms.

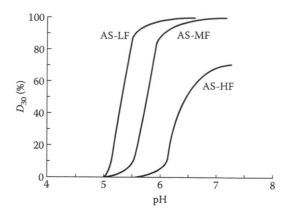

FIGURE 10.33 Dissolution profiles of cephalexin pellets coated with HPMCAS.

The pellets were then administered to human volunteers and urinary samples were collected to determine the urinary excretion rate of cephalexin. The results in Figure 10.34 represent the urinary excretion rate–time curve of the core pellets and coated pellets. AS-HF had the most delayed peak compared with the other grades. This type was then used as an enteric polymer for this dosage type.

The pellets were then administered to human volunteers, and urinary samples were collected to determine the urinary excretion rate of cephalexin. The results in Figure 10.35 represent the urinary excretion rate–time curve of the core pellets and coated pellets. AS-HF had the most delayed peak compared with the other grades. This type was then used as an enteric polymer for this dosage form. The coated pellets and core pellets were blended in a mixture at a ratio of 3:7. This form was next administered to the volunteers and the urinary excretion rate of cephalexin was determined. The results in Figure 10.35 indicate that sustained absorption was achieved. As a control, pH-independent sustained-release pellets were prepared to compare bioavailabilities. The pellets were prepared by coating the pellets with ethylcellulose. Three samples, having different release rates, were prepared and administered to healthy volunteers. Table 10.29 shows the T_{max} (time at the peak of blood drug concentration) and total urinary recovery of cephalexin from the pH-dependent type and the pH-independent type. The total urinary recovery was decreased as the T_{max} was delayed, but the pH-dependent type had significantly higher urinary recovery, which is

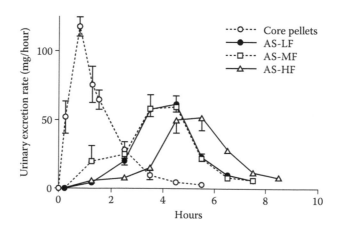

FIGURE 10.34 Urinary excretion rate–time curve of cephalexin pellets uncoated or coated with HPMCAS. Pellets (0.5 g cephalexin) were administered to healthy volunteers after a light meal. Data represent mean ± standard error (n = 3).

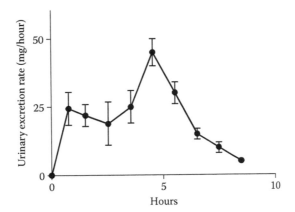

FIGURE 10.35 Urinary excretion rate–time curve of pH-dependent controlled release cephalexin pellets using HPMCAS (AS-HF). Uncoated:coated = 3:7. Pellets (0.5 g cephalexin) were administered to healthy volunteers after a light meal. Data represent mean ± standard error (n = 3).

TABLE 10.29
Total Urinary Recovery and T_{max} of Drug After Oral Administration of Cephalexin Pellets Coated with HPMCAS or Ethylcellulose

Coating Material	T_{max} (Hour)	Total Urinary Recovery (%)[a]
Uncoated	0.75	94
HPMCAS		
AS-LF	4.5	99
AS-MF	4.5	99
AS-HF	5.5	91
Ethylcellulose		
15%	1.5	98
20%	2.5	75
25%	3.5	40

[a] The amount of cephalexin in urinary samples was determined by high-performance liquid chromatography.

equivalent to bioavailability in this case, than the pH-independent type with a longer T_{max}. This difference is considered to be due to cephalexin having a narrow absorption window in the upper area of the small intestine [11], and time-dependent release pellets tend to pass the absorption window before drug release has been completed.

REFERENCES

1. Singiser RE. Japanese Patent 37-12294, 1962.
2. Rowe RC. Molecular weight and molecular weight distribution of hydroxylpropyl methylcellulose used in the film coating of tablets. *J Pharm Pharmacol* 1980; 32:116–119.
3. Kato T, Tokuya N, Takahashi A. Measurements of molecular weight and molecular weight distribution for water-soluble cellulose derivatives used in the film coating of tablets. Kobunshi Ronbunshu 1982; 39:293–298.
4. Callahan JC, Cleary GW, Elefant M et al. Equilibrium moisture content of pharmaceutical excipients. *Drug Dev Ind Pharm* 1982; 8:355–369.

5. Hawes MR. The effect of some commonly used excipients on the physical properties of film forming used in the aqueous coating of pharmaceutical tablets. Paper at the Panel of the Pharmaceutical Society of Great Britain, 1978.

6. Obara S, Muto H, Shigeno H et al. A three-month repeated oral administration study of a low viscosity grade of hydroxypropyl methylcellulose in rats. *J Toxicological Sci* 1999; 24:33–43.

7. Kokubo H, Obara S, Nishiyama Y. Application of extremely low viscosity methylcellulose (MC) for pellet film coating. *Chem Pharm Bull* 1998; 11:1803–1806.

8. Obara S, McGinity JW. Influence of processing variables on the properties of free films prepared from aqueous polymeric dispersions by a spray technique. *Int J Pharm* 1995; 126:1–10.

9. Obara S, McGinity JW. Properties of free films prepared from aqueous polymers by a spraying technique. *Pharm Res* 1994; 11:1562–1567.

10. Brunemann J, Nishiyama Y, Kokubo H. A plasticizer separation system for aqueous enteric coating using a concentric dual-feed spray nozzle. *Pharm Tech Europe* 2002; 14:41–48.

11. Maekawa H, Takagishi Y, Iwamoto K et al. Cephalexin preparation with prolonged activity. *Jpn J Antiblot* 1977; 30:631–638.

12. Shin-Etsu Chemical Co., Ltd., Internal document.

11 Applications of Formulated Systems for the Aqueous Film Coating of Pharmaceutical Oral Solid Dosage Forms

Ali R. Rajabi-Siahboomi and Thomas P. Farrell

CONTENTS

INTRODUCTION

Aqueous coating technology remains the main option for film coating of oral solid dosage forms. This is irrespective of the purpose of the film coating applications—that is, for conventional or modified-release film coatings. The main reasons for its continued popularity are the environmental limitations of organic solvents used and recent advances in the formulation of aqueous film coating materials as well as major improvements made in the coating machines and their ancillaries.

    This chapter reviews the use of formulated aqueous coating systems for the following:

- Conventional film-coating systems (immediate release)
- Enteric film-coating systems (delayed release)
- Barrier membrane controlled-release film-coating systems (extended release)

Prior editions of this book have included chapters covering aqueous film-coating formulations for immediate release (Opadry® and Opadry II), delayed release (Acryl-EZE® and Sureteric®), and sustained release (Surelease®) applications. This chapter serves to update the previous editions and provide a summary of recent advances made in the formulation and application of aqueous film coatings.

AQUEOUS FILM COATING FOR IMMEDIATE-RELEASE FORMULATIONS

As detailed in prior editions of this book [1,2], the original Opadry formulations, introduced in late 1970s, were comprised of low-viscosity hypromellose (HPMC), plasticizers, and pigments. These fully formulated Opadry formulations provided numerous advantages versus the use of individual raw materials, including the reduction of the number of raw materials for QC testing, reduced dispersion preparation time, consistent color-matched formulations, good processability, excellent appearance on tablets and good mechanical film properties. Opadry formulations have enjoyed widespread, successful use globally and are still found today on a great variety of marketed products. The solid levels of dispersions of HPMC-based Opadry formulations must be kept in the range of 10%–15% by weight in water to achieve processible dispersion viscosities of 300–600 centipoise. In order to increase productivity by decreasing coating time and/or increasing spray rate, the Opadry II family of products comprising HPMC and polysaccharides were introduced in the 1980s. With Opadry II, processible dispersions can be obtained at 20% rather than 10%–15% solids, which allows for the productivity increase. In addition, the inclusion of polysaccharides increased the adhesion of Opadry II coating systems to the substrates.

    Although HPMC-based Opadry II formulations have also been very successful commercially, unmet needs of the pharmaceutical industry emerged in the last two decades for which additional fully formulated film coating options were required. Given the advent of direct-to-consumer advertising for both OTC and prescription pharmaceutical products globally and especially in the United States, coatings providing enhanced aesthetic characteristics (e.g., clear logo definition, gloss, and pearlescence) were sought to establish unique brand identity. In addition, *functional* attributes (e.g., moisture and oxygen protection) were also sought within immediate-release film coatings in order to preserve labile active pharmaceutical ingredients. In response, film coatings were developed that improve aesthetic characteristics of dosage forms and provide functional benefits.

    Further advances and evolution in the development of fully formulated aqueous film coatings led to new film coatings based on polyvinyl alcohol (PVA) and sodium carboxymethycellulose

TABLE 11.1

Moisture and Oxygen Permeability of Pure Polymeric Films (Film Thickness = ~100 Microns)

Polymer	WVTR (25°C/80% RH) (Grams H_2O/100 in²/day)	O_2TR (at 25°C/60% RH) (cm³ O_2/100 in²/day)
NaCMC	96	0.04
HPMC	30	10.0
HPC	15	11.0
PVA	10	0.04

Note: HPC, hydroxypropylcellulose; HPMC, hypromellose; NaCMC, sodium carboxymethylcellulose; O_2TR, oxygen transmission rate; PVA, polyvinyl alcohol; WVTR, water vapor transmission rate.

(NaCMC). Film coatings comprising PVA offer the formulator the same or greater production conveniences afforded to them when using Opadry formulations containing hypromellose (HPMC) and also provide functionality previously unrealized. PVA-based films are known to have relatively low moisture vapor and oxygen permeability. On the other hand, NaCMC-based films have low oxygen permeability but relatively high water vapor permeability (Table 11.1).

PVA-BASED FILM COATINGS

Opadry amb and Opadry II 85 series are two different families of PVA-based products that were commercialized in the mid-to-late 1990s. The Opadry amb ("aqueous moisture barrier") formulation was optimized to provide the low water vapor transmission rate (WVTR) while still affording all the conveniences of fully formulated film-coating systems. It is supplied as a color-matched system and can be readily dispersed into water at the 20% solids level. Owing to the inherent tackiness of the PVA polymer, the maximum achievable spray rates obtained with Opadry amb are not as high as those of HPMC-based Opadry II film coatings. The Opadry II 85 series family of products was developed to provide high productivity (significantly higher spray rate than Opadry amb), ease of application, and moisture barrier properties.

FORMULATING AND COATING MOISTURE-SENSITIVE PRODUCTS

The fact that PVA-based films have inherently low WVTR does not guarantee that dosage forms coated with film coatings comprising PVA will be stable in moisture-rich environments. The ultimate stability of moisture-sensitive products is dependent on both formulation and processing variables including (a) core excipients and manufacturing conditions (e.g., relative humidity of the facility), (b) film-coating formulations, (c) film-coating process parameters, and (d) packaging materials [3]. Core excipients possessing low water activity are preferred in the development of moisture-sensitive formulations because they can preserve an active from hydrolytic degradation by tightly binding moisture in the core. One example of an excipient with low water activity is Starch 1500® (a partially pregelatinized maize starch), which has been shown to reduce the hydrolytic degradation rate of acetylsalicylic acid and ranitidine hydrochloride [4–6].

The preservation of dosage forms from the deleterious effects of water by coating with a PVA-based film coating also has been demonstrated. In one case, the hydrolytic degradation rate of acetylsalicylic acid was decreased [7]. In another, the stability of a powdered Echinacea extract, which normally liquefies in a period of a few hours at ambient conditions, was dramatically extended to 18 months by layering the Echinacea powder onto nonpareil beads and then coating with an Opadry II 85 series formula [8].

The ultimate moisture content of coated tablets is also significantly influenced by coating process parameters. Coating process parameters were systematically studied through a carefully constructed design of experiments. Spray rate, airflow, and inlet temperature were found to significantly influence both moisture content and aesthetics of coated products [9]. Most importantly, it was found that the moisture content of a dosage form could be maintained or even decreased during an aqueous film-coating process. Depending on the airflow and inlet temperatures utilized, the moisture content of the coated multivitamin varied between 0.6% and 2.7% versus the starting moisture content of 1.4% in the uncoated core. Therefore, it is possible to coat moisture-sensitive products using aqueous film coating processes and avoid the use of potentially hazardous organic solvents, which historically have been selected in this type of application. In a separate study, the use of a pre-drying step was also found to significantly impact the final moisture content of coated tablets [10].

COATING OXYGEN LABILE DOSAGE FORMS

PVA-based film coatings generally have significantly lower oxygen transmission rates than HPMC-based film coatings and usually provide sufficiently high oxygen barriers for most drug products. Still, even greater oxygen barriers are sometimes required. Although NaCMC-based Opaglos 2 and Opadry fx coating formulations were originally developed to enhance the aesthetic characteristics of dosage forms, both film coatings also possess excellent oxygen barrier properties as well. These properties have been demonstrated both in the measurement of oxygen transmission rates [11] and also in the preservation of ibuprofen under conditions known to result in oxidation of the active [12]. In the latter study, the ibuprofen assay of uncoated cores and cores coated with an HPMC-based Opadry coating (3% weight gain) was only about 80% after storage in a convection oven at 60°C for 11 weeks. In contrast, the ibuprofen assay of cores coated with Opadry fx (3% weight gain) was still about 99% under the same conditions (Figure 11.1). Additional considerations when using NaCMC-based film coatings are that NaCMC is an ionic polymer (low aqueous solubility at pH < 2, which may affect disintegration) and forms relatively viscous solutions. Therefore, it is generally recommended that NaCMC-based film coatings be applied as 7.5% w/w aqueous dispersions and to a relatively low weight gain of 0.5%–1.0% if they are being used for aesthetic characteristics alone.

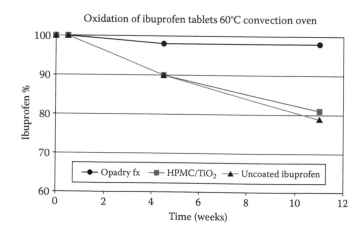

FIGURE 11.1 Oxidation of ibuprofen tablets. Ibuprofen tablets uncoated, coated with a standard Opadry film coating comprising hypromellose and titanium dioxide, or coated with Opadry fx. Film coatings applied to 3% weight gain.

FURTHER CONSIDERATIONS FOR DRUGS PRONE TO DEGRADATION

There are a significant number of small molecules that are susceptible to degradation by hydrolysis and oxidation and, for a subset of these molecules, coatings with still lower moisture and oxygen permeation rates are desired. In addition to the potential reaction of water and active oxygen with APIs, film-coating components or impurities within those components have been determined to react with drug molecules in some cases. For example, formic acid and formaldehyde, determined to be impurities in multiple excipients [13,14], have been shown to react with APIs. A degradation product was formed during the long-term stability studies (of the low-dose formulation of Avapro (irbesartan) film-coated tablet [15]. The degradant was identified as the hydroxymethyl derivative (formaldehyde adduct) of the drug substance, irbesartan, and the authors concluded that the formaldehyde was an impurity found in polyethylene glycol (PEG). The degradant was eliminated by removing PEG from the coating formulation.

Another research group determined that both formic acid and formaldehyde jointly participated in degradation chemistry [16]. Significant degradation of the amine-based smoking cessation drug varenicline tartrate in an early development phase osmotic controlled-release (CR) formulation yielded predominantly two products: N-methylvarenicline (NMV) and N-formylvarenicline (NFV). NMV was produced by reaction of the amine moiety with both formaldehyde and formic acid in an Eschweiler-Clarke reaction, and NFV was formed by reaction of formic acid alone with varenicline. It was postulated that both formaldehyde and formic acid were formed from oxidative degradation of PEG, which was used in the osmotic coating. The authors also concluded that the degradation chemistry was heavily dependent on the physical state of the PEG. When the concentration of PEG in the coating was sufficiently low, the PEG remained phase compatible with the other component of the coating (cellulose acetate) such that its degradation (and the resulting drug reactivity) was effectively eliminated. To further ensure that PEG oxidation would not occur to any significant extent, oxygen scavengers were added to the product packaging.

In light of the potential for API reactions with water, active oxygen species, and film coating components/impurities, an established need was the development of film coatings that have a low reactivity profile and also protect susceptible APIs from degradation by ambient moisture and oxygen. In response to this market need, a significantly enhanced moisture barrier film coating system was developed by Colorcon called Opadry amb II. It has excellent moisture barrier properties (Figure 11.2), good productivity (Table 11.2), and a low reactivity profile due to the absence of PEG [17]. The primary film former in Opadry amb II is PVA, and, ordinarily, PEG is required as a

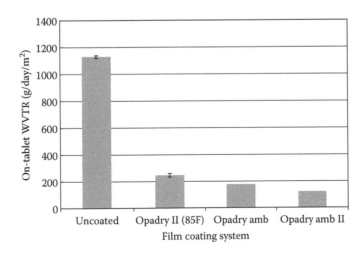

FIGURE 11.2 On-tablet water vapor transmission rates of uncoated and coated placebos (3% coating weight gain; 40°C, 75% humidity).

TABLE 11.2

Comparison of Typical Film Coating Spray Rates (Coatings Dispersed at 20% w/w in Water)

Formula	15″ Pan (g/min)	24″ Pan (g/min)	48″ Pan (g/min)
Opadry amb II	20	60	250–300
PVA-based Opadry II	20	60	260–430
Opadry amb	10	30	130–215

de-tackifier to achieve high spray rates. In the case of Opadry amb II, glyceryl monocaprylocaprate (GMCC) is used as a de-tackifier and has less potential for degradation versus PEG.

During the development of Opadry amb II, a refined on-tablet water vapor transmission rate (WVTR) methodology was also developed as a potentially more relevant method than the traditional measurement of WVTR on free films [18]. The moisture uptake of coated and uncoated placebos was evaluated in a dynamic vapor sorption–desorption (DVS) apparatus. The placebos were 10 mm, round, biconvex tablets containing lactose (69.4%), Starch 1500 partially pre-gelatinized maize starch (15.0%), microcrystalline cellulose (15.0%), magnesium stearate (0.5%), and fumed silica (0.1%). Both uncoated and coated tablets were initially dried at 40°C and 0% relative humidity (RH) until an equilibrium dry condition of less than 0.0002% weight loss per minute was obtained for 10 minutes. The tablets were then exposed to a 40°C and 75% RH environment, and the moisture uptake was recorded gravimetrically as a function of time until the rate of change in moisture uptake rate was less than 0.0002% over a 10-minute period. A steady-state moisture uptake rate was determined from the slope of the predominantly linear portion of the moisture uptake profile. The on-tablet water vapor transmission rate (WVTR) through the coating was determined by normalizing the steady state moisture uptake rate by the tablet surface area. The on-tablet WVTR of at least three tablets from each coating trial was determined, and the average value and standard deviation were reported as shown in Figure 11.2.

Within this study, it was also determined that the average on-tablet WVTR and variability decrease with increasing weight gain as shown in Figure 11.3. The decreases in average on-tablet WVTR were 40% and 27% when the weight gain was increased from 2% to 3% and 3% to 4%, respectively, suggesting that the reduction of WVTR would plateau if weight gain were increased further. Formic acid and formaldehyde impurities were also measured in Opadry amb II after storage for six months at accelerated storage conditions. Formic acid levels remained essentially unchanged

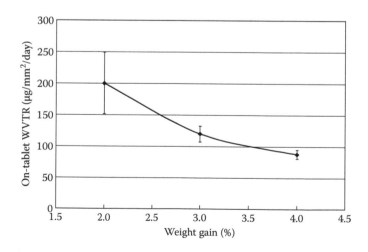

FIGURE 11.3 On-tablet WVTR vs. coating weight gain for Opadry amb II. DVS conditions: 40°C, 75% RH.

TABLE 11.3
Formic Acid and Formaldehyde Levels in Opadry amb II and PVA-Based Opadry II after Storage for Six Months at Accelerated Conditions

	Formic Acid Results (ppm)	
Storage Condition	Opadry amb II	PVA-Based Opadry II
Initial	28	21
After six months at 30°C/65% RH	22	62
After six months at 40°C/75% RH	25	72
	Formaldehyde Six Months Results (ppm)	
Storage Condition	Opadry amb II	PVA-Based Opadry II
Initial	7	3
After six months at 30°C/65% RH	<5	10
After six months at 30°C/65% RH	<5	7

for Opadry amb II over the six-month period, regardless of storage conditions whereas the formic acid content increased over time at accelerated storage conditions in PVA-based Opadry II due to PEG degradation (Table 11.3). Formaldehyde levels remained at relatively low levels in both coating types regardless of storage conditions. Taken together, these data indicate that Opadry amb II may offer protective advantages for drugs that may react with water, formic acid, and formaldehyde.

FILM COATING FORMULATIONS AND PROCESSES FOR HIGHER PRODUCTIVITY

Pharmaceutical dosage form manufacturers are continuing to strive for the development of the most efficient unit operations that allow maximum throughput and return on equipment investments. The film-coating process is no exception. The time it takes to apply a given amount of film coating on a dosage form is dependent on two main factors: the concentration of the film coating in the aqueous dispersion (often referred to as solids content) and the spray rate (Table 11.4). The maximum film-coating concentration in an aqueous dispersion is often governed by the viscosity of the dispersion. For most film-coating processes, the processible viscosity is about 300–600 centipoise. Depending on polymer type and use level in the film-coating formulation, this target viscosity is achieved at different solids levels. For many HPMC-based formulations (e.g., basic Opadry formulations), this target viscosity is reached at relatively low film-coating concentrations of 10%–15% solids. For HPMC-based Opadry II formulations, in which polysaccharides replace a portion of the HPMC, and for PVA-based Opadry II formulations, film-coating concentrations of at least 20% are possible. Some PVA-based film-coating

TABLE 11.4
Impact of Film-Coating Process Parameters on Coating Time (Example: 24″ Coating Pan; 14 kg Pan Load; 3% Coating Weight Gain)

Film Coating Concentration in Water (%)	Aqueous Dispersion to Spray (kg)	Spray Rate (g/min)	Total Coating Time (min)	Process Time Savings versus Base Case (%)
15 (base case)	2.80	60	46.7	–
20	2.10	60	35.0	25
20	2.10	45	46.7	–
25	1.68	60	28.0	40
30	1.40	60	23.3	50

formulations may be applied at concentrations of 25%–30% in standard perforated film-coating pans to achieve color uniformity with significant reductions in coating time compared to coating dispersions applied at lower solids concentrations; however, optimization of pan speed, spray rates, and spray distribution is essential when applying high solids coating dispersions [19]. It is stressed that color uniformity will usually be achieved in significantly less time than actual coating weight uniformity, and it is the latter that impacts the film-coating robustness and barrier properties. Therefore, the film-coating process parameters must be carefully optimized in view of the objectives of coating in the first place. In some cases, such as achieving maximum barrier properties or high gloss [20], applying a film coating at lower solids and a slower spray rate may be the best option.

Continuous-coating processes offer still greater potential improvements in film coating productivity. In one study, it was determined that a 3% weight gain of film coating could be applied in a standard 48″ perforated, batch-coating pan and a continuous coater in 40 and 10 minutes, respectively [21]. The reason for this is that the number of tablets per unit time presented to the spray zone is substantially higher in continuous-coating equipment than in typical batch coating pans. Both PVA- and HPMC-based film coatings have been successfully coated using continuous coaters, but film coatings that can be applied at higher concentrations and at faster spray rates are typically better suited for continuous coating. Optimized coating formulations based on an ethylene glycol and vinyl alcohol graft copolymer have been recently commercialized as Opadry QX by Colorcon that can be applied up to 35% solid levels [22], which makes such formulations highly desirable for use in continuous-coating machines. This significantly higher productivity of this novel film-coating system is postulated to be due to its low dispersion viscosity and surface activity of the polymer, enabling optimized droplet formation during spray and rapid spreading of droplets on surface of the substrate, forming a very smooth appearance film.

COATINGS FOR DIETARY SUPPLEMENTS

Dietary supplement manufacturers continue to request film-coating formulations specific to their needs. Both Opadry NS and Nutraficient® were designed as immediate-release coating systems specifically for nutritional, herbal, and dietary supplement products regulated as foods. The systems include food-approved, label-friendly raw materials. Taking this one step further, Opadry NutraPure™ is an aqueous, clear USDA Certified Organic coating developed in response to the dietary and food supplement market demand for even more natural and label-friendly coating alternatives. The coating creates an exceptionally glossy surface for tablets, similar to the finish of sugar coating, and contains only food-approved, plant-based ingredients. Opadry Nutrapure has no components derived from corn or soy, lessening any concerns over allergens, and also no components derived from genetically modified organisms (GMOs) for consumers who may have related concerns. As with most other clear coatings, it is recommended that Opadry Nutrapure be coated at relatively low solids (e.g., 12%) to achieve maximum gloss values.

CONCLUSIONS FOR IMMEDIATE RELEASE–FORMULATED FILM-COATING SYSTEMS

Fully formulated film-coating systems have been developed to meet evolving customer demands for active preservation, aesthetics, and productivity improvement. Regulatory requirements and consumer preferences in the dietary supplement market often necessitate the customization of formulas that have the broadest appeal possible while still retaining key technical and appearance attributes.

FILM COATING FOR MODIFIED-RELEASE FORMULATIONS

The applications of aqueous modified-release film-coating formulations depend on the type of polymers used, the majority of which are water-insoluble (at low pH media for enteric polymers).

Therefore, aqueous polymeric dispersions in the form of a latex or pseudolatex are used for pharmaceutical functional film-coating systems.

FORMULATED AQUEOUS ENTERIC FILM COATING SYSTEMS

Enteric coating systems are a subset of modified-release film coatings, which are intended to remain intact (and thus prevent any drug from being released or any acidic media being absorbed) for different periods of time after ingestion but ultimately to dissolve in order to permit the drug to be rapidly released thereafter. These formulations are also referred to as delayed-release systems, with which the delay of the onset of drug release, after ingestion, will depend on the type of the polymer used in the film coating and the transit of the dosage form through the gastrointestinal tract. Although USP has set forward specific disintegration and dissolution testing for enteric-coated tablets (USP monographs <701> and <711>, respectively), these methods do not quantitatively measure the acid media that may have been absorbed into the tablet core (i.e., acid uptake) while residing in the stomach. This is especially important to measure when an acid labile drug is in the core formulation; therefore, a reproducible method to measure percentage of acid taken up (absorbed) by an enteric-coated tablet is described.

METHOD FOR ACID UPTAKE MEASUREMENT

Accurately weigh six to 50 tablets (W_0) and expose to the acid media (0.1N HCl or pH 4.5 acetate buffer) for two hours at 37°C in a disintegration apparatus. The tablets should remain intact if enteric coating is successful. Then remove the tablets, pat dry to remove surface moisture and reweigh (W_t). From the differences in weights before and after exposure to acid media, the percent acid uptake may be calculated as shown in Equation 11.1:

$$\% \text{ acid uptake} = [(W_t - W_0)/W_0] \times 100 \qquad (11.1)$$

The lower the percentage of acid uptake, the more effective the enteric coating will be in protection of the drug in the core. However, values of up to 10% acid uptake have been shown to be acceptable for protecting highly acid labile drugs, such as proton pump inhibitors [23].

The highly functional nature of enteric coatings dictates that some film characteristics are extremely important, including possession of the following:

- Good mechanical properties that guarantee reproducibility and ruggedness (toughness) in performance.
- Permeability characteristics of the film that ensure the quantities of drug released through intact films are low and meet compendial requirements.
- Good polymer chemical stability, a characteristic that also helps to ensure that predictable product performance is achieved. Therefore, performance of the product does not change with time of storage.

COMMERCIALLY AVAILABLE AQUEOUS ENTERIC SYSTEMS

There are two formulated enteric systems available from Colorcon Inc. for aqueous enteric coatings of oral solid dosage forms: tablets, granules, pellets, and capsules. Enteric coating of capsules is more challenging due to potential migration of plasticizer leading to a brittle film; however, there are examples in which capsules have been successfully enteric coated [24]. One system uses polyvinylacetate phthalate, PVAP (Phthalavin®) as the enteric polymer and is called Sureteric, and another uses methacrylic acid copolymer type C (EUDRAGIT® L 100-55) called Acryl-EZE. Details of Sureteric coating systems have been discussed in a previous edition of this book [1].

Here in this chapter, the dispersion characteristics and applications of Acryl-EZE systems will be discussed.

Acryl-EZE Formulated Aqueous Enteric Systems

Acryl-EZE is a family of products developed and patented by Colorcon Inc. (US patent number 6,420,473 B1). These systems are all available in dry powder form, ready for dispersion in water and are formulated using the enteric polymer methacrylic acid copolymer type C (EUDRAGIT L100-55). Formulations may contain a plasticizer, neutralizing agent, flow aid, surfactant (wetting agent), de-tackifier, and pigment blends. The presence and level of each ingredient depends on the formulation, which is prepared on case-by-case basis to meet the requirements of specific application and specific regional regulatory compliance (for pigmented systems). These requirements may vary from color matching to pH trigger point that is the pH of the media at which the film starts to dissolve rendering the dosage form for disintegration, de-aggregation, and finally dissolution of the medicament.

Acryl-EZE powder is reconstituted to 20% w/w solids dispersion and applied on tablets at a weight gain range of 7% to 10% for enteric performance, depending on the size (surface area), mechanical, and other surface properties of the tablets.

The recommended storage conditions for the Acryl-EZE family of products are below 30°C/65% relative humidity, which provides a 12 month retest interval from the date of manufacturing.

Preparation of a Typical Acryl-EZE Dispersion

- Determine the amount and weigh the Acryl-EZE powder and water required to make a dispersion with 20% w/w solids based on the quantity of tablets to be coated and the target coating weight.
- Stir the water to form a vortex using a propeller stirrer and add the Acryl-EZE powder to the center of the liquid vortex in a slow steady stream, avoiding clumping and maintaining a vortex (similar to preparation of Opadry systems).
- Continue stirring for 20 minutes, and then pass the dispersion through a 250-micron sieve to remove any un-dispersed powder agglomerates or large particles prior to the coating process. This will prevent potential spray nozzle blockage during the coating process.
- Start the coating process, continue stirring the dispersion while the coating is ongoing to prevent settlement of suspended material.

Note: If a high-shear mixer is used to prepare the dispersion, only 10 minutes of stirring is required, and an antifoam emulsion, such as simethicone (at 0.5% w/w of the Acryl-EZE solids) should be added to water prior to the preparation.

Coating Process Recommendations

In order to maximize coating efficiency and prevent spray drying, a product bed temperature range of 30°C–32°C is recommended, keeping the atomization pressure low (typically 1.5–2 bar). Examples of typical coating-process parameters and enteric performance of Acryl-EZE applications are shown in the case studies below.

Case Study 1: Enteric Coating of 81-mg Aspirin Tablets Using Acryl-EZE

In this case study, tablet formulation, coating processes, and enteric performance of 81-mg aspirin tablets are demonstrated. The tablet formulation comprising 81-mg aspirin, partially pre-gelatinized

TABLE 11.5

Coating Process Conditions for Enteric Coating of 81-mg Aspirin Tablets with Acryl-EZE (Manesty AC-150)

Process Parameter	Values
Average inlet temperature (°C)	52.7
Average exhaust temperature (°C)	36.2
Average tablet bed temperature (°C)	30.0
Average spray rate (g/min)	343
Atomizing air pressure (bar)	3.0
Fan air pressure (bar)	2.5
Pan speed (rpm)	7
Airflow (m³/hour)	2600

TABLE 11.6

Enteric Performance of 81-mg Aspirin Tablets Coated with Acryl-EZE

Theoretical Weight Gain (%)	% Acid Uptake	% Released in 0.1N HCl after Two Hours	$T_{80\%}$ in Phosphate Buffer (pH = 6.8)
6	4.05	0.0	<30 minutes
7	2.76	0.0	<30 minutes
8	2.37	0.0	<30 minutes
9	2.56	0.0	<30 minutes
10	2.34	0.0	<30 minutes

maize starch (Starch 1500), microcrystalline cellulose, and stearic acid were prepared (170 mg tablet weight and 7.0 mm diameter, standard convex). The tablets were then enteric coated with pigmented Acryl-EZE 930 (20% coating dispersion), taking samples at 5%, 6%, 7%, 8%, 9%, and 10% theoretical weight gains. The tablet coating (pan load of 130 kg) was carried out in a 48″ side-vented pan (Accelacota-150, Manesty) equipped with four spray guns with individual peristaltic pumps. The coating process parameters used during the enteric coating are shown in Table 11.5. The total coating process time was 3.65 hours.

The tablets were tested for acid uptake and enteric performance after manufacture and three months storage at 40°C/75% RH (85 cc foil sealable HDPE bottles). Table 11.6 shows that the acid uptake for 81-mg aspirin enteric-coated tablets was less than 5% when the tablets had 6% or more coating weight gains. There was no detectable drug release in 0.1N HCl acid phase for two hours, and more than 75% of the drug released in pH 6.8 phosphate buffer phase within 90 minutes, meeting the USP 38 requirements for delayed-release aspirin tablets. Dissolution and free salicylic acid content testing on the coated tablets after three months of storage at 40°C/75% RH indicated excellent stability results and passed the USP delayed-release requirements [25].

ENTERIC COATING OF PROTON PUMP INHIBITORS

Proton pump inhibitors (PPIs) are used in the treatment of acid-related gastro-duodenal disorders by reducing gastric acid secretion. PPIs are substituted benzimidazoles, and each share a similar core structure and mode of action but differ in substituent groups [26,27]. The type of substituents affects the chemical properties of the compounds, which directly influence their rates of reactions and therefore their stability in different media [28]. The stability of PPIs in aqueous media is a function

TABLE 11.7
Delayed Release Solid Oral Dosage Forms of Proton Pump Inhibitors (PPIs) Available in the U.S. Market

Proton Pump Inhibitor (PPI)	Proprietary Name (U.S.)	Manufacturer	Solid Dosage Form	Strength (mg)
Omeprazole	Prilosec®	AstraZeneca	Capsule	10, 20, 40
Lansoprazole	Prevacid®	TAP Pharmaceuticals	Capsule, MUPS tablet	15, 30
Rabeprazole sodium	Aciphex®	Janssen	Tablet	20
Pantoprazole Sodium	Protonix®	Wyeth-Ayerst	Tablet	20, 40
Esomeprazole magnesium	Nexium®	AstraZeneca	Capsule	20, 40

of pH with an increased rate of degradation as the pH decreases. Consequently, most oral dosage forms of PPIs are formulated as enteric-coated granules, tablets, and multi-particulates. Table 11.7 shows the current PPIs available in tablet and capsule forms in the United States.

Multiple-dose treatment of patients with PPIs results in a decrease in their gastric acid secretion with a subsequent elevation in gastric pH [29,30]. Due to the rise in gastric pH, the enteric-coated dosage form when administered will be subjected to a higher pH environment than is typically found in a healthy, fasted stomach (simulated in vitro utilizing 0.1N HCl, USP). Therefore, in the following two case studies, different acid phases—0.1N HCl (pH 1.2) and/or pH 4.5 acetate buffer (USP)—have been investigated for acid uptake, enteric protection testing, and subsequent drug release in higher pH (phosphate buffers) to better simulate the gastric environment of patients who are administered multiple doses of this class of medication.

Case Study 2: Enteric Coating of 20-mg Rabeprazole Sodium Tablets with Acryl-EZE

Tablets of 20 mg rabeprazole sodium (Na) were prepared (total weight 146 mg, 6.3 mm diameter) by an organic wet granulation method and seal-coated with alcoholic ethylcellulose/magnesium oxide 1:1% w/w (theoretical weight gain of 1.37% or 1.5 mg/cm^2) as described by Saeki et al. [31]. The enteric coating was applied using Acryl-EZE 93F at various weight gains of 8.1%, 10.1%, 12.1%, and 14.1% in a partially perforated coating pan (LDCS5, Vector Corporation). Table 11.8 shows the coating process parameters used in this case study. The percentage of acid uptake for the enteric-coated

TABLE 11.8
Coating Process Parameters Used for Enteric Coating of Rabeprazole Tablets with Acryl-EZE (Vector LCDS5)

Parameter	Values
Pan volume (L)	1.3
Pan charge (kg)	1
Inlet temperature (°C)	63
Outlet temperature (°C)	35
Fluid delivery rate (g/min)	12
Process air flow (CFM/CMH)	40/68
Pan rotational speed (rpm)	25
Atomization air pressure (psi/bar)	18.5/1.3

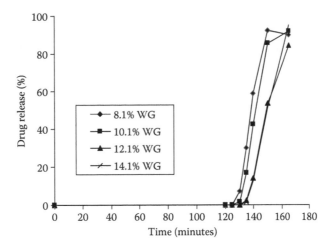

FIGURE 11.4 Drug release profiles of rabeprazole sodium tablets in 0.1N HCL, followed by phosphate buffer (pH 7.8).

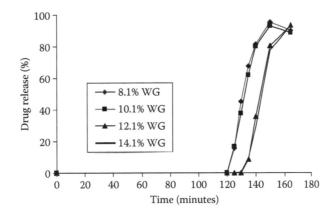

FIGURE 11.5 Drug release profiles of rabeprazole sodium tablets in acetate buffer (pH 4.5), followed by phosphate buffer (pH 7.8).

rabeprazole sodium tablets were 4.3% in 0.1N HCl and 5.4% in an intermediate pH 4.5 acetate buffer. Visual inspection of the tablets after two hours in each media indicated no signs of rabeprazole sodium degradation. Any degradation of rabeprazole Na would have led to a yellow or purple discoloration of the tablet, film layer, or dissolution medium.

Drug release profiles of Acryl-EZE coated rabeprazole Na tablets are shown in Figures 11.4 and 11.5 with less than 10% drug release in 0.1N HCl acid or pH 4.5 acetate buffer and more than 80% dissolved after 45 minutes in pH 7.8 phosphate buffer. Visual observation showed no signs of degradation in the dissolution vessel, and HPLC chromatograms did not indicate any degradant peaks for the assayed tablets [23]. Figures 11.4 and 11.5 also show that the release profiles of enteric-coated rabeprazole Na tablets could be modulated by varying the Acryl-EZE coating weight gain.

Scale-Up of Coating Processes

Successful acid resistance of enteric-coated solid dosage forms require careful selection of coating processes on small laboratory and large-scale coating facilities. A major challenge and

TABLE 11.9

Coating Process Parameters Used for Medium- to Large-Scale Acryl-EZE Enteric Coating of Tablets in Different Coating Machines

Coating Parameter	O'Hara 48″ Pan	Accela 150	Accela-60 DXL	GS-300	Glatt GC-1000	HCT 130-XL	Bamtri BGB-150E
Solids content (%w/w)	20	20	20	20	20	20	20
Theoretical weight gain (%)	10	10	10	10	10	10	10
Tablet charge (kg)[a]	140	120	360	180	75	245	140
Inlet air temperature (°C)	53	53	54	60	65	70	55
Drying air volume (m³/hr)[b]	2600	2600	6800	1800	1500	2040	N/A
Tablet bed temperature (°C)	29–32	29–35	29–36	35–37	32	34	34
Exhaust air temperature (°C)	34–38	36–38	37–40	32–37	40	32	37
Pre-warm tablet bed (°C)	34–36	34–36	34–36	38	38	N/A	N/A
Spray equipment	4 × SSVAU	4 × Manesty	5 × JAU	3 × Graco	3 × ABC	8 × Freund	3 × Bamtri
Fluid nozzle (mm)	1.5	1.2	1.5	1.4	1.2	1.2	1.2
Air cap (mm)	3.3	4	3.4	#4	N/A	3.0	N/A
Atomizing air pressure (bar)[c]	2.8	3.0	5.1	2.5	3.5	145 (slpm)	2.1
Pattern air pressure (bar)[c]	2.1	2.5	N/A	N/A	3.5 turns	60 (slpm)	N/A
Gun-to-bed distance (cm)	21–23	23–24	25–30	25	20	20	17
Spray rate (g/min)	350	340	500–600	230–330	120–180	600	250
Baffles	4	4	4	6	N/A	N/A	2
Pan speed (rpm)	6–8	7	3.5–4.5	10–11	13	6	6.5

[a] Maximize pan charge.

[b] Air volume may be decreased to prevent edge chipping.

[c] Adjust to maximize efficiency.

time-consuming process for new product development relates to successful transfer of technology from the laboratory scale to the production scale. A successful technology transfer will depend on (a) a robust product (for the core to be coated and the coating formulation to be applied) and (b) identification of critical processing parameters and tolerance ranges for those parameters. Table 11.9 shows the process parameters established for the applications of Acryl-EZE for typical coating machines from small to large scale [32].

Other Factors to Be Considered When Enteric Coating with Acryl-EZE

Seal Coat

Most robust core formulations with high mechanical strength do not require a seal coat. However, core formulations containing acid-sensitive drugs, such as PPIs, may require a seal coat to prevent degradation of the drug by the acidic polymer in the enteric film coat.

The seal coat should provide mechanical strength, be inert, and not interfere with the drug or the enteric polymer. For example, it has been reported [33] that polyethylene glycol (PEG), used as a plasticizer in the seal coat, may interact with the aspirin in an enteric-coated tablet and, on storage, lead to dissolution failure. However, when an Opadry seal coat formulated with triacetin was used, no such interaction was observed. Acryl-EZE and Opadry systems are compatible with no issues for transferring the liquid feed line from seal coat to Acryl-EZE dispersion during the coating process. For general applications of a seal coat, Opadry 03K, at 2%–3% weight gain, is recommended.

Use of Pumps

Acryl-EZE aqueous-enteric coatings can be applied successfully using various types of coating equipment; however, use of gear pumps should be avoided. A major limitation of gear pumps in the application of aqueous polymeric dispersions (including latices and pseudolatices) is the sensitivity of the dispersion to the shear generated inside gear pumps. This may result in agglomeration (or coagulation) of the polymer system due to significant wear of the pump mechanism, which, in turn, will lead to the dispersed material (polymer and pigments) penetrating between the gear surfaces and the pump housing, causing the pump to seize up.

Tablet Shape

Tablet shape may have a significant effect on the performance of applied functional films including enteric coatings. Shallower shapes are more prone to edge attrition and may result in a non-uniform film coverage on the edges of the tablet. If the core characteristics are such that a large weight gain of enteric coating is necessary, a 1%–2% seal coat can be used to strengthen and smooth the core edges. In addition to enhancing enteric protection, the seal coat may allow for a much-reduced level of enteric coating. This can result in product performance, time, and cost savings [34].

THE LATEST DEVELOPMENTS IN AQUEOUS ENTERIC COATINGS

Acryl-EZE II, which was introduced by Colorcon in 2014, is the latest class of formulations with higher polymer concentrations, which are suitable for tablets and multi-particulates, including PPI products.

Case Study 3: Enteric Coating of Lansoprazole Multiparticulates Using Acryl-EZE II

The coating performance of Acryl-EZE II versus a comparably formulated EUDRAGIT L30 D55 system was recently evaluated using lansoprazole multiparticulates as a model substrate on small and large coating scales [35]. Lansoprazole was layered onto 850- to 1000-μm sugar spheres (Suglets® PF011 Colorcon, Inc., USA) using an HPMC-based Opadry coating as a binder. After drug layering, an Opadry seal coat was applied to the multi-particulates. The multi-particulates were then enteric coated up to 40% weight gain with either Acryl-EZE II or a comparably formulated coating containing EUDRAGIT L 30 D-55 prepared by stepwise addition of plasticizer, de-tackifier, and pigments. The enteric-coated lansoprazole multi-particulates were prepared at small (2 kg) and large (50 kg) scale in a fluid bed coater using identical coating parameters for both coating systems as shown in Table 11.10. The enteric-coated multiparticulates were resistant to both pH 1.2 and pH 4.5 media for 60 minutes at weight gains as low as 20% while still allowing complete release of the drug at pH 6.8 (Figure 11.6). The release profiles were similar for the multi-particulates coated with both enteric coating types.

DELAYED-RELEASE COATING OF DIETARY SUPPLEMENT (NUTRITIONAL) PRODUCTS

There are additional challenges when considering delayed release coatings for dietary supplements, such as fish oil and garlic tablets. Enteric polymers commonly used in pharmaceutical delayed-release dosage forms, such as methacrylic acid copolymers, are not approved for use in products

TABLE 11.10

Coating Process Conditions for the Preparation of Enteric-Coated Lansoprazole Multi-Particulates

Fluid Bed Coater Type		Glatt GPCG-2			Vector VFC-60		
Initial charge (Kg)		2.33 (small)			46.63 (large)		
Wurster column diameter		18 cm			46 cm		
Distribution plate/screen		C/250 μm			P62/250 μm		
Process Parameters	**Drug Layer**	**Seal Coat**	**Enteric Coat**	**Drug Layer**	**Seal Coat**	**Enteric Coat**
Inlet temperature (°C)	65	65	55	75	75	62
Product temperature (°C)	45	45	35	45	45	35
Exhaust temperature (°C)	45	45	35	44	44	35
Air velocity (m³/hour)	130	130	130	1359	1359	1359
Atomization pressure (bar)	2	2	2	2.4	2.4	3
Flow rate (g/minute)	18	10	15	265	250	250

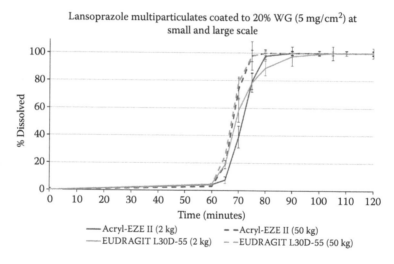

FIGURE 11.6 Dissolution profiles of enteric coated lansoprazole multiparticulates at 20% weight gain (5 mg/cm²) of enteric coating.

regulated as foods (e.g., dietary supplements) in most of the world. The Nutrateric® system was developed using only ingredients that are generally regarded as safe (GRAS) by the FDA. It is based on Surelease and NS Enteric® comprising sodium alginate, which both have GRAS status. The sodium alginate in NS Enteric is a pH-dependent pore former, which prevents release of active components in the acidic pH of the stomach but then dissolves in the intestines, allowing release of the active components there. Because several nutritional products are high volume, such as fish oil soft gelatin capsules and garlic tablets, and they are enteric coated for taste masking, it prompted the investigation of the coating performance of the Nutrateric system in a continuous coater [21]. It was determined that successful enteric performance was obtained at weight gains as low as 3.5% on placebo mineral oil soft gelatin tablets (Table 11.11), and several marketed fish oil and other dietary supplements products are coated with Nutrateric. The bed temperature was maintained at 33°C to prevent the soft gelatin capsules from melting and agglomerating. This bed temperature is lower than the 45°C–47°C ordinarily recommended for the coating of Surelease onto multi-particulates; however, other process parameters were adjusted to ensure good film formation.

TABLE 11.11

Continuous Coating Process Conditions and Coated Product Performance for Nutrateric (Substrate: 1000-mg Mineral Oil Soft Gelatin Tablets)

Process Parameter/Test Result	Trial 1	Trial 2	Trial 3
Weight gain (%)	4	3.5	4
Throughput rate (kg/hour)	130	200	200
Spray rate (g/minute)	865	1165	1335
Bed temperature (°C)	33	33	33
Minutes of resistance to simulated gastric fluid ($n = 6$ coated softgels taken after 30 minutes of continuous operation)	>60	>60	>60
Disintegration time in simulated intestinal fluid ($n = 6$ coated softgels taken after 30 minutes of continuous operation)	29.5 ± 4.8	31.5 ± 6.1	26.8 ± 5.2

EXTENDED-RELEASE AQUEOUS FILM-COATING SYSTEM: SURELEASE

Organic solvent–based coating of ethylcellulose has been employed in the formulation of extended-release oral solid dosage forms. However, the use of aqueous coating systems is preferred whenever possible due to environmental and operator safety considerations. The film-forming processes of organic and aqueous coatings of ethylcellulose are different with the latter being a result of coalescence of the ethylcellulose particles in the dispersion when sprayed on the surface of the substrate [36].

Surelease is a family of fully formulated, aqueous dispersion products, manufactured by Colorcon [37], designed specifically for modified drug release, such as extended-release, programmable-release, and taste-masking applications. Using ethylcellulose, a water-insoluble polymer, as the rate-controlling excipient in Surelease, there are major regulatory and technological benefits along with the reproducible release profiles that are achieved. The compositions of various types of Surelease are summarized in Table 11.12.

APPLICATIONS OF SURELEASE

Dispersion Preparation

Surelease is supplied as a 25% (w/w) solids dispersion, which is recommended to be diluted with water to 15% (w/w) solids before use. Prior to dilution, the container of Surelease is required to be agitated to ensure homogenization of solids in the dispersion. Then the dispersion is diluted by adding two parts of purified water to three parts of Surelease and stirred with a low shear mixer for

TABLE 11.12

Composition of Surelease Product Range (Surelease-E-7-x)

Ingredient	Function	19020	19030	19040
Ethylcellulose	Polymer	✓	✓	✓
Fractionated coconut oil	Plasticizer			✓
Dibutyl sebacate	Plasticizer	✓	✓	
Ammonium hydroxide (28%)	Stabilizer	✓	✓	✓
Oleic acid	Stabilizer/ Plasticizer	✓	✓	✓
Purified water	Vehicle	✓	✓	✓
Colloidal SiO_2	Flow aid		✓	

approximately 15 minutes. It is advisable to continue gentle agitation throughout the coating process to prevent potential sedimentation of solid particles.

Coating Process Recommendations

In order to maximize coalescence and prevent spray drying, a product bed temperature range of 45°C–47°C is generally recommended, keeping the atomization pressure around 1.5–2 bars. Some typical process conditions established with Glatt fluid-bed coating machines are shown in Table 11.13.

Extended-Release Coating of Multi-Particulates

Surelease is applied onto drug-layered nonpareils, extruded spheres, granules, drug crystals, and mini-tablets preferably using fluid-bed coating technology. Top spray coating may be used for small particulates, such as drug crystals; however, a Würster process (bottom spray) is generally recommended.

Drug release from Surelease-coated multiparticulates is mainly controlled by the coating film thickness (theoretical weight gain) as shown in Figure 11.7 for chlorpheniramine maleate layered on nonpareil beads.

TABLE 11.13

Typical Process Parameters Used for Application of Surelease to Drug-Layered Pellets for Bottom Spray Würster Systems

Process Parameter	Coating Process Conditions		
	Glatt GPCG-3	Glatt GPCG-60	Glatt GPCG-200
Batch size (kg)	3	70	200
Spray gun	Schlick 970	Schlick HS	Schlick 940
Fluidizing air volume (m³/hour)	83–107	800–900	N/A
Inlet air temperature (°C)	64–67	60–66	72–75
Exhaust air temperature (°C)	40–45	39–41	47–51
Product bed temperature (°C)	41–47	40–46	43–46
Atomizing air pressure (bar)	1.5	2.0	2.0
Spray rate (g/minute)	25–28	210–306	500–650

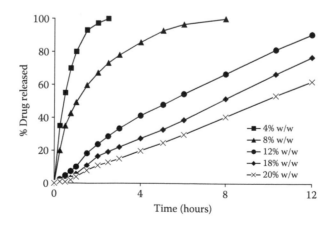

FIGURE 11.7 Effect of theoretical coating weight gain of Surelease E-7-19040 on chlorpheniramine maleate release profiles.

In addition, the aqueous solubility of the drug has major influence on drug release rate as shown in Figure 11.8 in which four drugs with different water solubilities have been layered on nonpareil pellets and then coated with Surelease E-7-19040 (16% theoretical weight gain). A highly water-soluble drug, such as guaifenesin, is released faster than less soluble drugs with no lag time. However, amlodipine, a poorly soluble drug, is released very slowly and with a considerable lag time.

In the case of poorly water-soluble drugs, a low theoretical weight gain (thin film) of Surelease may be sufficient to achieve the desired release profile. However, low weight gain on multi-particulate systems (very large surface area) may lead to batch-to-batch inconsistency. Altering the permeability of the Surelease film by incorporating a hydrophilic additive will enable the user to apply higher theoretical weight gain, reduce lag time, and ensure consistent faster release profiles for drugs with low aqueous solubilities [38]. Figure 11.9 shows the inclusion of hypromellose (METHOCEL™ E5) in Surelease E-7-19040 as a permeability enhancer (11% theoretical weight gain) and its effect on drug release profiles.

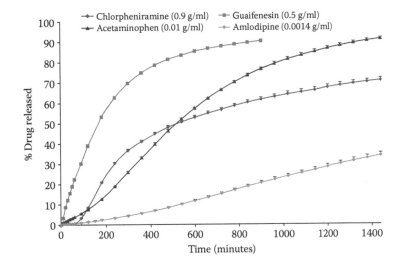

FIGURE 11.8 Effect of aqueous solubility on drug release rate from beads coated with 16% w/w Surelease E-7-19040.

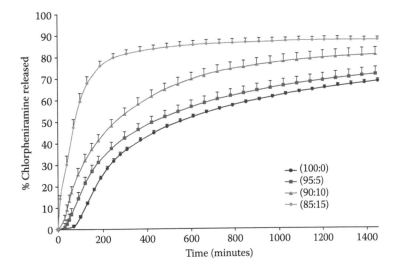

FIGURE 11.9 The influence of a permeability enhancer on chlorpheniramine maleate release from pellets coated with Surelease E-7-19040: Methocel E5 (theoretical weight gain 11%).

Extended-Release Coating of Tablets

Hydrophilic matrix tablets formulated with highly soluble drugs are often characterized by an initial rapid burst release of drug, prior to adequate gel layer formation. Surelease has been utilized to film coat the matrix tablets to inhibit the burst release as well as a method for modulating drug release rate [39]. Recent work has also shown that zero-order drug release may be achieved when coating Surelease onto hydrophilic matrices using glipizide as a model drug [40]. The steps to prepare this dosage form are much simpler and include aqueous film coating in contrast to the more involved preparation of bilayer tablets and coating with an organic solvent that are typically used to prepare push–pull osmotic pump systems.

Matrix Granulation

Undiluted Surelease may be used as a binder in high-shear, low-shear, or fluid-bed granulation. The granules are then compressed into tablets in order to generate an extended drug release profile.

Film Curing

The majority of ER barrier membrane-coating systems require thermal post-coating treatment ("curing") in order to achieve reproducible and storage-stable drug release characteristics. For example, FMC BioPolymer literature recommends that multi-particulates coated with Aquacoat ECD are incubated in a tray dryer or fluid bed column at 60°C for two hours post-coating to promote complete coalescence of polymer particles in the film [41].

The Surelease family of products is a fully formulated, optimally plasticized system, and as a consequence of plasticization of the polymer during manufacture, generally, Surelease films do not require a curing step. However, it is advisable to test for the occurrence of incomplete polymer coalescence during coating by placing the Surelease-coated products at 50°C–60°C for two, 12, and 24 hours and comparing the release profiles from these units with "uncured" beads. A curing effect may be noted if the elevated temperature incubation results in a decrease in the rate of drug release. The need for a curing step may be eliminated through optimization of the coating process.

SUMMARY OF AQUEOUS FILM COATING FOR MR FORMULATIONS

This section of the chapter provided examples of applications of Acryl-EZE and the Surelease family of products to achieve enteric (delayed) and extended-release profiles, respectively. Typical process conditions and performance of the products were highlighted as guidelines for the formulators and production personnel. The successful application of these products will require careful consideration of drug properties, condition of the equipment utilized, and in-depth understanding of the technology selected. Both Acryl-EZE and Surelease provide various product options for delayed- and extended-release formulation as well as ease of application to help the formulators develop their products in a timely manner and ensure consistent production performance when in the market.

REFERENCES

1. Porter, SC. Use of Opadry, Sureteric, and Surelease for the aqueous film coating of pharmaceutical oral dosage forms. In: Aqueous Polymeric Coatings for Pharmaceutical Dosage Forms, 2nd ed., Vol. 79, edited by JW McGinity, New York: Marcel Dekker, Inc., 1997;327–372.
2. Farrell T, Rajabi-Siahboomi A. The applications of formulated systems for the aqueous film coating of pharmaceutical oral solid dosage forms. In: Aqueous Polymeric Coatings for Pharmaceutical Dosage Forms, 3rd ed., edited by JW McGinity & L Felton, CRC Press, 2008;323–343.
3. Rajabi-Siahboomi A, Levina M, Upadhye S, Teckoe J. Excipient Selection in Oral Solid Dosage Formulations Containing Moisture Sensitive Drugs. In: Excipient Applications in Formulation Design and Drug Delivery, edited by A Narang & S Boddu, Springer, 2015 (in press).

4. Cunningham C, Scattergood L. Production-scale process and performance comparison of two fully formulated aqueous enteric coating systems. 2001 AAPS Annual Meeting—Online reprint (accessed 11 August 2015): http://www.colorcon.com/literature/marketing/mr/Delayed%20Release/Acryl-EZE /English/ads_acryleze_prod_scale_proc.pdf.
5. Levina M, Wan P. The influence of core formulation, film coating level and storage conditions on stability of ranitidine tablets. 2004 AAPS Annual Meeting.
6. Cunningham C, Kinsey B, Scattergood L. Formulation of acetylsalicylic acid tablets for aqueous enteric film coating. *Pharm Tech Europe*, May (2001).
7. Fegely K, Prusak B. Correlation of free salicylic acid content to the water vapor transmission properties of aqueous film coating systems. 2003 AAPS Annual Meeting.
8. Hughes K et al. Protection and processing of a highly hygroscopic herbal extract by drug layering and film coating. 2006 AAPS Annual Meeting—Online reprint (accessed 11 August 2015): http://www .colorcon.com/literature/marketing/fc/Opadry%20II/ads_opadry_II_prot_proc.pdf.
9. Cunningham C, Farrell T, Quiroga A. Coating moisture-sensitive products: Abstracts of Posters, Meeting of the Argentinean Association of Industrial Pharmacy and Biochemistry (SAFYBI), Buenos Aires, AR, Sep, (2005). Online reprint (accessed 11 August 2015): http://www.colorcon.com/literature /marketing/fc/Opadry%20II/ads_opadry_II_coat_moist.pdf.
10. Gulian S, Farrell T, Steffenino R. The effect of coating process conditions and coating formulation type on the quantity and location of water in film-coated tablets. 2005 AAPS Annual Meeting.
11. Gulian F, Steffenino R. Optical properties, film properties and stability of Opadry fx pearlescent film coating system. 2003 AAPS Annual Meeting.
12. Gulian F, Steffenino R, Ferrizzi D, Farrell T. Oxidative protection of ibuprofen using Opadry fx special effects film coating system. 2004 AAPS Annual Meeting.
13. Del Barrio MA, Hu J, Zhou P, Cauchon N. Simultaneous determination of formic acid and formaldehyde in pharmaceutical excipients using headspace GC/MS. *J Pharm Biomed Anal* 2006;41:738–743.
14. Farrell TP, Ferrizzi DF. Determination of trace formic acid and formaldehyde in film coatings comprising polyvinyl alcohol (PVA). 2008 AAPS Annual Meeting—Online reprint (accessed 9 July 2015): http://www.colorcon.com/literature/marketing/fc/Opadry%20II/ads_opadry_II_trace_formic_acid .pdf.
15. Wang G, Fiske JD, Jennings SP, Tomasella FP, Venkatapuram AP, Ray KL. Identification and control of a degradation product in Avapro™ film-coated tablet: Low dose formulation. *Pharm Dev Tech* 2008; 13(5):393–399.
16. Waterman KC, Arikpo WB, Fergione MB, Graul TW, Johnson BA, MacDonald BC, Roy MC, Timpano RJ. N-methylation and N-formylation of a secondary amine drug (varenicline) in an osmotic tablet. *J Pharm Sci* 2007;94(1):1–9.
17. Gimbel J, To D, Prusak B, Teckoe J, Rajabi-Siahboomi A. Evaluation of a novel, PEG-free, immediate release Opadry® aqueous moisture barrier film coating with high productivity. 2014 AAPS Annual Meeting—Online reprint (accessed 8 July 2015): http://www.colorcon.com/literature/marketing/fc /Opadry%20amb%20II/T3213.pdf.
18. To D, Teckoe J, Rajabi-Siahboomi A. A novel method to evaluate on-tablet moisture barrier performance of Opadry® film coating systems, 2014 AAPS Annual Meeting—Online reprint (accessed 9 July 2015): http://www.colorcon.com/literature/marketing/fc/Opadry%20amb%20II/T3214.pdf.
19. Cunningham CR, Neely CR. Film coating process considerations for the application of high productivity, high solids concentration film coating formulations. 2009 AAPS Annual Meeting—Online reprint (accessed 13 July 2015): http://www.colorcon.com/literature/marketing/fc/Opadry%20II/ads_fcprocess _hprod_hsolid.pdf.
20. Teckoe J, Mascaro T, Farrell TP, Rajabi-Siahboomi A. Process optimization of a novel immediate release film coating system using QbD principles. *AAPS PharmSciTech* 2013;14(2):531–540.
21. Cunningham C, Hansell J, Nuneviller III F, Rajabi-Siahboomi A. Evaluation of recent advances in continuous film coating processes. *Drug Dev Ind Pharm* 2010;36(2):227–233.
22. Cunningham C, Birkmire A, Rajabi-Siahboomi A. Application of a Developmental, High Productivity Film Coating in the GEA ConsiGma Coater. AAPS 2015 poster presentation (W4177).
23. Missaghi S, Fegely K, Ferrizzi D, Rajabi-Siahboomi AR. Application of a fully formulated aqueous enteric coating system on rabeprazole sodium 20 mg tablets. 2006 AAPS Annual Meeting—Online reprint (accessed 11 August 2015): http://www.colorcon.com/literature/marketing/mr/Delayed%20 Release/Acryl-EZE/English/ads_acryleze_app_full_form.pdf.

24. Fegely K, Simon BH, Rajabi-Siahboomi AR. Aqueous enteric coating application on non-banded hard gelatin capsules. 2006 AAPS Annual Meeting—Online reprint (accessed 11 August 2015): http://www .colorcon.com/literature/marketing/mr/Delayed%20Release/Acryl-EZE/English/ads_acryleze_aqu _ent_coat.pdf.

25. Cunningham C, Fegely K. One-step aqueous enteric coating systems: Scale-up evaluation, *Pharm Tech Europe*, October 2001.

26. Horn J. The proton-pump inhibitors: Similarities and differences. *Clin Ther* 2000;22(3):266–80.

27. PDR (Physician's Reference Desk), Montvale, NJ, Thomson PDR (2006).

28. Huber, R, Kohl B, Sachs G, Senn-bilfinger J, Simon W, Sturm E. Review Article: The continuing development of proton pump inhibitors with particular reference to pantoprazole. *Aliment Pharmacol Ther* 1995;9:363–378.

29. Miner P, Katz PO, Chen Y, Sostek M. Gastric acid control with esomeprazole, lansoprazole, omeprazole, pantoprazole, and rabeprazole: A five way crossover study. *Am J Gastroenterol* 2003;98(12):2616–20.

30. Rohss K, Lind T, Wilder-Smith C. Esomeprazole 40 mg provides more effective intragastric acid control than lansoprazole 30mg, omeprazole 20mg, pantoprazole 40mg, and rabeprazole 20 mg in patients with gastro-oesophogeal reflux symptoms. *Eur J Clin Pharmacol* 2004;60(8):531–9.

31. Saeki Y, Koyama N, Watanabe S, Aoki S. US Patent No. 5,035,889, Eisai Co., Ltd.

32. Coating Parameters for the Use of Acryl-EZE (Colorcon website document)—Accessed 11 August 2015 http://www.colorcon.com/literature/marketing/mr/Delayed%20Release/Acryl-EZE/English/coating _params.pdf.

33. Tsang LW. Enteric coating of oral solid dosage forms: Practical aspects of formulation and applications, presentation given at the 2007 Colorcon Modified Release Forum.

34. Cunningham CR, Kinsey BR, Scattergood LK, Turnbull N. The effect of tablet shape on the application of an enteric film coating. 2002 AAPS Annual Meeting.

35. Young C, Reyes G, Teckoe J, Rajabi-Siahboomi A. Evaluation of a Novel Fully Formulated Acryl-EZE® Enteric Coating System vs EUDRAGIT® L30D-55 on Lansoprazole Multiparticulates at Small and Large Coating Scale 2014 AAPS Annual Meeting—Online reprint (accessed 14 July 2015): http://www .colorcon.com/literature/marketing/mr/Delayed%20Release/Acryl-EZE%20II/W5161.pdf.

36. Binschaedler C, Gurny R, Doelker E. Theoretical concepts regarding the formation of films from aqueous microdispersions and application to coating. *Labo-Pharma-Probl Tech* 31/331, 389 (1983).

37. US Patents 4,123,403 and 4,502,888.

38. Ong K, Rege PR, Rajabi-Siahboomi AR. Hypromellose as a pore-former in aqueous ethylcellulose dispersion: Stability and film properties, 2006 AAPS Annual Meeting—Online reprint (accessed 11 August 2015): http://www.colorcon.com/literature/marketing/mr/Extended%20Release/Surelease/English/surelease _pore_former.pdf.

39. Dias VD, Gothoskar AV, Fegely KA, Rajabi-Siahboomi AR. Modulation of drug release from hypromellose (HPMC) matrices: Suppression of the initial burst effect. 2006 AAPS Annual Meeting—Online reprint (accessed 11 August 2015): http://www.colorcon.com/literature/marketing/mr/Extended%20 Release/METHOCEL/English/ads_methocel_mod_drug_rel.pdf.

40. Mehta RY, Missaghi S, Tiwari, S Farrell TP, Rajabi-Siahboomi AR. Barrier membrane coating of hydrophilic matrices: A simplified strategy to attain zero order drug release. 2013 CRS Annual Meeting—Online reprint (accessed 11 August 2015): http://www.colorcon.com/literature/marketing/mr /Extended%20Release/Surelease/English/CRS_2013_Mehta_BM_SURE.pdf.

41. Aquacoat ECD Brochure (FMC BioPolymer) p. 6 (accessed online 12 August 2015): http://www.fmcbio polymer.com/Portals/bio/content/Docs/AquaCoat%20ECD%207706%20.pdf.

12 Substrate Considerations when Coating Solid Dosage Forms

Elena Macchi and Linda A. Felton

CONTENTS

INTRODUCTION

The application of polymeric films onto solid dosage forms (e.g., tablets, pellets, and capsules) is widely employed in the pharmaceutical field for decorative, protective, and functional purposes [1–5]. Although dry coating has gained interest as a solvent-free technique [6–8], the majority of coatings are applied by a spray-atomization technique, with which polymers, either dissolved or dispersed in an aqueous or organic solvent, are sprayed onto substrates. Such coating processes are highly complex, involving coating formulation variables, processing variables, and substrate variables, all of which can impact the performance of the film as shown in Figure 12.1. This chapter describes physical and chemical properties of solid substrates that can impact the coating process and ultimately influence product performance.

MOVEMENT OF SUBSTRATE IN THE COATING APPARATUS

Various equipment is available for polymeric coating of solid dosage forms, and it is classified into three categories: pans, perforated pans, and fluid-bed equipment. More detailed information on coating equipment can be found in Chapter 3. All equipment relies on the substrate being repeatedly cycled through the spray zone until the desired coating mass and/or uniformity are reached [9]. When substrates pass into this spray zone, the atomized polymer-containing liquid droplets impinge on the surface and spread across it. In some cases, the liquid can also penetrate the surface of the substrate and, in more extreme cases, cause surface dissolution, resulting in physical mixing at the interface [10,11]. The cores are removed temporarily from the spraying zone, exposing them to the heated air, which facilitates solvent evaporation and film formation.

Obviously, the substrate to be coated must be sufficiently strong to withstand the stresses induced during movement in the apparatus. These stresses include not only the physical movement within the coating apparatus as the product hits the walls of the machinery but also the weight of all other products in the pan or fluid bed. Thus, highly friable substrates are unsuitable for coating. Moreover, a marginally strong tablet that is successfully coated at the laboratory scale could fail during scale-up, and hence, a robust substrate is needed [12]. If a substrate is friable, particles detached from

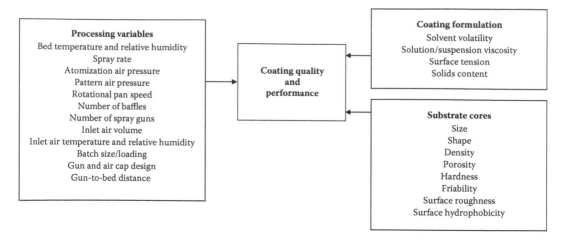

FIGURE 12.1   Schematic of variables that influence film formation, coating uniformity, and overall product performance in pan-coating processes.

its surface, especially at the edges, could be entrapped in the forming polymeric layer, resulting in coated products showing poor reproducibility in release.

The size, shape, and density of the substrate impact coating in several different ways. First, the size often dictates which type of coating equipment will be used. Pan coating is generally used to coat larger products (i.e., tablets and capsules) whereas pellets and mini-tabs are often coated in the fluid-bed apparatus. Pan-coating equipment relies on drum rotation to move the substrates in the pan. Tablets that are particularly dense tend to slide in the pan rather than tumble and this can create non-uniform movement through the spray zone, which ultimately adversely affects coating uniformity within the batch. Because baffles in the coating pan facilitate substrate tumbling, increasing the number of baffles may create better movement [13]. Density also plays a role in fluid bed–coating processes, which utilize an airstream to move substrates. Highly dense pellets may require higher fluidization velocity to achieve fluid movement of the bed [14].

Various experimental [13,15] and modeling techniques [16] have been used to investigate substrate movement within coating equipment. Although many of the published studies have focused on coating parameters, such as rotational pan speed, several works have investigated tablet properties that influence movement in coating pans. Pandey and Turton [17], for example, used a video-imaging technique to compare spherical particle movement to tablet movement in pan-coating processes. Parameters investigated included circulation time (time between successive tablet sightings at the bed surface) and surface time (time the substrate stays in the spray zone). Although the average circulation time between the spheres and tablets was not statistically different, the surface time for tablets was lower than the spheres when low bed fill levels were used. The authors also noted that the tablets moved faster than the spherical particles.

In addition to circulation time and surface time, the surface area of the substrate exposed to the spray nozzle can impact coating uniformity. For example, Wilson and Crossman [18] evaluated the coating thickness on the face, edge, and end/belly band of differently shaped tablets and demonstrated that better film uniformity is achieved with a more spherical geometry as it has no preferred orientation in the spray zone. More recently, Freireich et al. [19] combined discrete element and Monte Carlo modeling to investigate the intra-tablet coating variability of six tablet shapes in relation to their movement in the pan. Almond, bullet, half moon, shield, standard round convex, and sphere tablets were considered. For all shapes, the coating thickness on the tablet face was greater than at the tablet edges. Although some limitations of this approach were found, simulations also demonstrated that coating thickness variability of the non-spherical tablet shapes was larger for tablets showing a greater degree of preferred orientation in the spray zone.

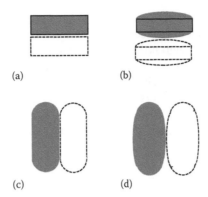

FIGURE 12.2 Schematic of various tablet shapes. (a) Flat-faced tablets, (b) convex tablets, (c) caplet, and (d) modified caplet. Modification to the shape can reduce surface area for tablet-to-tablet sticking.

Measurements by Freireich et al. [19], although consistent with works from other authors, showed some inconsistencies with results of Ho et al. [20], who reported that the thickness of the coating was increased from the tablets' center toward the edge. However, it is important to note that many parameters could have led to these opposing findings, for example, conditions of the process and differences in tablets radius of curvature and mobility as well as possible contacts with other tablets, which were not considered by the authors.

The shape of the substrate also influences its tendency to stick or agglomerate to other substrates during coating. For example, flat-faced tablets and caplets are likely to stick to some extent, depending on coating and processing conditions, simply due to the presence of the flat surfaces as shown in Figure 12.2a and c. Changing the coating formulation or processing conditions can reduce or, in some cases, eliminate this defect. Adding an anti-adherent, such as talc, to the coating formulation (or increasing the concentration of the anti-adherent), for example, has been shown to reduce the incidence of sticking. Wesseling et al. [21] investigated the effectiveness of talc and glyceryl mono-stearate (GMS) on the tackiness of free films based on EUDRAGIT® L 30D and EUDRAGIT® RS 30D. The authors found that GMS was able to reduce film tackiness at lower concentrations than talc; moreover, such reduction in tackiness could not be achieved by increasing talc concentration in the film. Adding a slight degree of curvature to the shape of the solid substrate can also minimize the extent of sticking by reducing the flat areas, as shown in Figure 12.2b and d.

PHYSICAL AND CHEMICAL STABILITY

The substrate, the active pharmaceutical ingredient (API), and any other excipient in the substrate core should be stable during processing and subsequent storage. As mentioned earlier, the drug products must be physically strong enough to withstand the tumbling/movement in the coating apparatus. In addition, the dosage forms are exposed to the elevated temperatures used to facilitate solvent evaporation and film formation. Bed temperatures must be higher than the minimum film-forming temperature of the polymer coating system for aqueous-based dispersions to ensure film formation occurs [22]. Moreover, for aqueous-based dispersions, a post-coating curing or drying step is often required to achieve coalescence of the polymer chains in the film. Thus, it should be readily apparent that the API and excipients should be thermally stable and, for aqueous-based coatings, should be stable in the presence of water.

One of the reasons for applying polymeric film coatings to solid substrates is to protect moisture-sensitive drug products. These films slow the rate of water vapor penetration from ambient air into the substrate, acting as a barrier between the core and the environment. Bley et al. [23], for example, coated freeze-dried garlic powder–containing tablets with aqueous solutions or dispersions of hydroxypropyl methylcellulose (HPMC), poly(vinyl alcohol), ethyl cellulose, and poly(methacrylate-methylmethacrylates) and studied the influence of coating and curing conditions on the stability of the active ingredient allicin. Although no degradation was detected immediately after the coating process, stability of the product was improved as the polymer films slowed the rate of water vapor penetration for tablets coated with poly(vinyl alcohol) and poly(methacrylate-methylmethacrylates). For drugs susceptible to hydrolysis, an organic-based barrier coating may be necessary to protect the API from exposure to the solvent during aqueous coating processes. Another potential option is to use high solids coatings that require less solvent (water) and have an added benefit of shorter processing time [24–26].

Conditions in the pan to which the substrates are exposed during coating have also been shown to impact long-term stability of drugs susceptible to hydrolysis. Kestur et al. [27] used PyroButton® data logging devices to monitor the micro-environmental conditions (temperature and relative humidity) within the coating bed. Expanding on the findings by Badawy et al. [28] for the improvement in the stability of coated tablets containing the moisture-sensitive drug brivanib alaninate, the authors found that the dryer the conditions in the coating pan, the lower the content of free water (i.e., water activity) in the tablets after coating, which resulted in a higher drug stability after six months storage under accelerated conditions.

In addition, proper selection of the core excipients is critical for the stability of moisture-sensitive drugs. In preformulation compatibility studies, excipients possessing low water activity should be preferred as they are able to bind moisture in the core preferentially over the drug. Hence, the effects of changes in relative humidity on drug stability could be limited, resulting in an extension of product shelf life. For example, Cunningham et al. [29] reported the use of Starch 1500® to decrease hydrolytic degradation of acetylsalicylic acid in enteric-coated tablets after three months storage at 40°C/75% RH.

It goes without saying that compatibility of ingredients with the polymer formulation is an obvious issue but one that should not be overlooked. Interactions and incompatibilities between APIs and/or excipients with polymers have been reported in the literature [30]. For example, Bodmeier and Paeratakul [31] found that flocculation or coagulation of ethylcellulose suspensions occurred in the presence of salts of basic drugs. Although interactions between the drugs and the latices occurred when anionic surfactants were employed, stabilization of the suspension could be achieved by the replacement with a nonionic surfactant. In another study, Murthy et al. [32] reported some incompatibilities for hard gelatin capsules coated with Aquateric® upon storage. Coated capsules showed a slower release of their content after three months storage at room temperature as a consequence of a reaction between gelatin and the cellulose acetate phthalate polymer (or its hydrolysis products), which made the gelatin insoluble. Wang et al. [33] studied the stability of irbesartan in the core of commercially available tablets coated with Opadry™ II White. The authors demonstrated that upon storage, formaldehyde was liberated from polyethylene glycol contained in the coating formulation, resulting in the formation of the formaldehyde adduct of irbesartan (hydroxymethyl derivative). Enteric polymers, used to protect acid-labile drugs from exposure to the acidic environment of the stomach as well as to protect the gastric mucosa from irritation due to the contact with drugs [34], are themselves weak acids and can adversely affect the chemical stability and thus bioavailability of an acid-labile API. More information on polymer interactions with drugs and excipients can be found in Chapter 13.

To prevent or minimize drug–polymer or excipient–polymer interactions, the application of an inert, water-soluble sub-coat can act as a barrier to physically separate the core from the polymeric film [30]. Suitable materials for sub-coats include sucrose, polyethylene glycol, povidone, polyvinylalcohol, amylopectin, HPMC, and hydroxypropylcellulose (HPC). For example, Bozdag et al. [35] used sub-coatings to protect the acid-labile API omeprazole in the substrate from enteric polymers of the coating. Other potential strategies to prevent such interactions between acid-labile drugs and enteric polymers include the use of basic excipients in the core (e.g., sodium, potassium, calcium, magnesium, and aluminum salts of phosphoric acid and weak inorganic or organic acids) and the addition of pH-buffering substances (e.g., aluminium, calcium, and magnesium hydroxides; magnesium oxides; and trihydroxymethylaminomethane) to the tablet core to create a more neutral microenvironment around drug particles.

Aqueous-based coating of soft gelatin capsules is further complicated because of the liquid fill material. Felton et al. [36] reported that a 30- to 60-minute pre-warming of the cores before coating was necessary to avoid the formation of bubbles in an EUDRAGIT® L film, attributed to non-uniform drying of the polymer. Increasing the temperature of the fill liquid during the pre-warming stage allowed more homogeneous drying of the film, thus achieving a smooth and more uniform coating. In this same work, the fill liquid in conjunction with the plasticizer in the polymeric coating was shown to influence the adhesive and enteric performance of the coated capsules. The impact of the hydrophilicity of the plasticizer used for the coating suspension on the enteric protection of softgels filled with either the hydrophilic polyethylene glycol 400 or the hydrophobic Miglyol® 812 was investigated. When the hydrophilic plasticizer triethyl citrate (TEC) was used in the coating formulation, enteric protection was observed for soft gelatin capsules irrespective of fill liquid. In contrast, the hydrophobic tributyl citrate (TBC) plasticizer in the coating formulation afforded gastric protection only for substrates filled with Miglyol® 812.

SURFACE INTERACTIONS AND SUBSTRATE WETTING

Wetting of a substrate can be described as replacing a solid–vapor interface with a solid–liquid interface. For film-coating processes, there are three types of wetting: adhesional, immersional, and spreading as shown in Figure 12.3.

Adhesional wetting involves the loss of solid–vapor and liquid–vapor interfaces as the atomized polymer-containing droplets initially impinge on the substrate surface. Immersional wetting involves a loss in the solid–vapor interface as the liquid penetrates into the substrate. Last, spreading wetting is a loss of the solid–vapor interface with an increase in both solid–liquid and liquid–vapor interfaces as the polymer-containing droplets spread across the substrate surface. Such spreading is critical for film formation as the polymer chains in close proximity intertwine to form the film.

In reference to types of wetting, adhesional wetting is always spontaneous whereas immersional wetting is spontaneous when the contact angle between the polymer-containing liquid droplet and the substrate surface is less than 90°. In contrast, spreading wetting is never spontaneous and requires energy to occur. Energy for spreading wetting is typically gained from the momentum of the sprayed droplets hitting the substrate surface during coating, and obviously coating formulation variables can impact this type of wetting. Substrate formulation variables can also be manipulated to improve wetting, predominately adhesional and immersional wetting.

It should be readily apparent that the quantity and type of excipients compressed into tablets influence the chemical properties of the substrate surface, which, in turn, affects interfacial bond formation and wetting. Good spreadability of the aqueous polymer-containing droplets on the substrates during coating can be ensured by tablets showing a hydrophilic surface as a high number of interfacial hydrogen bonds are formed. The interaction between the tablet surface and the droplet allows the latter to spread and droplets to coalesce/fuse with other polymer chains in close

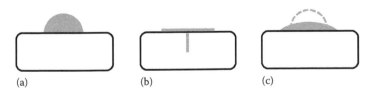

(a) (b) (c)

FIGURE 12.3 Types of substrate wetting in film coating processes. (a) adhesional, (b) immersional, and (c) spreading.

proximity, thus leading to the formation of a continuous film. On the contrary, in the case of spraying an aqueous suspension onto a hydrophobic substrate, the poor interfacial interactions limit wetting and spreading. Felton and McGinity [37] demonstrated the contact angle of an aqueous-based acrylic dispersion onto tablet surfaces increased as the substrate became more hydrophobic through the addition of a wax excipient to the tablet formulation. Porter [38] studied the coating of sustained-release wax matrix tablets with aqueous dispersions and confirmed hurdles in the formation of a good-quality film due to poor wettability of the substrate surface.

As expected, wetting of the substrate is directly related to adhesion of the polymeric film. Aulton [39] extensively discussed the importance of wetting and spreading in order to have good film adhesion. Ingredients at the substrate surface, however, could partially dissolve during the coating process and ultimately migrate into the coating. For example, Felton and Perry [10] were able to quantify the extent of physical mixing at the film–tablet interface. In this study, X-ray photoelectron spectroscopy combined with ion-sputtering profiling was able to detect phosphorous from the tablet core and chlorine from the polymer film, and the film–tablet interfacial region was reported as the area where both elements were found simultaneously. Such physical mixing and migration into the film can potentially compromise the quality of the applied coating and, more importantly, the performance of the coated dosage forms.

Curing time and temperature have been reported to influence migration of the drug to the applied film. Bodmeier and Paeratakul [40] showed that an increase of ibuprofen migration to the bead surface was promoted by increasing curing temperature and time, resulting in faster drug release.

Adhesion of the polymer film to the substrate is affected by two major forces: the strength of the interfacial bond (mainly hydrogen bonds but also dipole–dipole and dipole-induced dipole interactions) and the internal stress in the applied coating (resulting from the sum of thermal, volumetric, and shrinkage stresses). More detail about polymer adhesion can be found in Chapter 6. Here, however, we briefly discuss substrate surface properties that impact adhesion.

Polymer adhesion is influenced by the chemical and physical properties of both the surface and of the whole substrate (e.g., shape). Surface roughness relates to adhesion in terms of wettability and the extent of interfacial interactions with a rougher surface providing more interfacial interactions and thus better film–substrate adhesion as reported by Nadkarni et al. [41]. Although film application onto a tablet compressed above a critical pressure could result in insufficient adhesion because of a highly smooth surface, below a critical compression pressure, cohesive failure could occur, and the tablet laminates during adhesion testing due to the formation of intermolecular bonds between the substrate and the film that are stronger than the bonding forces of the compressed powders of the tablets [37,42]. The authors showed that tablet hardness influenced the adhesion of EUDRAGIT® L films applied from an aqueous dispersion with a lower tablet hardness producing a rougher surface with an increased total area available for film–substrate interactions.

Tablets with a more porous structure allow for greater immersional wetting. Fisher and Rowe [43], for example, studied the influence of the rate and depth of fluid penetration on the interfacial contact between a polymer and a tablet. If the rate of water penetration into the tablet core during coating is too high, the surface of the coated product may appear uneven where the surface of the tablet is more porous (generally corresponding to the crown or the curved portion of a convex tablet) or areas of the tablet surface with locally high amounts of superdisintegrants [44]. Adjustments to processing parameters to create more efficient drying or a slower application of the sprayed formulation can diminished this defect. Alternatively, increasing the solids content of the coating formulation to make a more viscous solution may decrease the extent of immersional wetting. With super-disintegrants, better mixing prior to tableting may reduce the incidence of cratering. In extreme cases, it may be necessary to replace these excipients.

Substrate swelling is another issue to consider when applying polymeric films. Swelling of the substrate can occur upon contact with water contained in the sprayed polymeric formulation and/or because of thermal expansion from the heated drying air. Differences in thermal expansion of the substrate and the coating contribute to the development of higher internal stresses in the film, which adversely affects film–substrate adhesion. Swelling and film adhesion issues have also been reported with gelatin capsules [45]. Careful control of processing conditions, particularly spray rate and bed temperature, is required when coating gelatin capsules because gelatin itself is soluble in warm/hot water, and residual water in the capsule shell is necessary for its mechanical strength. Swelling has also been reported when HPC-based capsules prepared by injection molding technique [46,47] and HPMC capsules [48] were coated with an aqueous-based polymer although no issues related to substrate adhesion were noted.

Tablet shape is an important consideration in film-coating operations. As mentioned earlier, shape affects movement in the coating apparatus and the exposed surface area in the spray zone. Shape also impacts adhesion. Rowe [49], for example, investigated the adhesion of polymeric films to tablet surfaces and attributed the lack of adhesion of an applied film at the edges of a tablet to the formation of high internal stresses in the film. Cunningham et al. [50] investigated the influence of tablet shape on enteric performance and demonstrated that high polymer weight gains and a sub-coat were required to achieve enteric protection for shallow convex tablets whereas lower amounts of polymer and no sub-coating were required when standard convex tablets were used.

As mentioned, excipients in the core impact wetting and spreading and thus also affect film–substrate adhesion. Excipients with hydroxyl groups tend to provide stronger adhesion with polymeric aqueous dispersions. Rowe [51] measured the adhesion of HPMC films (which has both primary and secondary hydroxyl groups in its structure) onto tablets containing different excipients (e.g., microcrystalline cellulose [MCC], sucrose, and anhydrous as well as spray-dried lactose). HPMC films applied onto MCC-based tablets exhibited the strongest adhesion due to the higher number of hydroxyl groups in the MCC structure in comparison to the other excipients studied. Similar results were found by Lehtola et al. [52]. In this latter study, the authors also investigated the effect of increasing amounts of the hydrophobic lubricant magnesium stearate (MgSt) to the investigated tablet formulations and showed that as the tablet surface presented a greater number of nonpolar hydrocarbon groups, interfacial interactions with the applied polymer suspensions were limited, which resulted in decreased adhesion with the greatest decrease observed with the MCC tablets. More recently, Pandey et al. [53] investigated the use of sodium lauryl sulfate (SLS, 0% and 1% w/w) and MgSt (0.5% and 1.75% w/w) in the preparation of tablets from blends prepared by direct compression, wet granulation, and dry granulation. The authors demonstrated the important role of these excipients on both compaction (on which these ingredients had a negative effect, in particular in dry granulation) and coating performance (indicated by the incidence of logo bridging or infilling). Focusing on the quality of coated tablets, the authors concluded that the use of SLS in tablets caused a lack of film–tablet adhesion, which was more pronounced when MgSt was added.

OTHER SUBSTRATE PROPERTIES TO CONSIDER

All prescription drug products in the United States are required to have some type of identification, typically a unique number or logo printed on the substrate surface or debossed directly into the tablet face. Coating debossed tablets, however, presents additional challenges for the formulation scientist. The debossing should not be placed across the crown or curved area of the tablet as shown in Figure 12.4, as this location tends to be mechanically weak and thus is prone to surface erosion as the tablets tumbling during coating as well as surface dissolution, both of which can result in obscuring of the debossed image.

Bridging or infilling of the debossed area, referred to as bridging of the intagliations, is a defect with which the coating does not adequately adhere to the tablet but rather forms a bridge across the edges of the logo, also obscuring to varying degrees the debossed image. Studies on factors important to consider when coating substrates with intagliations go back to the 1980s. Kim et al. [54] demonstrated that the incidence of bridging increased with increasing film thickness. The shape of the intagliation also influences the incidence of logo bridging. Rowe and Forse [55] showed that when the width and depth of intagliations was doubled, keeping the angle and the width:depth

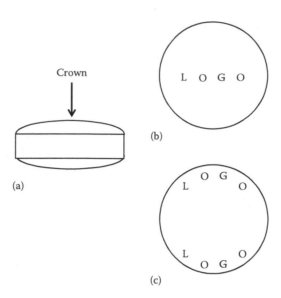

FIGURE 12.4 Schematic representation of (a) the crown/curved area of a convex tablet and of the logo placed (b) across the crown or (c) at the edge of the tablet (which ensures a better duration of the debossed image).

ratio constant, the incidence of bridging decreased from ~85% to ~55%. The increase of the area of intagliations would indeed expose greater interfacial area of contact, thus leading to a decrease in such defect.

For hard gelatin and HPMC capsules, poor deposition of the coating at the cap and body closure area has been reported. This issue, which can cause a failure in the desired release performance, can be resolved by banding the closed capsule at the joint of caps and bodies. Felton et al. [48] and Macchi et al. [46] have demonstrated the necessity of banding the closure of capsules before coating with the enteric EUDRAGIT® L 30 D-55 in order to obtain effective modified release dosage forms. Huyghebaert et al. [56] suggested a method to avoid the banding step. In that study, HPMC-based caps and bodies were coated separately with aqueous dispersions based on different polymers (i.e., EUDRAGIT® FS 30 D, EUDRAGIT® L 30 D-55, Aqoat® AS-HF and Sureteric®) in a fluid-bed apparatus, where the coating was also applied to the area of the body where the cap overlaps. Although the method was found suitable for obtaining enteric-coated HPMC capsules with no need for band sealing, differences with respect to both the substrates' behavior during the coating process and the quality of the coated products were noticed based on the polymer used for the coating suspension. Interestingly, because of the different configuration of the closing area, no need for sealing prior to coating was demonstrated on starch [57] and HPC-based capsules [46].

CONCLUSIONS

Because the substrate can significantly influence the coating process and product performance, the formulation scientist should devote some thought to its design. The substrate should be sufficiently robust, both physically and chemically, to withstand the coating process. Starting with a robust substrate will help minimize the potential for coating defects, especially during scale up. The size, shape, and density of the substrate impact movement within the coating equipment and can affect the intra-tablet coating uniformity. Physical and chemical properties of the substrate impact wettability and film adhesion. Gelatin capsules and debossed tablets present additional challenges during coating.

REFERENCES

1. Felton, L. A., and C. J. Wiley. 2003. Blinding controlled-release tablets for clinical trials. *Drug Dev Ind Pharm* 29(1): 9–18.
2. Béchard, S. R., O. Quraishi, and E. Kwong. 1992. Film coating: Effect of titanium dioxide concentration and film thickness on the photostability of nifedipine. *Int J Pharm* 87: 133–139.
3. Felton, L. A., and G. S. Timmins. 2006. A nondestructive technique to determine the rate of oxygen permeation into solid dosage forms. *Pharm Dev Technol* 11(1): 141–147.
4. Pearnchob, N., J. Siepmann, and R. Bodmeier. 2003. Pharmaceutical applications of shellac: Moisture-protective and taste-masking coatings and extended-release matrix tablets. *Drug Dev Ind Pharm* 29(8): 925–938.
5. Palugan, L., M. Cerea, L. Zema et al. 2015. Coated pellets for oral colon delivery. *J Drug Deliv Sci Technol* 25: 1–15.
6. Obara, S., N. Maruyama, Y. Nishiyama et al. 1999. Dry coating: An innovative enteric coating method using a cellulose derivative. *Eur J Pharm Biopharm* 47(1): 51–59.
7. Pearnchob, N., and R. Bodmeier. 2003. Coating of pellets with micronized ethylcellulose particles by a dry powder coating technique. *Int J Pharm* 268(1–2): 1–11.
8. Cerea, M., W. Zheng, C. R. Young et al. 2004. A novel powder coating process for attaining taste masking and moisture protective films applied to tablets. *Int J Pharm* 279: 127–139.
9. Turton, R. 2008. Challenges in the modeling and prediction of coating of pharmaceutical dosage forms. *Powder Technol* 181(2): 186–194.
10. Felton, L. A., and W. L. Perry. 2002. A novel technique to quantify film-table interfacial thickness. *Pharm Dev Technol* 7(1): 43–47.
11. Barbash, D., J. E. Fulghum, J. Yang et al. 2009. A novel imaging technique to investigate the influence of atomization air pressure on film-tablet interfacial thickness. *Drug Dev Ind Pharm* 35(4): 480–486.
12. Porter, S. C., G. Sackett, and L. Liu. 2009. Development, optimization, and scale-up of process parameters: Pan coating. In: *Developing solid oral dosage forms: Pharmaceutical theory and practice*, ed. Y. Qui, Y. Chen, G. G. Z. Zhang, 761–805. Burlington: Academic Press.
13. Smith, G. W., G. S. Macleod, and J. T. Fell. 2003. Mixing efficiency in side-vented coating equipment. *AAPS PharmSciTech* 4(3): 1–5.
14. Toschkoff, G., and J. G. Khinast. 2013. Mathematical modeling of the coating process. *Int J Pharm* 457: 407–422.
15. Shelukar, S., J. Ho, J. Zega et al. 2000. Identification and characterization of factors controlling tablet coating uniformity in a Wurster coating process. *Powder Technol* 110: 29–36.
16. Turton, R. 2010. The application of modeling techniques to film-coating processes. *Drug Dev Ind Pharm* 36(2): 143–151.
17. Pandey, P., and R. Turton. 2005. Movement of different-shaped particles in a pan-coating device using novel video-imaging techniques. *AAPS PharmSciTech* 6(2) Article 34.
18. Wilson, K. E., and E. Crossman. 1997. The influence of tablet shape and pan speed on intra-tablet film coating uniformity. *Drug Dev Ind Pharm* 23(12): 1239–1243.
19. Freireich, B., W. R. Ketterhagen, and C. Wassgren. 2011. Intra-tablet coating variability for several pharmaceutical tablet shapes. *Chem Eng Sci* 66: 2535–2544.
20. Ho, L., R. Müller, M. Römer et al. 2007. Analysis of sustained-release tablet film coats using terahertz pulsed imaging. *J Control Release* 119(3): 253–261.
21. Wesseling, M., F. Kuppler, and R. Bodmeier. 1999. Tackiness of acrylic and cellulosic polymer films used in the coating of solid dosage forms. *Eur J Pharm Biopharm* 47: 73–78.
22. Felton, L. A. 2013. Mechanisms of polymeric film formation. *Int J Pharm* 457(2): 423–427.
23. Bley, O., J. Siepmann, and R. Bodmeier. 2009. Protection of moisture-sensitive drugs with aqueous polymer coatings: Importance of coating and curing conditions. *Int J Pharm* 378: 59–65.
24. Porter, S. C. 2009. Coating with high solids. *Manuf Chemist* 80(6): 44–46.
25. Rogers, T., P. Sheskey, H. Furukawa et al. 2008. Investigation of a novel hypromellose polymer for high-productivity tablet coating applications. In: *American Association of Pharmaceutical Scientists*, Atlanta, Georgia.
26. Porter, S. C., P. Hadfield, L. Pryce et al. 2011. An examination of the use of high-solids film-coating systems as a means of reducing the cost of aqueous coating processes. In: *American Association of Pharmaceutical Scientists*, Washington, District of Columbia.

27. Kestur, U., P. Pandey, S. Badawy et al. 2014. Controlling the chemical stability of a moisture-sensitive drug product through monitoring and identification of coating process microenvironment. *Int J Pharm* 476: 93–98.

28. Badawy, S. I. F., J. Lin, M. Gokhale et al. 2014. Quality by design development of brivanib alaninate tablets: Degradant and moisture control strategy. *Int J Pharm* 469(1): 111–120.

29. Cunningham, C.R., B. R. Kinsey, and L. K. Scattergood. 2001. Formulation of acetylsalicylic acid tablets for aqueous enteric film coating. *Pharm Tech Europe* 13(5): 44–53.

30. Bruce, D. L., and J. W. McGinity. 2008. Polymer interactions with drugs and excipients. In: *Aqueous polymeric coatings for pharmaceutical dosage forms*, 3rd ed., eds. J. W. McGinity, and L. A. Felton, 369–408. New York: Informa Healthcare.

31. Bodmeier, R., and O. Paeratakul. 1989. Evaluation of drug-containing polymer films prepared from aqueous latexes. *Pharmaceut Res* 6(8): 725–730.

32. Murthy, K. S., N. A. Enders, M. Mahjour et al. 1986. A comparative evaluation of aqueous enteric polymers in capsule coatings. *Pharm Technol* 10: 36–46.

33. Wang, G., J. D. Fiske, S. P. Jennings et al. 2008. Identification and control of a degradation product in Avapro™ film-coated tablet: Low dose formulation. *Pharm Dev Technol* 13: 393–399.

34. Dulin, W. 2010. Oral targeted drug delivery systems: Enteric coating. In: *Oral controlled release formulation design and drug delivery: Theory to practice*, ed. H. Wen, and K. Park, 205–224. New Jersey: John Wiley & Sons.

35. Bozdag, S., S. Calis, and M. Sumnu. 1999. Formulation and stability evaluation of enteric coated omeprazole formulations. *STP Pharma Sci* 9: 321–327.

36. Felton, L. A., M. M. Haase, N. H. Shah et al. 1995. Physical and enteric properties of soft gelatin capsules coated with Eudragit® L 30 D-55. *Int J Pharm* 113: 17–24.

37. Felton, L. A., and J. W. McGinity. 1996. Influence of tablet hardness and hydrophobicity on the adhesive properties of an acrylic resin copolymer. *Pharm Dev Technol* 1(4): 381–389.

38. Porter, S. C. 1997. Use of Opadry, Sureteric, and Surelease for the aqueous film coating of pharmaceutical oral dosage forms. In: *Aqueous polymeric coatings for pharmaceutical dosage forms*, 2nd ed., ed. J. W. McGinity, 327–372. New York: Marcel Dekker, Inc.

39. Aulton, M. E. 1995. Surface effects in film coating. In: *Pharmaceutical Coating Technology*, ed. G. Cole, 118–151. Bristol: Taylor & Francis Ltd.

40. Bodmeier, R., and O. Paeratakul. 1994. The effect of curing on drug release and morphological properties of ethylcellulose pseudolatex-coated beads. *Drug Dev Ind Pharm* 20(9): 1517–1533.

41. Nadkarni, P. D., D. O. Kildsig, P. A. Kramer et al. 1975. Effects of surface roughness and coating solvent on film adhesion to tablets. *J Pharm Sci* 64(9): 1554–1557.

42. Felton, L. A., and J. W. McGinity. 1999. Adhesion of polymeric films to pharmaceutical solids. *Eur J Pharm Biopharm* 47: 3–14.

43. Fisher, D. G., and R. C. Rowe. 1976. The adhesion of film coatings to tablets surfaces-instrumentation and preliminary evaluation. *J Pharm Pharmacol* 28: 886–889.

44. Levina, M., and C. R. Cunningham. 2005. The effect of core design and formulation on the quality of film coated tablets. *Pharm Technol Eur* 17(4): 29–37.

45. Thoma, K., and K. Bechtold. 1992. Enteric coated hard gelatin capsules. *Capsugel Technical Bulletin*.

46. Macchi, E., L. Zema, A. Maroni et al. 2015. Enteric-coating of pulsatile-release HPC capsules prepared by injection molding. *Eur J Pharm Sci* 70: 1–11.

47. Macchi, E., L. Zema, P. Pandey et al. 2016. Influence of temperature and relative humidity conditions on the pan coating of hydroxypropyl cellulose molded capsules. *Eur J Pharm Biopharm* 100: 47–57.

48. Felton, L. A., C. J. Wiley, and A. L. Friar. 2002. Enteric coating of gelatin and cellulosic capsules using an aqueous-based acrylic polymer. In: *American Association of Pharmaceutical Scientists*, Toronto, Canada.

49. Rowe, R. C. 1981. The adhesion of film coatings to tablet surface—A problem of stress distribution. *J Pharm Pharmacol* 33: 610–612.

50. Cunningham, C. R., B. R. Kinsey, L. K. Scattergood et al. 2002. The effect of tablet shape on the application of an enteric film coating. In: *American Association of Pharmaceutical Scientists*, Toronto, Canada.

51. Rowe, R. C. 1977. The adhesion of film coatings to tablet surfaces—The effect of some direct compression excipients and lubricants. *J Pharm Pharmacol* 29: 723–726.

52. Lehtola, V.-M., J. T. Heinämäki, P. Nikupaavo et al. 1995. Effect of some excipients and compression pressure on the adhesion of aqueous-based hydroxypropyl methylcellulose film coatings to tablet surface. *Drug Dev Ind Pharm* 21(12): 1365–1375.

53. Pandey, P., D. S. Bindra, S. Gour et al. 2014. Excipient-process interactions and their impact on tablet compaction and film coating. *J Pharm Sci* 103: 3666–3674.

54. Kim, S., A. Mankad, and P. Sheen. 1986. The effect of application rate of coating suspension on the incidence of the bridging of monograms on aqueous film-coated tablets. *Drug Dev Ind Pharm* 12(6): 801–809.

55. Rowe, R. C., and S. F. Forse. 1981. The effect of intagliation shape on the incidence of bridging on film-coated tablets. *J Pharm Pharmacol* 33: 412.

56. Huyghebaert, N., A. Vermeire, J. P. Remon. 2004. Alternative method for enteric coating of HPMC capsules resulting in ready-to-use enteric-coated capsules. *Eur J Pharm Sci* 21: 617–623.

57. Vilivalam, V. D., L. Illum, and K. Iqbal. 2000. Starch capsules: An alternative system for oral drug delivery. *Pharm Sci Technol To* 3(2): 64–69.

13 Polymer Interactions with Drugs and Excipients

L. Diane Bruce and James W. McGinity

CONTENTS

INTRODUCTION

Drug and excipient interactions with polymeric film-forming agents influence the properties, functionality, and permeability of the applied film. Interactions can take place in the solid state at the substrate and polymer film interface, during preparation of a coating solution, or during dissolution. Some drug–polymer and excipient–polymer interactions are important in the polymeric film-formation process, and other interactions have deleterious effects on the functionality of the film.

Until the 1950s sugar coating was the only coating option for pharmaceutical preparations to provide elegance, mask taste or unpleasant odor, and protect light-sensitive or hygroscopic drugs. Sugar coating was a lengthy, time-consuming process with the outcome dependent on the skill of the operator [1]. Film coating was introduced and eventually replaced sugar coating. Film coating consisted of polymers dissolved in organic solutions and sprayed onto a substrate. Polymeric film coatings impart a variety of functions, including targeted drug delivery and sustained drug release. Over the years, however, film coating using organic solvents decreased largely due to governmental and environmental regulations limiting use due to toxicity and safety concerns. FDA regulations have also increasingly limited residual solvent content in pharmaceutical drug products, making solvent coating less desirable. As a result, aqueous-based polymeric film coatings are more commonly used in the industry. Aqueous polymeric-based coatings, although less toxic and safer to handle, have disadvantages. Problems arise due to interactions between the functional coating and the substrate, migration of water-soluble drugs into the film, and polymer aging affecting drug release and stability. In the last decade, solvent-free polymeric coating or dry coating has emerged as an alternative film coating technology. Dry coatings can be applied by atomization of molten materials, softened powder layering, and electrostatic adhesion. Many of the issues with aqueous-based and solvent-based film coatings are alleviated with this solvent-free technology [1,2].

Aqueous-based polymeric coating systems are comprised of polymers that are readily soluble in water or consist of aqueous colloidal dispersions of insoluble polymer particles. Both types of aqueous coatings differ in their film-forming mechanisms, each having advantages and disadvantages of use. Aqueous polymer solutions when prepared with high solids content can be highly viscous, leading to potential clogging of spray nozzles. Film formation in aqueous polymer solutions occurs as the polymer chains become entangled once sprayed droplets hit the substrate surface and water begins to evaporate. As drying occurs, a viscous, gelled, three-dimensional polymer network transforms into a continuous film. This process of film formation is also similar in an organic-based solvent system. The strength and durability of the resultant film is a function of the polymeric molecular weight and concentration of polymer solids in solution. Alternatively, aqueous polymeric colloidal dispersions consist of water-insoluble polymer particles suspended in an aqueous medium. The solids content of the dispersion can be increased without a significant increase in viscosity, thus reducing the potential for spray nozzle clogging. Higher solids content limits water penetration into the substrate and requires less water removal during the coating and drying process. This reduces processing energy requirements and overall time required for coating.

Film formation for aqueous polymeric colloidal dispersions is a complex mechanism in which water evaporates from between spherical polymer particles after deposition onto the substrate. The polymer particles begin to coalesce as they are drawn closer together as a result of viscous flow and surface tension. Frenkel's equation describes this phenomenon through a relationship between the half angle of contact, θ, as a function of surface tension, σ, time, t, polymer viscosity, η, and particle radius, r [3]. The degree of particle coalescence is characterized by the angle θ and improves as surface tension or interfacial tension increases and as the viscosity of the polymer spheres decrease. Plasticizing agents added to the dispersion are necessary for softening and deformation of the polymer spheres by swelling the polymer particles and reducing the minimal film-forming temperature, thus enhancing polymer coalescence and entanglement, which lead to film formation. The balance of forces required to maintain a stable disperse system make colloidal dispersions particularly sensitive to shear forces, additives, and temperature, requiring special care during preparation and processing.

Dry powder polymer coating offers the advantages of a solvent-free system, a necessity for water- or solvent-sensitive actives. The film formation mechanism is similar to solvent-based coatings and occurs through evaporation, polymer coalescence and sintering [2]. Polymer interactions with plasticizers, low-melting primers, or subcoat polymers, and the substrate are necessary for smooth functional film formation [1,2].

Although much of the discussion in this chapter concerning polymer interactions with drugs and excipients are focused on the aqueous-based polymeric film coatings, the same principles also apply to solvent and solvent-free polymer coating systems. The physicochemical properties of the film-forming polymer, excipients, and substrate must be considered for the desired results.

NON-COVALENT INTERACTIONS

Many of the interactions that occur between drugs, excipients, and polymers can be categorized as non-covalent. Non-covalent interactions include van der Waals attractions (or dispersion forces), hydrogen bonding, and electrostatic interactions (also called ionic bonding). All of these interactions involve an electrical charge due to temporary dipoles or ion formation. Dipoles occur when electrons are not shared equally between two atoms having different electro-negativities. The atom with the greater electro-negativity draws the shared electron pair closer to it, resulting in partial positively charged and negatively charged ends of the molecule.

Electrostatic interactions and hydrogen bonding are the strongest non-covalent type interactions. A hydrogen bond is the result of very strong dipole–dipole attraction between hydrogen atoms bonded to small, strongly electronegative atoms that have at least one unshared pair of electrons (typically nitrogen, oxygen, and fluorine). Electrostatic interactions or ionic bonding occur commonly in inorganic molecules and salts of organic molecules due to the attraction between negatively and positively charged atoms.

The interactions between drugs, excipients, and polymers can involve a single interaction or a combination of these non-covalent interactions. The results of these interactions can produce a wide variety of outcomes that may be bad (poor dissolution of poor bioavailability of the coated product due to film failure) or good (enhancement in film properties due to an interaction). Several studies have been conducted to determine how drug and excipient interactions with polymers affect the polymer system.

DRUG INTERACTIONS WITH EUDRAGIT® POLYMERS

Eudragit® acrylic polymer dispersions and resins have a variety of pharmaceutical applications, and interactions have been characterized for these polymers with several drug substances [4–15]. The functional groups on the Eudragit polymer backbones (and in some cases, the charges associated with the Eudragit polymers) make them readily reactive with drug substances.

Eudragit polymers used in the referenced studies were divided into anionic (Eudragit L and Eudragit S), cationic (Eudragit E), and zwitterionic (Eudragit RS and Eudragit RL) categories. In most of the case studies, complexes or co-precipitates were prepared between different drug substances and Eudragit polymer dispersions using a technique in which the dispersions were diluted and neutralized prior to introduction of an aqueous solution of a drug. Solid dispersions were also prepared by dissolving the Eudragit polymers and a drug in a solvent mixture and evaporating the solvent to collect the dried precipitate. The techniques used to study the interactions included carbon nuclear magnetic resonance (c-NMR) or proton nuclear magnetic resonance (h-NMR) spectroscopy, Fourier transform–infrared (FT-IR) spectroscopy, and differential scanning calorimetry (DSC) analysis of physical mixtures and formed complexes of the drugs and polymers, and dissolution studies. In all of the investigations, when an interaction between the drug substances and the Eudragit polymer was indicated, the interaction was found to be due to either a saline bond formation (ionic) or hydrogen bonding between the drug and the polymer.

In one study, a complex was formed between carteolol HCl and Eudragit L. Analyses of the complex using NMR spectroscopy demonstrated that the drug carteolol HCl was in the ammonium salt form. During preparation of the complex, the Eudragit polymer was neutralized with NaOH to form (R-COONa) subunits. Carteolol HCl, when added to the neutralized polymer, interacted to form an ionic bond with the carboxylate anions on the polymer. The carteolol ammonium salt interaction with the polymer carboxylate ion groups is illustrated in Figure 13.1 [4]. A similar salt (ionic bonding) was also reported between propranolol HCl and Eudragit L in another study. The complex between propranolol HCl and Eudragit L polymer was prepared similarly by neutralizing the polymer with NaOH prior to addition of the drug. The authors of this study confirmed the presence of a salt or ionic bond between propranolol and Eudragit L by FT-IR analysis of the complexes [5]. Additional findings of ionic bonding between the ammonium salts of drug substances and the anionic group of the Eudragit L polymers are reported in the literature [10,15].

Free-base drug substances are also known to interact with Eudragit polymers. When morphine HCl was added to neutralized Eudragit L polymer to form a complex, analysis using NMR

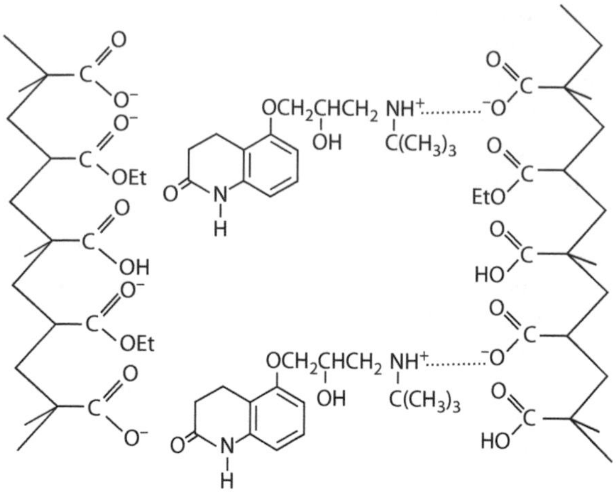

FIGURE 13.1 Types of bonds between drugs and Eudragit® L: saline bond corresponding to carteolol. (From Holgado MA et al., *Int J Pharm* 1995; 114:13–21.)

spectroscopy indicated the drug was in its free-base form. Morphine free–base was shown to interact with Eudragit L by hydrogen bonding with the polar groups of the polymer [6,7]. Two possibilities of hydrogen bonding are shown in Figure 13.2 between morphine free–base hydroxy and amine nitrogen groups with the polymer carboxylic functional groups. Similar hydrogen bonding between the nitrogen groups of dipyridamole free–base and Eudragit S, and naltrexone free–base and Eudragit L have also been observed [8,9].

Drug molecular structures directly affect the type of interaction and strength of the bond that will form with Eudragit polymers as shown in the two preceding examples. The secondary amine functional group in the carteolol is more reactive and less stable than the tertiary amine in morphine, resulting in an ionic bonding between the drug and polymer instead of hydrogen bonding. The co-precipitates prepared in the interaction between morphine HCl and Eudragit L were high in drug content. The high yield and efficiency of the process was attributed to the ability of the morphine to react with the polymer at multiple sites via hydrogen bonding.

In some cases, drug substances show signs of interaction with one category of Eudragit polymers but no interaction with others. Such an example occurred in co-precipitates prepared from

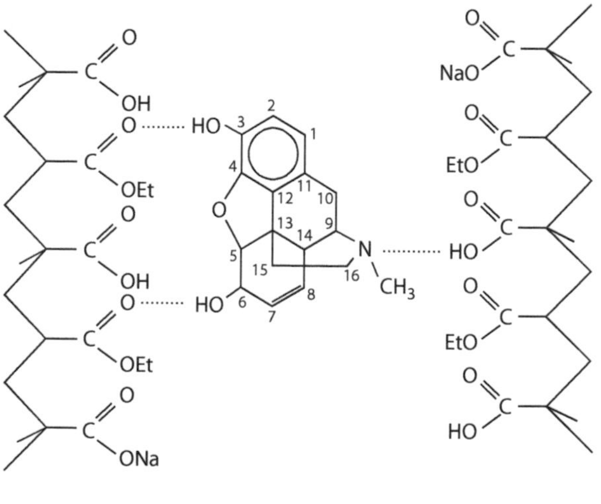

FIGURE 13.2 Hydrogen bonds between morphine and Eudragit® L. (From Alvarez-Fuentes J et al., *Drug Dev Ind Pharm* 1994; 20(15):2409–2424.)

ibuprofen and anionic (Eudragit S 100) or zwitterionic (Eudragit RL/RS) polymers. No interaction was found between ibuprofen and these polymers using FT-IR, X-ray diffraction, or DSC analysis of the co-precipitates [11]. However, when the drug was co-precipitated with a blend of Eudragit E 100 and Eudragit S 100, a significant interaction was found due to the addition of the cationic polymer (Eudragit E). Dissolution studies of co-precipitates of ibuprofen and Eudragit RS and RL showed a delayed release of the drug; however, this was explained to be due to the swelling of the zwitterionic polymers rather than due to an interaction with ibuprofen. The release rate of the drug continued to decrease with increased polymer concentration as a result of increased thickness of the diffusion layer surrounding the drug.

Several studies were conducted by Lin et al. using a variety of drug substances, including indomethacin, warfarin, and piroxicam combined with one or all categories of the previously described Eudragit polymers, primarily Eudragit E, RL, and S [12–14]. The drug substances were found to interact with the Eudragit E polymer but not with the other Eudragits. Indomethacin and warfarin were observed to increase in solubility when a saturated amount of the drug was combined with Eudragit E in acetone. The solubility of the drug substances was also shown to increase when the drugs were combined in solution with Eudragit S and RL but not to the extent observed with Eudragit E. The enhanced solubility was attributed to an interaction or binding between the drug substance and the polymer, thereby increasing the drug solubility in the medium. With each of the three drug substances, there was a shift in FT-IR spectra when dissolved in acetone with Eudragit E, but no shift was observed when the drug substances were combined with Eudragit RL or S. The shift in spectra was substantial evidence of a drug–polymer interaction due predominately to hydrogen bonding between the drug substances and the Eudragit polymer. The release of warfarin and indomethacin from tablets prepared from physical mixtures, ground mixtures, and films of Eudragit polymers containing drug supported this conclusion. As shown in Figure 13.3, the release of warfarin in pH 7.4 phosphate buffer from tablets prepared from warfarin–Eudragit cast film or ground mixture is significantly delayed due to the molecular interaction between warfarin and Eudragit E [14]. Results were similar for indomethocin. A delayed-release profile between these drug substances and Eudragit RL was also observed and was not attributed to an interaction, but to the insolubility of the polymers in pH 7.4 phosphate buffer. No indication of an interaction was found using FT-IR or DSC analysis of physical mixtures, ground mixtures, cast films, and preheated samples of drug substance and Eudragit RL combinations.

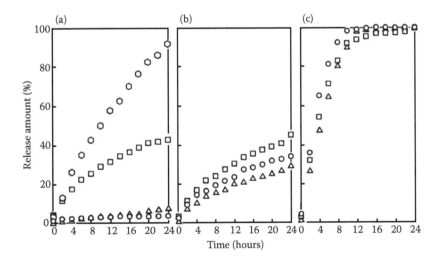

FIGURE 13.3   Release behavior of warfarin from tablets prepared by warfarin–Eudragit® resins: (a) Eudragit E; (b) Eudragit RL; (c) Eudragit S. (○) Tablets prepared from warfarin/spray-dried lactose; (□) tablets prepared from physical mixture; (△) tablets prepared from the cast film; (◔) tablets prepared from 24 h ground mixture. (From Lin SY et al., *Eur J Pharm Sci* 1994; 1:313–322.)

Drug–polymer interactions can also occur during the dissolution process as observed in prepared films of Eudragit RS and Eudragit RL containing salicylic acid [16,17]. Jenquin et al. [16] found the rate of drug release from Eudragit films to be related to the properties of the drug and the type of polymer in the film. Both salicylic acid and chlorpheniramine maleate (CPM) were incorporated into Eudragit RS and RL films with increasing drug concentrations. For CPM, the release of drug from the film increased with increasing drug loading. However, when salicylic acid was incorporated into Eudragit RL film, a trend (at 2%, 5%, and 10% drug concentration in the film) of a peak in the concentration of the salicylic acid released into the media followed by a sharp decline in concentration occurred (Figure 13.4) [16]. This decline in drug concentration in the dissolution media was later shown using Langmuir adsorption isotherms to be due to adsorption of salicylic acid onto the Eudragit polymers. It was theorized that the salicylic acid after diffusing through the film, ionized (became negatively charged) in the neutral pH of the dissolution media and interacted with the positively charged quaternary ammonium groups on the polymer via electrostatic binding. This adsorption was greater for Eudragit RL than for Eudragit RS due to the greater ratio of quaternary ammonium groups in the Eudragit RL available to bind with the salicylic acid. The adsorption isotherms clearly showed adsorption of salicylic acid to RL and RS polymers and no adsorption occurring for CPM (Figure 13.5). Furthermore, salicylic acid release decreased with increasing ionic strength of the dissolution media. This effect was due to a decrease in adsorption of the drug to the polymer by inhibition of electrostatic binding in the presence of Cl ions in the Tris buffers (Figure 13.6). The chloride ions in the media interacted with the quaternary ammonium groups on the polymer and buffered the electrostatic charge so that the negatively charged groups of the salicylic acid were unable to bind. Some adsorption was maintained even with the increase in ionic strength of the media, which may have been due to hydrogen bonding and van der Waals forces of attraction, referred to by the authors as nonelectrostatic interactions. When CPM was incorporated into the Eudragit RS/RL film, the same dissolution phenomenon did not occur. CPM remained ionized (positively charged) in the dissolution media. The positively charged CPM and the positively charged quaternary ammonium groups on the Eudragit polymers were not electrostatically attracted, resulting in constant concentration of CPM in the dissolution medium.

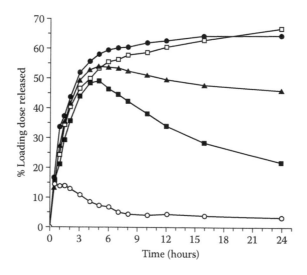

FIGURE 13.4 Effect of drug concentration on the release of salicylic acid from Eudragit® RL PM films. (□) 20% drug; (●) 15% drug; (▲) 10% drug; (■) 5% drug; (○) 2% drug. (From Jenquin MR et al., *J Pharm Sci* 1990; 79(9):811–816.)

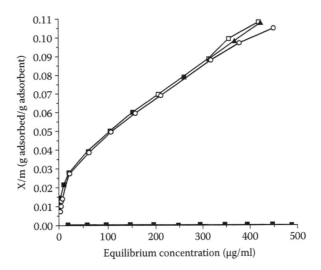

FIGURE 13.5 Isotherms for the adsorption of SA and CM to Eudragit® RL 100 particles of varying sizes. (□) SA, polymer size 106–125 μm; (▲) SA, polymer size 180–250 μm; (○) SA, polymer size 420–600 μm; (■) CM, polymer size 420–600 μm. CM, chlorpheniramine maleate; SA, salicylic acid. (From Jenquin MR et al., *J Pharm Sci* 1990; 79(9):811–816.)

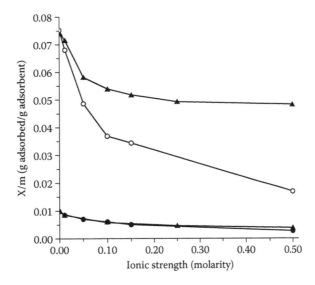

FIGURE 13.6 Effect of ionic strength on the adsorption of SA to Eudragit® RL 100. Two initial drug concentrations and two ionic strength systems were evaluated: (△) 0.1 mg/mL SA in NaCl; (▲) 1.0 mg/mL SA in NaCl; (●) 0.1 mg/mL SA in Tris–HCl; (○) 1.0 mg/mL SA in Tris–HCl. SA, salicylic acid. (From Jenquin MR et al., *J Pharm Sci* 1990; 79(9):811–816.)

INTERACTIONS WITH COLLOIDAL DISPERSIONS

An interaction between a drug or excipient and a polymer can be the result of a disturbance in the electrostatic balance of the polymeric dispersion medium. When fine particles are dispersed in a liquid medium, collisions occur between particles due to Brownian movement. When particles collide, they will either remain in permanent contact or rebound to remain freely suspended. The status of the particles is dependent upon the attractive and repulsive forces of interaction between the particles. There are five possible types of forces that influence particle stability: electrostatic forces of repulsion, van der Waals forces or electromagnetic forces of attraction, Born forces or short-range repulsive forces, steric forces dependent on the geometry and conformation of molecules at the particle interfaces, and solvation forces due to changes in quantities of adsorbed solvent on approach of neighboring particles [18]. According to the DLVO (Derjaguin, Landau, Verwey, Overbeek) theory, electrostatic charge stabilization of dispersed particles is a result of the balance between electrostatic repulsive and van der Waals attractive forces between particles [19]. An electrostatic repulsive charge around each dispersed polymer particle prevents aggregation or flocculation. A double layer of charge exists at the surface of the particle: one that is tightly bound to the particle surface and another that is more diffuse. Ions on the particle surface attract counter-ions from the media and form the tightly bound layer. As distance increases from the particle surface, the counter-ion concentration becomes more diffuse until eventually the concentrations of anions and cations result in electric neutrality. The zeta potential is the difference in the charge potential between the surface of the tightly bound layer and the neutral region in a solution [19]. If the zeta potential of a colloidal dispersion is reduced below a certain value (which is different for every system), the attractive van der Waals forces will exceed the repulsive forces, and the particles will flocculate. This effect is a disadvantage of polymeric colloidal dispersions that results in a sensitivity to additives, such as electrolytes, changes in pH, temperature, and shear forces. All of these factors can lead to changes in the thickness of the diffuse double layer between adjacent suspended polymer particles, leading to coagulation or flocculation. Numerous drug–polymer or excipient–polymer interactions may result from the disturbance of polymer colloidal particle electrostatic balance and compression of the electric double layer between polymer particles.

INTERACTIONS WITH COLLOIDAL DISPERSIONS DUE TO CHANGES IN THE DIFFUSE DOUBLE LAYER

Goodman and Banker were familiar with polymeric colloidal dispersion flocculation in the presence of drug substances and used this phenomenon as a novel approach to formulate a sustained-release dosage form. The acid salts of cationic nitrogen–containing drugs that were found to flocculate acrylic anionic copolymer emulsions are listed in Table 13.1 [20]. These drugs readily caused flocculation of the dispersions with flocculation values of 10 to 20. The flocculation value is the concentration of drug in mmoles/L required to cause complete flocculation within two hours. The flocculation or drug entrapment procedure called for slowly adding the acrylic copolymer emulsion to a constantly mixed solution of drug (methapyrilene hydrochloride) in distilled water. This resulted in immediate flocculation and precipitation of the system. The mixture was vacuum filtered, collected, and dried for four hours at 50°C and then comminuted and screened. The interaction between the drug and polymer was studied by plotting the relationship between drug concentration and acrylic copolymer drug entrapment ratio, determining flocculation values at various polymeric emulsion pH levels, and determining in vitro release rates of precipitates and tablet preparations. A linear relationship was observed between the initial amount of drug in solution and the amount of drug entrapped in the solid product. Additionally, the pH of the anionic polymeric copolymer emulsion affected the amount of drug entrapped. Raising the pH of the polymer emulsion improved the stability of the polymer against electrolyte-induced flocculation by increasing the hydrophilicity and thus the solvation of the polymer. Decreasing the

TABLE 13.1
Flocculation of an Acrylic Copolymer Emulsion System by Acid Salts of Various Cationic Nitrogen-Containing Medicinals

Compound	Molecular Weight	Type of Amine	Flocculation Value
d-Amphetamine sulfate	368.5	Primary	10
Chlorpromazine HCl	355.3	Tertiary	10
Atropine sulfate	694.9	Tertiary	10
Homatropine methyl bromide	370.3	Quaternary ammonium	20
Ephedrine HCl	201.7	Secondary	20–25
Phenylephrine HCl	203.7	Tertiary	40–50
Morphine sulfate	758.9	Tertiary	10–15
Dihydrocodeinone bitartrate	494.5	Tertiary	20
Methapyrilene HCl	297.9	Tertiary	10
Pyrilamine maleate	401.5	Tertiary	10
Chlorpheniramine maleate	390.9	Tertiary	10

Source: H. Goodman, G. S. Banker, *J Pharm Sci*, 59, 8, 1131–1137, 1970.

pH of the polymer emulsion had the opposite effect, increasing the hydrophobicity of the polymer particles and making them more prone to flocculating in the presence of methapyrilene HCl. Goodman and Banker [20] theorized the cause of flocculation to be due to the added electrolyte (methapyrilene HCl) decreasing the thickness of the diffuse ionic layer between polymer particles in the latex emulsion. Analyses of the drug–polymer precipitate indicated that the drug was still in its chloride salt form. The authors concluded that the drug did not chemically interact with the polymer by forming an ionic bond, and that the interaction was therefore due to the ability of the polymeric particles to enclose or spatially surround the drug molecules. This interaction phenomenon resulted in the formation of precipitates that, when pressed into tablets, yielded suitable sustained-release properties and reduced toxicity when tested in animal models. In another study, Rhodes et al. demonstrated that entrapment and flocculation of an acrylic copolymer dispersion can also be influenced by drug salt type [21]. In this study, carboxylic acid anion salts were found to facilitate and enhance the amount of drug (chlorpheniramine) bound or entrapped by the polymer for sustained-release action.

Bruce et al. observed chlorpheniramine maleate to flocculate Eudragit L 30 D-55 dispersions [22]. This interaction was theorized to be responsible for poor film formation and premature drug release from the enteric-coated CPM pellets in acidic media. Additional weight gain of the enteric polymer on the CPM pellets was required to delay drug release in acidic media. Adsorption isotherms were used to study the drug–polymer interaction by plotting the log of the amount of drug adsorbed per unit mass adsorbent as a function of log CPM concentration remaining in solution. A linear relationship was observed that best correlated with the Freundlich model, indicating that adsorption of CPM to Eudragit L 30 D-55 takes place in multiple layers rather than by a specific chemical interaction between the drug and the polymer. FT-IR scans of the drug and polymer precipitate and physical mixture were identical, verification that no specific chemical bonding occurred between the drug and polymer system. The interaction between drug and polymer was attributed to the change in the thickness of the diffuse ionic layer around the polymer particles by CPM and resultant entrapment of CPM in the polymer floccules.

The incompatibilities between the drug substances and the colloidal dispersions observed by these authors could affect the film-formation process and explain premature drug leakage and dose dumping from modified or sustained-release film-coated dosage forms. A flocculation interaction at the substrate polymer film interface can also occur during film application. A film may not form properly if the drug and polymer interact, as demonstrated by Bodemeier and Paeratakul [23]. They were unable to prepare films from ethylcellulose pseudolatex (Aquacoat® and Surelease®) and Eudragit NE 30 D dispersions incorporating the drugs propanol HCl and CPM due to flocculation of the colloidal dispersions. The authors postulated the flocculation was due to interaction of the cationic drug salts with the anionic surfactants (sodium lauryl sulfate or ammonium oleate) used to stabilize the polymer emulsion systems. This type of electrostatic interaction with the surfactant changed the thickness of the double layer stabilizing the dispersed polymer particles. To prove that the cationic drug salts interacted with the anionic surfactant, the authors prepared a plasticized ethylcellulose pseudolatex by replacing the anionic surfactants with the nonionic surfactant Pluronic P103. The same cationic drug salts were added to the newly processed polymeric dispersions (Pluronic P103 stabilizing the disperse system) and flocculation did not occur. As a result, films were successfully prepared.

INTERACTIONS WITH COLLOIDAL DISPERSIONS DUE TO BRIDGING OR DEPLETION FLOCCULATION

Hydrophilic, water-soluble polymers, such as hydroxypropyl methyl cellulose (HPMC), hydroxy propyl cellulose (HPC), and polyethylene glycol (PEG), are commonly used to increase or accelerate drug release from coatings or films prepared from water-insoluble polymers [24–27]. During dissolution, the incorporated water-soluble polymer dissolves, allowing pores to form that can increase film permeability. Addition of these pore-forming polymers to polymeric colloidal dispersions must be done with caution and at appropriate concentrations to avoid flocculation. The added polymers can adsorb onto the surface of the polymer colloidal particles, and the chains of the adsorbed polymer extend into the aqueous media. This leads to two possible occurrences. First, the colloidal particles may collide and the interpenetrating polymer chains create a zone of osmotic pressure differential and diffusion of the medium into itself, driving particles apart. Second, polymer adsorption onto the colloidal particles may cause destabilization of the colloidal system by means of bridging flocculation. Bridging flocculation occurs when the adsorbed polymer chain extends to another particle and forms a bridge, typically occurring at low polymer concentrations. At higher polymer concentrations, the particles may become coated with the adsorbed polymer and repel one another, potentially re-stabilizing the polymeric colloidal dispersion [28]. Flocculation may also occur with non-adsorbing polymers. Free non-adsorbing polymers added to the dispersion can reach a critical flocculation concentration and cause a phase separation through a sudden increase in the viscosity of the dispersion. This phenomenon is known as depletion flocculation.

Depletion flocculation was observed when HPMC was added as a film pore-forming agent to an ethylcellulose dispersion (Aquacoat) [29]. The extent of flocculation was determined by measuring the sedimentation volume F of the flocculated dispersion. F is defined as the ratio of flocculated sediment to original dispersion volume. The addition of HPMC to the ethylcellulose dispersion flocculated the colloidal polymer particles above a critical HPMC concentration. The minimum HPMC concentration and viscosity grade necessary to cause flocculation are shown in Figure 13.7. Flocculation was observed to occur at ranges of 3% to 10% HPMC concentration (based on polymer content). The higher-molecular-weight grades of HPMC were more efficient flocculants. The solids content of the ethylcellulose dispersion influenced flocculation, and the HPMC concentration necessary to cause flocculation decreased with increasing solids content of the dispersion (Figure 13.8). According to this study, low-molecular-weight pore-forming agents should be added to the colloidal dispersion and prepared at low solids concentration to prevent flocculation and interaction.

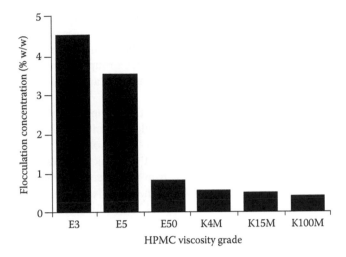

FIGURE 13.7   Flocculation concentration (based on solids content of Aquacoat®) of different HPMC viscosity grades (Aquacoat, 15% w/w solids content). HPMC, hydroxypropyl methyl cellulose. (From Wong D, Bodemeier R, *Eur J Pharm Biopharm* 1996; 42(1):12–15.)

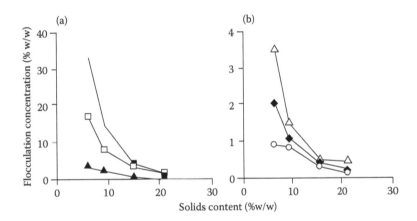

FIGURE 13.8   Flocculation concentration (based on solids content of Aquacoat®) as a function of pseudolatex solids content with different HPMC viscosity grades: (a) (■) E3; (□) E5; (▲) E50; (b) (△) K4M; (◆) K15M; (o) K100M. HPMC, hydroxypropyl methyl cellulose. (From Wong D, Bodemeier R, *Eur J Pharm Biopharm* 1996; 42(1):12–15.)

PLASTICIZATION OF POLYMERIC FILMS BY DRUGS AND EXCIPIENTS

INTERACTIONS BETWEEN POLYMERS AND PLASTICIZERS

Film-forming polymers typically require the addition of excipients or adjuvants in order to enhance the film-forming process and properties of the film. Plasticizers are added to impart flexibility and distensibility, increase toughness, improve strength, and reduce brittleness of the polymer. The plasticizer is usually a small compound with low molecular weight and functional groups that interact with the polymer to decrease the intermolecular cohesive forces between polymer chains, thereby increasing polymer segmental mobility and free volume. To be an effective plasticizer, the compound must be compatible and miscible with the polymer. The size, shape, and the nature of the functional groups will determine the functionality of the plasticizer. The degree of interaction between plasticizer and polymer will affect film mechanical properties, glass transition temperature (T_g), drug release, and permeability of the film. In general, the lower the molecular weight of the compound, the greater the plasticizing action it will exhibit on the polymer. Plasticizers commonly used in film-coating applications can be divided into three groups: polyols consisting of propylene and polyethylene glycols, organic esters including the phthalate and citrate esters, and vegetable oils and glycerides, such as acetylated monoglycerides.

Several methods may be employed to determine the compatibility or degree of interaction between a plasticizer and polymer material and include determination of the change in T_g of the plasticized polymer, comparison of the plasticizer–polymer solubility parameters, measurement of the intrinsic viscosity of the polymer dissolved in the plasticizer, examination of a plasticized polymer film (either visually or by scanning electron microscopy [SEM] for transparency), and mechanical testing of the plasticized polymer film. These are typical methods employed to determine plasticizer–polymer compatibility; other methods can be found in the literature [30].

Glass Transition Temperature

The plasticizing effect of a compound on a polymer will result in a change in the polymer's T_g. This is the temperature at which the polymer changes from a hard glassy material to a softer, rubbery state. A differential scanning calorimeter or modulated differential scanning calorimeter is employed to measure the T_g of the polymer–plasticizer combination. A polymer will exhibit a lowering of T_g when combined with a compatible plasticizer if sufficient interaction occurs to result in plasticization. A plasticizer with a high degree of interaction will cause a greater lowering in the glass transition temperature of the polymer than an equivalent concentration of one having a poor interaction. If the T_g of the polymer increases rather than decreases, the compound has an anti-plasticizing effect on the polymer, due to immobilization of the polymer chains by hydrogen bonding, van der Waals attractions, and steric hindrance between the polymer and plasticizer, thereby increasing the stiffness of the polymer chains [31]. Jackson and Caldwell explained the occurrence of antiplasticization as a result of a reduction in the free volume of the polymer, interaction between the polar groups of the polymer and the anti-plasticizer, and a physical stiffening action due to the presence of rigid anti-plasticizer molecules adjacent to the polar groups of the polymer [32].

Solubility Parameter

The solubility parameter is calculated from the heat of mixing of two components and predicts component interaction and compatibility as demonstrated by Okhamafe and York from the relationship defined by Hildebrand and Scott [30] in Equation 13.1:

$$\Delta H = V_m[(\Delta E^1/V^1)^{1/2} - (\Delta E^2/V^2)^{1/2}]^2 \, \varphi^1 \cdot \varphi^2 \qquad (13.1)$$

where V_m is the total volume of the mixture, ΔE is the energy of vaporization, V is the molar volume, and φ is the volume fraction of the components. $\Delta E/V$ is generally referred to as the cohesive energy density and its square root as the solubility parameter δ. Thus, the above equation can take the following form (Equation 13.2):

$$\Delta H = V_m(\delta_1 - \delta_2)^2 \varphi_1 \cdot \varphi_2 \qquad (13.2)$$

If δ_1 and δ_2 are equivalent, the heat of mixing is zero, indicating maximum interaction, solubility, and compatibility. Therefore, equivalency of polymer and plasticizer calculated solubility parameters dictates the compatibility and miscibility of the two compounds. Wang et al. calculated the solubility parameters for Eudragit RS 30 D and plasticizer combinations including acetyl tributyl citrate (ATBC), acetyl triethyl citrate (ATEC), diethyl phthalate (DEP), triacetin, and triethyl citrate (TEC), and found all these plasticizers to differ from the polymer by less than 6.3 $(J/cm^3)^{1/2}$, indicating miscibility and compatibility with Eudragit RS 30 D [33]. Solubility parameters were used by Wu and McGinity to predict the compatibility and effectiveness of CPM, ibuprofen, and methylparaben and theophylline as plasticizers for the polymer Eudragit RS 30 D [34]. Compounds whose solubility parameters differed from the polymer by 9.4 $(J/cm^3)^{1/2}$ demonstrated no plasticization effect on Eudragit RS 30 D.

Intrinsic Viscosity

Polymer–plasticizer interactions can be studied by determining the intrinsic viscosity of the polymer dissolved in the plasticizer solution using the following relationship (Equation 13.3)

$$\eta_{sp}/c = [\eta] + k'[\eta]^2 c \tag{13.3}$$

where η_{sp}/c is reduced viscosity, $[\eta]$ is intrinsic viscosity, defined as the limit of the reduced viscosity (η_{sp}/c) as the concentration approaches zero, c is the concentration of the solution, and k' is an interaction constant, also called the Huggins constant [35]. A high intrinsic viscosity value indicates a greater degree of polymer–plasticizer interaction, reflecting the tendency of the polymer to uncoil and associate with the plasticizer solvent. Shah and Zatz measured intrinsic viscosity to assess the plasticizer–polymer interaction between cellulose ester polymers and dimethyl phthalate, diethyl phthalate, dibutyl phthalate, and glyceryl triacetate [36]. The authors found that intrinsic viscosity dropped as the phthalate hydrocarbon chain length increased, but the results did not directly correspond with mechanical testing observations. Assessing polymer–plasticizer interactions by the intrinsic viscosity method has limitations because not all polymers can be dissolved in a plasticizer. This method also assumes that the concentration of polymer in the system is low and thus cannot accurately reflect the levels of polymer–polymer interaction in the system. Hutchings et al. found that differences for η (intrinsic viscosity) and k' (interaction constant) for various plasticizers evaluated did not reflect differences between the plasticizers with respect to their interactions with ethylcellulose. Determination of solubility parameters indicated a rank order with methanol > PEG > citrate esters and triacetin > diesters > oleic acid/oleyl alcohol, but no differences in intrinsic viscosity or interaction constant could be identified. This was attributed to the rigidity of the ethylcellulose molecules, where influences of the solvent molecules (plasticizer) on the polymer chains are less significant, and changes in intrinsic viscosity from one solvent system (plasticizer) to another are more subtle. Due to this, the authors suggested that evaluation of polymer–plasticizer interactions via intrinsic viscosity and interaction constant is not an ideal method for determining plasticizer suitability for film-coating additives [37].

Mechanical Properties

Interactions between plasticizer molecules and polymer molecules have an effect on the mechanical properties of the polymer film. An ideal film should be hard and tough without being brittle. Mechanically, this translates into having a high tensile strength and large elongation or strain before breaking. These properties can be quantified through tensile testing and plotting the resulting stress (σ) versus strain (ε) curve. The tensile strength, also referred to as ultimate tensile strength, is the maximum tensile stress sustained by a test specimen during a tensile test. The area under the stress–strain curve is a measure of the material's toughness [38].

Hutchings et al. illustrated the relationship between plasticizer functional groups and molecular structure and free film mechanical properties [39]. The authors prepared free films with 10 different plasticizers from the following classifications: branched esters, di-acid esters, and fatty acids/alcohols. Stress, strain, and elastic modulus were plotted as a function of plasticizer type or class and concentration. A rank order for influence on mechanical properties can be seen within each class (Figure 13.9). In general, increasing the amount of plasticizer in the films lead to a reduction of free film modulus and stress values while strain at rupture values increased. Values obtained for stress for the di-acid esters and the fatty acid/alcohol were generally lower than those obtained for citrate esters and triacetin. The authors attributed this to the long-chain molecular structure of the di-acid esters and fatty acid/alcohols. In the branched ester class, the plasticizers with hydroxyl groups, TEC and tributyl citrate (TBC), demonstrated the lowest modulus values as a result of

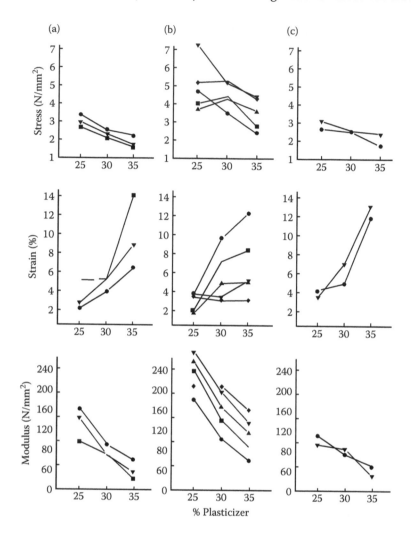

FIGURE 13.9 Effect of plasticizer type and amount on free film stress (N/mm²), strain (%), and modulus (N/mm²). (a) Di-acid esters: (●) DBS; (▼) DBA; (■) DMS; (b) branched esters: (●) TEC; (▼) ATEC; (■) TBC; (▲) ATBC; (◆) TRI; (c) fatty acid/alcohols: (●) OALC; (▼) OLAC. ATBC, acetyl tributyl citrate; ATEC, acetyl triethyl citrate; DBA, dibutyl adipate; DBS, dibutyl sebacate; DEP, diethyl phthalate; DMS, dimethyl sebacate; OALC, oleyl alcohol; OLAC, acid oleic; TBC, tributyl citrate; TEC, triethyl citrate; TRI, triacetin. (From Hutchings D et al., *Int J Pharm* 1994; 104:203–213.)

hydrogen bonding interactions with the polymer. The esterified compounds ATEC, ATBC, and triacetin are unable to interact with the polymer as effectively.

Gutiérrez-Rocca and McGinity were also able to explain differences in mechanical properties of polymeric films by considering the chemical structure of the plasticizer [40]. The plasticizers used in this study were TEC, ATEC, TBC, and ATBC incorporated in Eudragit L 100–55. All plasticizers lowered the glass transition temperature of the polymer with increasing concentrations except for TBC and ATBC, which plateaued at 10% w/w (based on dry polymer weight). TBC and ATBC are water-insoluble plasticizers that are poorly miscible with the polymer. All plasticizers added to the film resulted in a reduction of tensile strength. The water-soluble plasticizers resulted in a significant increase in percentage elongation of the film. The elastic modulus (Young's modulus) continually decreased with increasing concentration of water-soluble plasticizers. There were no statistically significant changes in elastic modulus when greater than 10% concentration of the water-insoluble plasticizers was incorporated into the film. These results correlated with the findings for glass transition temperature. Eudragit L 100–55 contains carboxylic functional groups capable of interacting with the plasticizer molecules by hydrogen bonding, electrostatic interactions, and dispersion forces. The results of the mechanical and thermal testing can be explained by examining the chemical structures and functional groups of both the polymer and the plasticizers (Figure 13.10). The smaller plasticizers triacetin and TEC have the greatest ability to interact with the polymer. In triacetin, the carbonyl oxygens are readily available to interact through hydrogen bonding with the carboxyl hydrogens of the copolymer. This is also true for TEC, but the presence of the ethyl group may reduce the accessibility of the carbonyl oxygen for hydrogen bonding. Side chains on the other compounds hinder the availability of the functional groups for bonding. The results of the study indicate that smaller water-soluble plasticizers have a higher affinity to diffuse into, and interact with, the polymer, increasing the molecular mobility of the polymer chains. These results are in agreement with the findings of Bodemeier and Paeratakul, who also examined the differences that water-soluble and water-insoluble plasticizers have on the mechanical properties of both dry and wet films [41]. Dry Eudragit RS 30 D films plasticized with the water-soluble plasticizers TEC and triacetin had higher elongation and lower puncture strength, corresponding to a lower elastic modulus compared to the water-insoluble plasticizers. This indicated that the water-soluble plasticizers were more efficient plasticizers. Corresponding wet films were significantly more flexible than the dry films plasticized with the water-soluble plasticizers. However, a

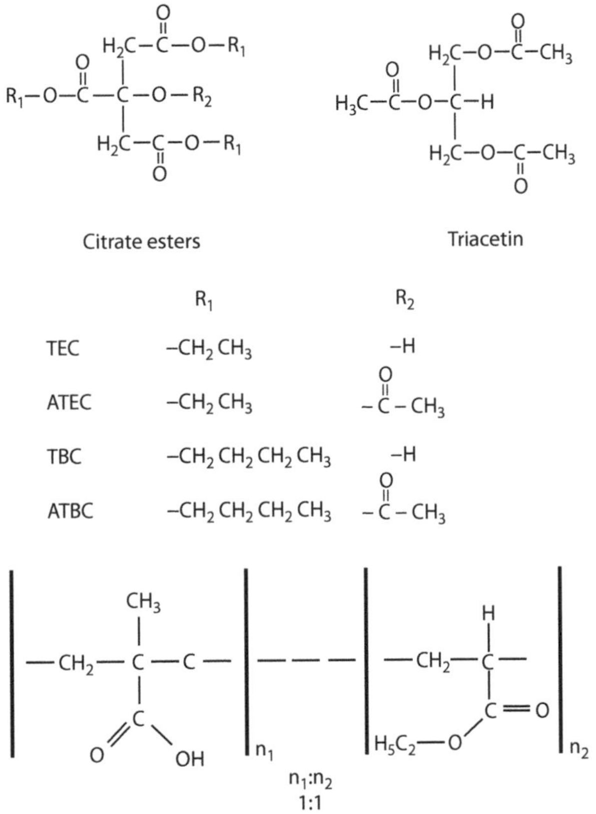

FIGURE 13.10 Molecular structure of triacetin and the citrate ester plasticizers and structural characteristics of poly(methacrylic acid ethyl acrylate), commercially available as Eudragit® L 30 D and L 100–55. ATBC, acetyl tributyl citrate; ATEC, acetyl triethyl citrate; TBC, tributyl citrate; TEC, triethyl citrate. (From Gutiérrez-Rocca JC, McGinity JW, *Int J Pharm* 1994; 103:293–301.)

disadvantage in using water-soluble plasticizers is the tendency for the compounds to leach from the films (as determined amount of plasticizer remaining shown in Table 13.2), impacting the film's mechanical properties.

A compatible and efficient plasticizer will function to make a softer, tougher polymer and will reduce brittleness. Using an indentation test, Alton and Abdul-Razzak observed that HPMC films plasticized with PEG 600, 1500, 4000, and 6000 became softer and more viscoelastic with increasing plasticizer content [42]. Mechanical testing of the same films showed a reduction in tensile strength, an increase in elongation, and lowering of elastic modulus as the molecular weight of the PEG decreased (Figure 13.11). The lower-molecular-weight PEG proved to be the most efficient plasticizer and produced the best films. The authors used the gel theory to explain the interaction between plasticizer and polymers as a competition between a plasticizer, the polymer, and solvent (water) molecules for polymer active sites. According to this theory, there will be more of the lower

TABLE 13.2

Mechanical Properties of Dry and Wet Eudragit® RS 30 D Films Plasticized with Water-Soluble and Water-Insoluble Plasticizers (20% w/w) (Standard Deviation in Parentheses)

Plasticizer (Film	Puncture Strength (MPa)		Elongation (%)		Plasticizer
Thickness μm)	Dry	Wet	Dry	Wet	Remaining (%)
TEC (309)	1.99 (0.22)	0.93 (0.05)	142.8 (1.3)	38.4 (4.6)	56.29 (1.79)
Triacetin (302)	1.82 (0.38)	0.61 (0.07)	120.9 (6.0)	6.8 (0.6)	35.92 (1.06)
ATBC (314)	4.30 (0.09)	1.11 (0.13)	77.8 (7.6)	85.2 (3.6)	101.84 (1.67)
ATEC (323)	4.01 (0.18)	1.01 (0.02)	86.9 (5.5)	64.3 (8.5)	90.38 (0.05)
DBP (327)	3.18 (0.47)	0.88 (0.19)	93.2 (12.6)	106.9 (9.2)	99.95 (1.88)
DBS (324)	2.37 (0.09)	0.79 (0.04)	91.8 (2.0)	59.7 (3.6)	88.34 (0.66)
DEP (324)	2.47 (0.40)	0.91 (0.03)	91.1 (3.2)	51.0 (3.8)	95.27 (1.53)
TBC (319)	2.37 (0.40)	0.86 (0.03)	113.5 (1.8)	86.6 (3.4)	97.79 (2.06)

Source: R. Bodemeier, O. Paeratakul, *Pharm Res*, 11, 6, 882–888, 1994.

Note: ATBC, acetyl tributyl citrate; ATEC, acetyl triethyl citrate; DBP, dibutyl phthalate; DBS, dibutyl sebacate; DEP, diethyl phthalate; TBC, tributyl citrate; TEC, triethyl citrate.

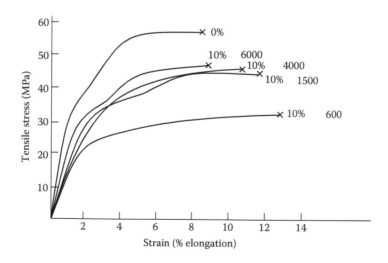

FIGURE 13.11 Stress–strain curves for HPMC films containing 10% of different grades of polyethylene glycol. HPMC, hydroxypropyl methyl cellulose. (From Aulton ME et al., *Drug Dev Ind Pharm* 1981; 7(6):649–668.)

molecular weight PEG molecules to compete for and interact with polymer active sites. The plasticizer thus reduces the number of active sites available for polymer–polymer contact and thereby reduces the rigidity of the polymer.

Film Permeability/Dissolution

Addition of a plasticizer not only changes the mechanical properties of a film, but also changes the permeability, adhesion, and drug-release characteristics. Typically, plasticizing agents added to aqueous colloidal polymeric dispersions lower the rate of drug release from the resultant film due to enhanced coalescence of colloidal polymer particles, decreased brittleness, and an increase in flexibility, toughness, and strength. These effects contribute to improved film performance.

Plasticizing agents can also have the opposite effect and can increase the rate of drug release and film permeability due to increased polymer segmental mobility during processing. After processing, the polymer chains remain in a loosened state upon cooling and are thus more permeable. Additionally, an excess of plasticizing agent incorporated into a film can migrate to the film surface, accelerating the rate of drug release due to picking and sticking or attraction of moisture [43,44]. Drug-release rate can also be influenced by the solubility of the plasticizing agent [45]. TEC, a water-soluble plasticizer, was incorporated into Eudragit RS films containing the drug propanolol HCl. Release rates of propanolol HCl were plotted versus TEC concentration in the films. The release rate constant was high at low plasticizer content, dropped to a minimum plateau, and then increased again with higher plasticizer concentrations (Figure 13.12). The observed increase in release rate with higher TEC concentrations was explained as leaching of the water-soluble plasticizer from the film. The high release rate constant at lower plasticizer levels was attributed to incomplete film formation and polymer coalescence due to insufficient plasticizer content. In contrast, the release rate of Eudragit L 30 D-55 was delayed in pH 5.5 media by the incorporation of high-molecular-weight PEG, a water-soluble plasticizer [46]. The drug-release rate from the films decreased with the addition of PEG, as the molecular weight of the PEG increased. Films containing PEG 8000 had the slowest rate of release in pH 5.5 due to the formation of a film that was not permeable or soluble at a pH where Eudragit L 30 D-55 normally dissolves.

NONTRADITIONAL PLASTICIZERS

Compounds other than those commonly recognized as plasticizers can produce plasticization effects in film coatings. In certain applications, excipients and drug compounds have the potential

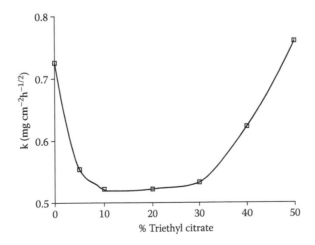

FIGURE 13.12 Effect of triethyl citrate concentration on release rate constant of propranolol HCl (50 mg)—Eudragit® RS films. (From Bodmeier R, Paeratakul O, *Int J Pharm* 1990; 59:197–204.)

to act as nontraditional plasticizers. Drug compounds have been reported in the literature to function as polymer-plasticizing agents [34,47,48]. This is important to keep in mind during formulation development of a film-coated system. Drug compounds and excipients can migrate during the application of an aqueous film coating, depositing in the film and influencing the mechanical and drug-release properties. Lidocaine HCl, as an example, was found to function as a plasticizer in Eudragit E 100 for both extruded and solvent-cast films [47]. The polymer T_g was lowered and the elongation at failure was increased with the addition of lidocaine HCl either alone or in combination with the plasticizing agent TEC. Lidocaine was a more effective plasticizer in solution-cast films than extruded films due to better intermolecular mixing in solution than in the melt.

Other compounds that have been shown to act as film-coating plasticizers include citric acid and urea. Okhamafe and York incorporated citric acid and urea in HPMC and polyvinyl alcohol (PVA) films, and then used thermomechanical analysis (TMA) and DSC analysis to measure plasticizer effectiveness [49]. The authors found the two compounds to be effective plasticizers for both polymers based on a significant lowering of the polymer T_g. Urea and citric acid were shown to be better plasticizers than PEG 1000 when the T_g data were compared, indicating that additives with many hydrogen bonding groups strongly interact with the polymers to enhance segmental mobility. Citric acid was also found to plasticize Eudragit L 30 D-55 films in another study by Bruce et al. [50]. Citric acid reduced polymer T_g by 26°C when incorporated at 10% w/w based on polymer dry weight. The T_g of the films was reduced by an additional 15°C when 15% w/w (based on polymer dry weight) TEC was also incorporated into the film along with the citric acid (Table 13.3). This data helped explain an observed reduction in drug release from sodium valproate enteric-coated pellets when citric acid was included in the pellet matrix. The citric acid not only lowered pellet microenvironmental pH to prevent premature ionization of the polymer and reduced drug solubility by conversion of the sodium valproate to valproic acid, but also plasticized the film and lowered the film elastic modulus.

Methyl paraben, ibuprofen, chlorpheniramine maleate (CPM), and theophylline were investigated as nontraditional plasticizers for Eudragit RS 30 D by Wu and McGinity [34]. All the compounds except theophylline resulted in a decrease in the polymer T_g when incorporated into the film. Methyl paraben, CPM, and ibuprofen were also shown to decrease the Young's modulus of Eudragit RS 30 D when compression testing of coated beads was conducted using a Chatillon digital force gauge. A decrease in Young's modulus was also observed for films containing increasing concentrations of methyl paraben. These compounds are able to interact with the ammonium and ester groups of the polymer by hydrogen bonding and electrostatic forces, which weaken the inter-chain bonding within the polymer. This explanation was supported by X-ray diffraction analysis, which demonstrated that the Eudragit RS 30 D dispersion changed from a crystalline to an amorphous pattern when methylparaben, CPM, and ibuprofen were included in the film.

TABLE 13.3

Average Glass Transition Temperature Values of Eudragit® L 30 D-55–Cast Films Containing Varying Levels of Citric Acid ($n = 3$)

Citric Acid (% Based on Dry Polymer)	Film T_g (°C)	
	Without Triethyl Citrate	**With 15% Triethyl Citrate**
0	110 ± 4.2	84 ± 1.0
5	112 ± 2.0	76 ± 2.2
10	84 ± 2.5	69 ± 1.8
20	88 ± 4.0	62 ± 2.8
40	80 ± 2.3	65 ± 1.3

Source: L. D. Bruce et al., *Int J Pharm*, 264, 85–96, 2003.

ANTI-PLASTICIZATION

Functional groups on a plasticizer that strongly interact with a polymer can result in an increase in the polymer T_g or produce an anti-plasticization effect [51]. When a compound anti-plasticizes a polymer, there is an opposite effect to plasticization on thermal and mechanical properties; T_g and Young's modulus can both increase rather than decrease, and elongation decreases. This was the case when streptomycin sulfate was incorporated and cast into Eudragit E 30 D and Scopacryl D340 films. Streptomycin sulfate made the films brittle with decreased elongation at the failure [52]. In the same study, Dittgen [52] also found that when plasticizers (such as glycol monoethyl ether, propylene glycol, glycerol, and ethylene glycol) were added to films, the molecular weight of the plasticizers as well as the hydroxyl groups and position of hydroxyl groups available to interact with the polymers affected the plasticization of the films and the elongation at break. In general, the glycol with the lower molecular weight had the greatest plasticizing action. The monoethyl ether of glycol was the most effective plasticizer, demonstrating the greatest percentage elongation on the films, which was attributed to the ability of the functional groups to interact with the polymer and reduce internal polymer–polymer interactions, thus increasing polymer molecular mobility. Alternatively, anti-plasticization can occur from interactions between the plasticizer and polymer, resulting in a decrease in polymer mobility. Jackson and Caldwell concluded that anti-plasticization is the result of a combination of factors that include the reduction of polymer-free volume, interaction between the polar groups of the polymer and anti-plasticizer, and physical stiffening resulting from the presence of the rigid anti-plasticizer molecules adjacent to the polar groups of the polymer [32]. Anti-plasticization yields an increase in both tensile strength and Young's modulus. Guo found that anti-plasticization occurred at low levels of plasticizer concentration (for triacetin and PEG) in cellulose acetate films [53]. However, when the plasticizer content was increased above 5% w/w, plasticization occurred, resulting in a concurrent increase in film creep compliance with increasing plasticizer concentration. A plasticizing effect was demonstrated when the T_g of cellulose acetate films decreased with increasing PEG and triacetin content in the films (Figure 13.13). Although a decrease in film T_g at a 5% w/w concentration of plasticizer was observed, an associated drop in creep compliance at 37°C was also noted, which is indicative of anti-plasticization. The lower molecular weight plasticizers demonstrated the greatest effect on reduction of T_g similar to the

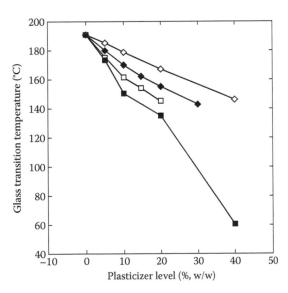

FIGURE 13.13 The effects of plasticizers on the glass transition temperature of cellulose acetate–free films. (◇) PEG 8000; (◆) PEG 4000; (□) PEG 600; (■) triacetin. (From Guo J, *Drug Dev Ind Pharm* 1993; 19(13):1541–1555.)

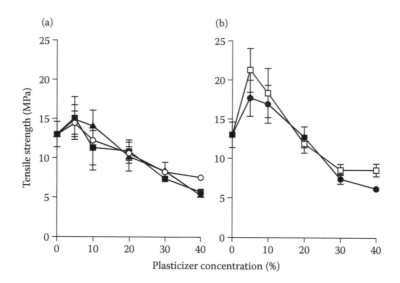

FIGURE 13.14 Tensile strength of pellets as a function of plasticizer content: (a) (■) ATEC; (○) TEC; (▲) ATBC; (b) (●) DEP; (□) triacetin. Concentrations of plasticizers are based on the dry weight of Eudragit® RS polymer. ATBC, acetyl tributyl citrate; ATEC, acetyl triethyl citrate; DEP, diethyl phthalate; TEC, triethyl citrate. (From Wang CC et al., *Int J Pharm* 1997; 152:153–163.)

findings of Dittgen for glycols. The relationship between the molecular weight and T_g between a polymer and a second component (e.g., plasticizer) can be explained in terms of the Couchman and Karasz equation (Equation 13.4) [54]:

$$T_g = (T_{g1} \cdot M_1) + (T_{g2} \cdot M_2) \qquad (13.4)$$

where T_g is the glass transition temperature of the plasticized polymer, T_{g1} and T_{g2} are the glass transition temperatures of the respective pure components, and M_1 and M_2 are the mass fractions. According to this relationship, films that are plasticized with a higher-molecular-weight plasticizer should have a higher T_g than films plasticized with a lower-molecular-weight compound. Wang et al. observed a significant increase in the tensile strength of pellet formulations containing Eudragit RS 30 D plasticized with low levels of TEC, ATBC, ATEC, DEP, and triacetin [33]. In this study, a plasticization threshold was evident once a critical plasticizer concentration was reached, similar to the findings by Guo. At higher concentrations, a plasticization effect occurred, as demonstrated by plotting tensile strength and Young's modulus as a function of plasticizer concentration (Figure 13.14). The glass transition temperature of the plasticized polymer demonstrated a continual decrease with plasticizer concentration.

EFFECT OF INSOLUBLE EXCIPIENTS ON POLYMER PROPERTIES

INSOLUBLE ADDITIVES IN FILM COATINGS

Insoluble additives included in aqueous film coatings can provide both color and photolytic protection, enhance appearance, and act as processing aids. Additives of this type commonly used in film coatings include pigments, surfactants, and anti-tack agents. The size, shape, concentration, and surface chemistry of insoluble additives can significantly influence polymer properties. The quality and surface chemistry of each insoluble additive must be considered for possible interactions.

Pigments used in film coatings include aluminum lakes of water-soluble dyes and opacifying agents, such as titanium dioxide and iron oxides. These additives typically result in an increase in

Young's modulus with a corresponding decrease in tensile strength, leading to detrimental effects on the mechanical properties of the film [55–59]. Interactions have been associated with the resulting size, shape, volume concentration, orientation, and chemical and physical bonding of the additive with the polymer. Moreover, insoluble additives may also influence water vapor permeability, crushing force, and dissolution of films [43,60].

Interactions between insoluble additives and polymer coating solutions can result in coagulation or agglomeration. In the case of polymethacrylate latices, this may be due to polymer sensitivity to electrolytes and pH change, described previously in this chapter. The quality of the additive can influence the stability of the polymer, as in the case of low-quality colored lakes associated with small amounts of water-soluble dyes containing strong electrolytes. This phenomenon was observed when the adjuvant Sicopharm rot 30, a red iron oxide pigment, was added to Eudragit L 30 D-55 dispersion [60]. The resulting Eudragit dispersion coagulated during the preparation of a coating suspension containing high pigment concentration. This interaction was caused by the high conductivity of the pigment due to adhering electrolytes (Table 13.4). To stabilize the dispersion, Sicopharm rot 30 was soxhlet-extracted in distilled water to remove adhering electrolytes, and sodium carboxymethylcellulose (NaCMC) was added to sterically stabilize the pigment particles. Compression of the electrical double layer between the polymer particles by the pigment disturbed the electrostatic balance of the polymer particles in the colloidal dispersion leading to polymer coagulation. Coagulation may also occur due to binding of fine pigment particles with a polymer-stabilizing

TABLE 13.4
Characterization of Pigments

Pigments	Particle Size (μm)	Morphology	Chemical Composition	Density (g/cm³)	Surface Area (m²/cm³)	pH of Suspension	Supernatant Conductivity (μS)
Titanium dioxide Kronos A	<15	Spherical	Titanium dioxide	4.02	36.98	8.29	380
Iriodin 110	<15	Platelets	Titanium dioxide–coated mica	2.85	34.20	9.57	75.4
Iriodin 100	10–60	Platelets	Titanium dioxide–coated mica	3.01	22.73	9.97	87.6
Mica M	<15	Platelets	Mica	2.49	16.48	8.29	32.2
Talkum IT extra	<15	Platelets	Talc	2.76	22.35	9.66	77.5
Sicopharm rot 30	<15	Platelets	Hematite (red iron oxide)	5.08	49.28	5.31	34.6[a]
EM 140662	<15	Platelets	Hematite-coated mica	3.90	30.58	4.30	343
Iriodin 502	10–60	Platelets	Hematite-coated mica	3.49	17.80	5.06	204

Source: K. A. Maul, P. C. Schmidt, *Int J Pharm*, 118, 103–112, 1995.

Note: Density determined in a gas comparison pycnometer, surface area determined by Brunauer, Emmet, and Teller (BET) method, pH of suspension measured with a pH-meter, and conductivity of the suspensions' supernatant determined using a conductometer.

[a] Soxhlet-extracted.

emulsifier or if surface charges associated with the pigment are opposite to those on the surface of a latex polymer. To avoid these interactions, electrolytes, acids, and bases are often added as solutions after first being diluted as much as possible. Good compatibility of latexes is generally found with talc, titanium dioxide, aluminum oxide, calcium phosphate, aluminum silicate, and ferrous oxides.

The degree of a polymer–filler interaction influences the internal stress of a polymer system. Additives lead to discontinuities in the polymer matrix network when adjacent polymer hydrogen bonds are broken. The bonding interaction occurring between the pigment and polymer, being a weaker interaction (usually dipole–dipole), constitutes an overall weakening in the structure and can result in a stress concentration. As filler concentration rises, internal stresses of the resultant films can also increase, leading to a fall in tensile strength, as explained by Okhamafe and York [57]. Polymer–filler interactions such as adsorption of polymer onto the surface of solid particles decrease polymer chain mobility, often resulting in a rise in T_g, a decrease in the deformation capacity of the film, and a corresponding decrease in polymer elongation. The polymer–filler interaction can influence the permeability of films because voids become more extensive as the degree of interaction increases [61,62]. As the amount of filler increases, the critical pigment volume concentration (CPVC) will eventually be exceeded. CPVC is the concentration at which the amount of polymer is not sufficient to bind the additive particles [61]. This concentration must be determined experimentally for each additive and polymer system. Furthermore, localized thermal stresses can develop due to the differences between the thermal expansion coefficients of the polymer and added solids, resulting in film cracking.

Porter studied the effect of titanium dioxide and lake additives on both free films and films applied to tablets by evaluating stress–strain relationships, diametrical crushing strength, and water vapor permeability [59]. The tensile properties of HPMC films were found to decrease with increasing concentration of titanium dioxide and FD & C yellow No. 5 aluminum lake, or a combination of both. Similar values were obtained for films prepared with either additive or in combination. The author also performed diametrical crushing tests on film-coated tablets containing the same additives. The crushing strength of film-coated tablets decreased as pigment volume concentration increased. The reduction in tensile strength was explained by the interaction of pigment with the polymer particles. This yielded discontinuities in the polymer matrix and internal polymer stress concentrations. Internal stress tends to increase as hydrogen bonds between adjacent polymer particles are broken and the pigments interact with polymer molecules. An increase in pigment concentration in the HPMC films resulted in a significant net decrease in water vapor permeability, particularly when titanium dioxide was added. The pigments served as a barrier to moisture diffusing through the film. However, when the concentration of titanium dioxide was raised above 10%, the permeability of the film increased. At higher concentrations (CPVC), the polymer cannot bind the pigment particles together, allowing pores to form in the film.

Pigments may differ significantly in their shapes and sizes, depending on the manufacturer. Particle morphology can have a greater influence than surface chemistry on film properties. Morphology influences particle packing, orientation, and interaction with a polymer as well as moisture and media penetration through a film. Maul and Schmidt compared the effect of pigment morphologies (platelets, spheres, and needles) on the drug-release properties of Eudragit L 30 D films [60]. The authors found that platelet-shaped pigments reduced drug release from enteric-coated pellets regardless of the surface activity or chemical constitution of the additives. For example, titanium dioxide platelets demonstrated more of a sustained release effect than titanium dioxide spheres or iron oxide needles when incorporated into film coatings applied to pellets. Although a difference in surface polarity of the compared pigments existed, the shape of the pigments played the most influential role on drug release properties of the films. The titanium dioxide spheres and the iron oxide needles were observed to form aggregates less than 1 µ in size within the films, and acted to wick-in the dissolution medium leading to faster drug release. When drug release was compared from film coated pellets containing various pigments of comparable shape and size having chemically different constitution, in all cases the platelet-shaped pigments reduced drug release. In a study by Gibson et al., particle shape was responsible for a greater increase in the internal stress and Young's modulus of HPMC films [63]. This was also observed by Okhamafe and York [57], supporting the theory that particle shape rather than chemistry has a greater influence on drug release rate and film permeability. Additional discussions of the influence of insoluble excipients on the properties of film coatings can be found in the literature [64].

SUBSTRATE EFFECTS ON POLYMER FILM FUNCTIONALITY

During aqueous film coating, highly water-soluble drug substances can dissolve or migrate in the aqueous dispersion and deposit in the coating layer. Changes in film permeability brought on by an interaction between the drug and polymer may be due to excessive plasticization of the film, leading to premature dissolution of the drug through polymer channels or pores, or premature ionization of a polymer film. Migration and deposition of the drug substance in the film coating could occur due to the affinity of the drug to the polymer or drug solubility in the polymer. In one example, Bodemeier and Paeratukul exposed CPM and ibuprofen beads coated with ethylcellulose to thermal treatment. The treatment led to a retardation of drug release for the highly water-soluble CPM, as compared to an increase in release of the ibuprofen compound [65]. Typically, the compound with the greater water solubility migrates into the film. In this case, release was enhanced due to migration of the poorly water-soluble drug substance (ibuprofen) into the film. This study showed that ibuprofen had a greater affinity for, and solubility in, the ethylcellulose polymer than CPM. Crystals of ibuprofen were observed in the film surface, confirming that drug migration had occurred. In another study by the same authors, crystals of propranolol HCl were observed in propranolol HCl–Eudragit NE films. Drug release from the film increased with increasing drug loading due to release of propranolol HCl through fluid-filled pores in the film. Permeability was enhanced due to pores or voids created by dissolution of drug crystals that migrated and dispersed throughout the film [66].

Drug substances in a substrate can interact with polymer or ionic surfactants that stabilize a coating, resulting in agglomeration of polymer particles as the coating is deposited on the substrate surface. One explanation for the agglomeration of particles leading to a discontinuous and porous film is adsorption of molecules to the surface of the polymer particle. The resulting stearic hindrance prevents polymer particle coalescence [67]. As discussed earlier in this chapter, anionic surfactants used to stabilize colloidal dispersions (e.g., ethylcellulose and Eudragit NE 30 D dispersions) can react with ionic salts of drug substances [23]. If the substrate is a strong electrolyte or ionic salt and is also very water soluble or hydrophilic, the substrate can wick away water needed to power the film coalescence. This will result in a discontinuous film and require greater amounts of polymer coating application to prevent film failure during dissolution. In addition, highly water-soluble substrates deposited in films during application, prematurely release drug through water-filled channels in the coating and can increase the internal osmotic pressure of the film [68]. This can result in undesirable osmotic pumping of drug through the membrane.

SUBSTRATE pH EFFECTS ON ENTERIC POLYMER FUNCTIONALITY

Alkaline substrates can prematurely ionize enteric polymer coatings as water and media penetrate into the substrate core. Several studies have documented the effect of acidic and alkaline substrates and the associated microenvironmental pH on release properties of enteric film coatings [22,69–73]. Dangel et al. conducted a series of studies to determine the effect of drug acidity or alkalinity on the enteric polymer dispersion Kollicoat MAE 30 DP [69,70]. To study this effect, the authors determined differences in the weight increase of tablets and pellets containing acetylsalicylic acid, indomethacin, and diclofenac sodium exposed to 0.1N HCl for two hours. The drug-release profiles of tablets and pellets in phosphate-buffered media were examined. Acid resistance was improved when an acidic drug such as acetylsalicylic acid or indomethacin comprised the tablet or pellet core. When the release of indomethacin and diclofenac sodium pellets and tablets was compared in simulated intestinal fluid, release from the substrates containing diclofenac sodium occurred much more rapidly. The authors attributed the higher release and the lower acid resistance of the pellets and tablet cores to the de-protonation of the film former by the alkaline drug. Increasing the thickness of the enteric film coating from 3 to 4 mg/cm^2 improved the acid resistance of both the pellets and the tablets containing diclofenac sodium. The weight increase observed during the acid resistance test as a result of gastric fluid (0.1N HCl) absorption or ingress was greater for pellets containing diclofenac sodium than for the tablets. This was attributed to the greater surface area of the pellets. Upon analysis, the amount of diclofenac sodium released into the acidic media was found to be relatively low, a finding attributed to the low solubility of this drug in acid.

Ozturk et al. modeled the release kinetics from enteric-coated dosage forms in buffered media [72]. This model predicted that tablet core pH would influence the pH profile in the coating layer, thereby affecting the dissolution rate of the polymer. The authors explained that dissolution for enteric-coated dosage forms may be modified by interaction of the polymer with drug and excipients in the core. The model was tested with three sample groups of polyvinyl acetate phthalate (PVAP) enteric-coated tablets to investigate the effect of the core pH on the enteric coating dissolution time. The enteric-coated tablets contained aspirin to obtain an acid core, citrate salts to maintain a core pH of 6.5, and a placebo control. According to this model, the presence of an acidic drug in the core formulation was predicted to lower the pH in the coating layer relative to that of the bulk dissolution media. Near the tablet surface (tablet/polymer interface), the H$^+$ concentration should be higher, suppressing the ionization of the polymer. At the polymer/boundary layer (polymer/dissolution media layer) interface, H$^+$ concentration decreases and ionization of the polymer proceeds, leading to an eventual increase in dissolution of the drug. A high surface pH would ionize the enteric polymer, leading to even faster dissolution rates of the tablet and drug. Tablets containing placebo and citrate salts had faster disintegration rates, consistent with model predictions. These data also explain the findings of Dressman and Amidon, who reported a significant effect on the disintegration times of enteric-coated tablets administered to dogs when tablet core pH was varied [73]. In their studies, mean disintegration times were greater for tablets with a core pH of 5 than for tablets with a core pH of 3 or 4. These data suggest that lower tablet micro-environmental pH delayed the dissolution of the enteric polymer.

Delayed dissolution in buffered media resulting from incorporation of an acid into a tablet core was also reported by Crotts et al., evidence that acidic tablet or pellet cores suppress ionization of enteric polymer functional groups [71]. Doherty and York theorized that the pH of the diffusion layer at the surface of a dosage form resembles that of a saturated solution of drug and excipients in the dissolution media and represents the micro-environmental pH of the system [74]. Doherty and York were able to actually measure the micro-environmental pH and surface pH using a micro-pH probe. In this study, they found an excellent correlation between the saturated solution pH and the measured surface pH during dissolution for a pure furosemide compact, further supporting the theory that acidic or basic drugs can influence the ionization and release of enteric polymers.

Pellet core pH was also shown to correlate with the release rate for CPM enteric-coated pellets in studies performed by Bruce et al. [22]. Pellets comprised of 10% w/w CPM and 55% w/w Emcompress® released 27% w/w CPM in acidic media after two hours whereas formulations comprised of the same level of CPM and 20% w/w or 40% w/w citric acid passed the enteric test with less than 10% w/w CPM release under the same conditions (Figure 13.15). The pellets were coated with a 10% weight gain of Eudragit L 30 D-55. Pellet pH was measured by grinding a sample of pellets, combining them with water, and measuring the pH of the resultant slurry. A trend of increasing drug release with increasing pellet pH was observed (Figure 13.16). This finding supports the theory that an increase in pellet substrate micro-environmental pH (in this case, using 55% w/w Emcompress, dibasic calcium phosphate) can potentially lead to the ionization of an enteric polymer coating, resulting in premature drug release.

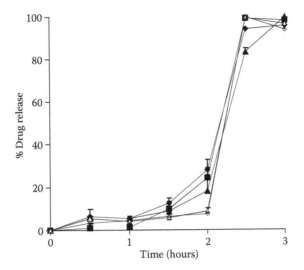

FIGURE 13.15 Influence of pellet composition on drug-release properties of pellets comprised of 10% w/w CPM and varying levels of Avicel®, Emcompress®, or citric acid and coated with 10% weight gain Eudragit® L 30 D-55: (◆) formulation C, 55% Emcompress; (■) formulation A, 47% Avicel and 40% lactose; (▲) formulation B, 87% Avicel; (△) formulation D, 20% citric acid; (◇) formulation E, 40% citric acid. Dissolution media consisting of 0.1N HCl, pH 1.2 from zero to two hours, and 0.05M phosphate buffer from two to three hours, 37°C, 100 rpm (*n* = 3). CPM, chlorpheniramine maleate. (From Bruce LD et al., *Drug Dev Ind Pharm* 2003; 29(8):909–924.)

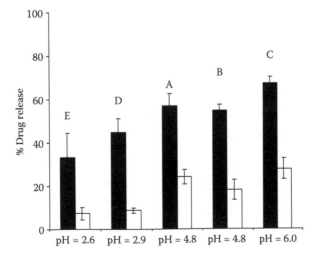

FIGURE 13.16 Comparison of drug-release properties of enteric-coated pellet formulations (A to E) after two hours in 0.1N HCl, pH 1.2, as a function of uncoated pellet core pH: (■) pellets coated with 7% weight gain Eudragit® L 30 D-55; (□) pellets coated with 10% weight gain Eudragit L 30 D-55. Dissolution conditions: 2 h, 750 mL 0.1N HCl, 37°C, 100 rpm (*n* = 3). (From Bruce LD et al., *Drug Dev Ind Pharm* 2003; 29(8):909–924.)

PREVENTING AND FACILITATING DRUG-POLYMER AND EXCIPIENT-POLYMER INTERACTIONS

SUBCOATING TO PREVENT INTERACTIONS

The most widely used method to prevent drug–polymer interactions is by the application of a seal coat or subcoat between the film-forming polymer and the substrate. The subcoat is typically a 2% to 3% weight gain of HPMC or some other nonfunctional or soluble polymer. A subcoat seals off the substrate from the functional polymer coating to prevent comingling. A subcoat may also be applied to prevent migration of a drug substance through the film of a functional polymer coating as required for highly water-soluble drug substances.

Yang and Ghebre-Sellassie used X-ray microprobe analysis of the cross-section of a diphenhydramine pellet core coated with Aquacoat to visually see the migration of drug into the film coating layer [75]. Chlorine atoms of the incorporated HCl salt were used as a probe to monitor drug distribution into the film. Samples were bombarded with an electron beam, producing X-ray emissions specific to chlorine. The resulting emissions were detected by an X-ray spectrometer. At lower coating bed temperatures, migration of drug into the film coating was enhanced. Applying a HPC subcoat prior to coating the pellets with Aquacoat limited drug migration into the film layer. This effect was shown on SEM photomicrograph and X-ray dot maps as a sharply defined boundary layer between the pellet core and the film layer. The release-rate constants calculated for pellets coated with Aquacoat at various coating temperatures were lowest when a subcoat was applied (Tables 13.5 and 13.6). Subcoats can also act as a barrier between the substrate and functional polymer. A well-known use of subcoating as a barrier is described in patent 4,786,505, or the "505" patent by Astra for the product Prilosec [76]. This patent claimed an oral pharmaceutical preparation comprising (a) a core region of omeprazole plus an alkaline-reacting compound, alkaline omeprazole salt, and/or combinations thereof; (b) an inert rapidly soluble subcoat; and (c) an outer layer deposited onto the subcoat comprising an enteric coating. The subcoat was necessary to prevent the acid labile omeprazole from contacting acidic functional groups of the enteric polymer, thereby preventing drug degradation.

In addition to acting as a barrier, subcoats may also increase the diffusional path length that drug substances must travel for dissolution to occur. Less of the functional polymeric coating may therefore be required, providing a cost savings in manufacturing. Several authors have reported improved acid resistance and the need for less enteric polymer to pass either the enteric dissolution or disintegration test when a subcoat was applied to a tablet or pellet core [22,49,71,77–79]. The

---

TABLE 13.5

**Release Rate Constant for Pellets Coated with Aquacoat®
at Different Bed Temperatures**

Bed Temperatures (°C)	k_1 (hr^{-1})[a]	k_2 (hr^{-1})[b]
22	1.045	
25	0.402	0.201
30	0.300	0.109
35	0.262	0.101
40	0.312	0.123
45	0.335	0.264
50	0.450	0.381

Source: S. T. Yang, I. Ghebre-Sellassie, *Int J Pharm*, 60, 109–124, 1990.

[a] First-order rate constant for the initial phase of dissolution (up to two hours).

[b] First-order rate constant for the final phase of dissolution (after two hours).

TABLE 13.6

Release Rate Constants for Subcoated Pellets Subsequently Coated with Aquacoat®

Bed Temperatures (°C)	k_1 (hr^{-1})[a]	k_2 (hr^{-1})[b]
22	0.536	–
25	0.308	0.185
30	0.295	0.104
35	0.261	0.136
40	0.267	0.103
45	0.290	0.217
50	0.418	0.301

Source: S. T. Yang, I. Ghebre-Sellassie, *Int J Pharm*, 60, 109–124, 1990.

[a] First-order rate constant for the initial phase of dissolution (up to two hours).

[b] First-order rate constant for the final phase of dissolution (after two hours).

subcoat in these cases acted as a barrier to prevent an interaction between the substrate and enteric polymer or increased the diffusional path length of the drug. More recently, hot-melt subcoating using a low melting wax or stearic acid has been used in combination with traditional polymer coating to function as a barrier and increase the drug diffusional path length. Less overall polymer coating was required to provide a sustained drug release profile using this subcoat application [80,81].

When selecting a suitable subcoat as a barrier coating, the solubility of the substrate or core, properties of the polymeric subcoat, and required coating weight gain must all be considered in order to select the appropriate subcoat polymer to apply for the appropriate release kinetics. The effect of core solubility on drug release from enteric-coated pellets was demonstrated in studies performed by Bruce et al. [22]. A weight gain of only 7% w/w Eudragit L 30 D-55 was required for pellets containing 30% w/w theophylline (a poorly water-soluble compound) to pass the enteric test. In contrast, pellets containing 10% w/w CPM (a highly water-soluble compound) required greater than 10% w/w weight gain of the same enteric polymer.

The authors experimented with three polymeric subcoats: Eudragit RD 100, an immediate-release coating; Eudragit RS 30 D, a sustained-release polymer; and Opadry AMB, a moisture barrier consisting of an immediate-release polymer. All subcoats were applied to pellets at a 3% w/w polymer weight gain followed by a 7% w/w polymer weight gain of Eudragit L 30 D-55. None of the subcoated pellets passed the enteric test at this low level of enteric coating; however, drug-release rates were found to correlate with both the wettability or contact angle and the water vapor transmission rate of the subcoat material. Subcoat materials with higher contact angle values and lower water vapor transmission rates (Eudragit RS 30 D and Opadry AMB) were more effective in delaying drug release in acid. Because CPM interacted with the Eudragit L 30 D-55 polymer, subcoating the pellet also prevented contact between drug and polymer, allowing a functional enteric film to form.

SUBCOATING TO FACILITATE INTERACTIONS

Subcoats or priming coats were shown to be necessary for the adhesion and spreading of dry powder polymer coatings [2]. Both low melting hydrophilic polymers and amphiphilic materials have been investigated as subcoats and substrate priming agents for dry powder film coating [82]. During powder coating, application of a primer layer over tablet cores reduces tablet surface tension and the work of adhesion. In one study, PEG 3350 was utilized as a priming agent and also integrated in Eudragit L100-55 powder coating to promote polymer adhesion and enhance the film formation process [83]. The application of Eudragit E PO and Eudragit RL PO subcoats assisted with adhesion of Eudragit L100-55 onto sodium valproate tablet cores, reducing the amount of Eudragit L100-55 of required enteric polymer to pass the enteric test [83].

The measurement of film contact angle with water was used to determine the most appropriate subcoat and priming agent for dry powder coating to facilitate adhesion interactions [83]. The best primer material had contact angle values similar to the applied powder polymer film material. In dry powder coating operations, the subcoat or prime coat must interact favorably with the molten polymer to enhance and promote polymer surface adhesion. Hydrophilic polymers, for example, interact best with hydrophilic subcoating or priming materials.

SUMMARY

Interactions between drugs, excipients, and polymeric coatings may be favorable or unfavorable as the examples and literature references given in this chapter have shown. Whether trying to avoid detrimental interactions or accomplish favorable interactions, it is necessary to consider the chemical makeup of a drug substance and any excipients used in the preparation of a substrate or film former. Considerations that should be made during formulation development include overall solubility of a drug in water or dissolution media, drug solubility in the polymeric film former, functional and ionizable groups available in the drug and excipients that could interact with the polymer, acidity or alkalinity of the drug substance, and plasticizing effects that both drugs and excipients can have on a polymer. The application of a barrier subcoat has proven effective in preventing most unwanted interactions between drugs or excipients in a substrate and an overlying functional polymeric film but may also promote adhesion and film formation in some dry powder coating applications.

REFERENCES

1. Gaur P, Mishra S, Rohit G, Singh A, Yasir M. Film coating technology: Past, present and future. *J Pharm Sci Pharmacology* 2014; 1:57–67.
2. Sauer D, Cerea M, DiNunzio J, McGinity JW. Dry powder coating of pharmaceuticals: A review. *Int J Pharm* 2013; 457:488–502.
3. Brown GL. Formation of films from polymer dispersions. *J Polym Sci* 1956; 22:423–434.
4. Holgado MA, Fernandez-Arevalo M, Alvarez-Fuentes J et al. Physical characterization of carteolol: Eudragit L binding interaction. *Int J Pharm* 1995; 114:13–21.
5. Lee H, Hajdu J, McGoff P. Propranolol: Methacrylic acid copolymer binding interaction. *J Pharm Sci* 1991; 80(2):178–180.
6. Alvarez-Fuentes J, Fernandez-Arevalo M, Holgado MA et al. Characterization of morphine polymeric coprecipitates. A biopharmaceutical study. *Pharmazie* 1994; 49:834–839.
7. Alvarez-Fuentes J, Fernandez-Arevalo M, Holgado MA et al. Morphine polymeric coprecipitates for controlled release: Elaboration and characterization. *Drug Dev Ind Pharm* 1994; 20(15):2409–2424.
8. Beten DB, Gelbcke M, Diallo B et al. Interaction between dipyridamole and Eudragit S. *Int J Pharm* 1992; 88:31–37.
9. Alvarez-Fuentes J, Caraball I, Boza A et al. Study of a complexation process between naltrexone and Eudragit L as an oral controlled release system. *Int J Pharm* 1997; 148:219–230.
10. Badawi AA, Fouli AM, El-Sayed AA. Drug release from matrices made of polymers with reacting sites. *Int J Pharm* 1980; 6:55–62.
11. Kislalioglu MS, Khan MA, Blount C et al. Physical characterization and dissolution properties of ibuprofen: Eudragit coprecipitates. *J Pharm Sci* 1991; 80(8):799–804.
12. Lin S, Perng R, Cheng C. Solid state interaction studies between drugs and polymers: Piroxicam-eudragit E, RL, or S resins. *Eur J Pharm Biopharm* 1996; 42(1):62–66.
13. Lin SY, Perng RI. Solid-state interaction studies of drugs/polymers I. indomethacin/Eudragit E, RL, or S Resins. *STP Pharm Sci* 1993; 3(6):465–471.
14. Lin SY, Cheng CL, Perng RI. Solid state interaction studies of drug-polymers (II): Warfarin-Eudragit E, RL, or S resins. *Eur J Pharm Sci* 1994; 1:313–322.

15. Sarisuta N, Kumpugdee M, Muller BW et al. Physico-chemical characterization of interactions between erythromycin and various film polymers. *Int J Pharm* 1999; 186:109–118.

16. Jenquin MR, Liebowitz SM, Sarabia RE et al. Physical and chemical factors influencing the release of drugs from acrylic resin films. *J Pharm Sci* 1990; 79(9):811–816.

17. Jenquin MR, Sarabia RE, Liebowitz SM et al. Relationship of film properties to drug release from monolithic films containing adjuvants. *J Pharm Sci* 1992; 81(10):983–989.

18. Florence AT, Attwood D. Disperse systems (Ch 7). In: Florence AT, Attwood D, eds. *Physicochemical Principles of Pharmacy*, 2nd ed. MacMillan Press Ltd., Great Britain, 1988, pp. 229–279.

19. Martin A. Colloids (Ch 15). In: Martin A, ed. *Physical Pharmacy*, 4th ed. Baltimore, MD: Williams & Wilkins, 1993, pp. 393–422.

20. Goodman H, Banker GS. Molecular-scale drug entrapment as a precise method of controlled drug release I: Entrapment of cationic drugs by polymeric flocculation. *J Pharm Sci* 1970; 59(8):1131–1137.

21. Rhodes CT, Wai K, Banker GS. Molecular scale drug entrapment as a precise method of controlled drug release II: Facilitated drug entrapment to polymeric colloidal dispersions. *J Pharm Sci* 1970; 59(11):1578–1581.

22. Bruce LD, Koleng JJ, McGinity JW. The influence of polymeric subcoats and pellet formulation on the release of chlorpheniramine maleate from enteric coated pellets. *Drug Dev Ind Pharm* 2003; 29(8):909–924.

23. Bodemeier R, Paeratakul O. Evaluation of drug-containing polymer films prepared from aqueous latexes. *Pharm Res* 1989; 6(8):725–730.

24. Borodkin S, Tucker FE. Drug release from hydroxyopropylcellulose-polyvinyl acetate films. *J Pharm Sci* 1974; 63:1359–1364.

25. Donbrow M, Friedman M. Enhancement of permeability of ethylcelullose films for drug penetration. *J Pharm Pharmacol* 1975; 27:633–646.

26. Donbrow M, Samuelov Y. Zero order drug release from double layered porous films: Release rate profiles from ethylcellulose, hydroxypropyl cellulose and polyethylene glycol mixtures. *J Pharm Pharmacol* 1980; 32:463–470.

27. Frohoff-Hulsmann MA, Schmitz A, Lippold BC. Aqueous ethyl cellulose dispersions containing plasticizers of different water solubility and hydroxypropyl methylcellulose as coating material for diffusion pellets I. Drug release rates from coated pellets. *Int J Pharm* 1999; 177:69–82.

28. Dickinson E, Eriksson L. Particle flocculation by adsorbing polymers. *Adv Colloid Interface Sci* 1991; 34:1–29.

29. Wong D, Bodemeier R. Flocculation of an aqueous colloidal ethyl cellulose dispersion (Aquacoat) with a water-soluble polymer, hydroxypropyl methylcellulose. *Eur J Pharm Biopharm* 1996; 42(1):12–15.

30. Okhamafe AO, York P. Interaction phenomena in pharmaceutical film coatings and testing methods. *Int J Pharm* 1987; 39:1–21.

31. Sears JK, Darby JR. The Technology of Plasticizers. New York: John Wiley & Sons, Inc., 1982, pp. 847–1085.

32. Jackson WJ, Caldwell JR. Antiplasticization. III. Characteristics and properties of antiplasticizable polymers. *J Appl Polym Sci* 1967; 11:227–244.

33. Wang CC, Zhang G, Shah NH et al. Influence of plasticizers on the mechanical properties of pellets containing Eudragit RS30D. *Int J Pharm* 1997; 152:153–163.

34. Wu C, McGinity JW. Non-traditional plasticization of polymeric films. *Int J Pharm* 1999; 177:15–27.

35. Allcock HR, Lampe FW. Secondary methods for molecular-weight determination (Ch 15). In: Allcock HR, Lampe FW, eds. *Contemporary Polymer Chemistry*, 2nd ed. Englewood Cliffs, NJ: Prentice Hall, 1990, p. 385.

36. Shah PS, Zatz JL. Plasticization of cellulose esters used in the coating of sustained release solid dosage forms. *Drug Dev Ind Pharm* 1992; 18(16):1759–1772.

37. Hutchings D, Nicklasson M, Sakr A. An evaluation of ethylcellulose-plasticizer interactions using intrinsic viscosity and interaction constant. *Pharmazie* 1993; 48:912–914.

38. Aulton ME, Mechanical properties of film coats (Ch 12). In: Cole G, ed. *Pharmaceutical Coating Technology*. Taylor & Francis, Inc., 1995, pp. 288–362.

39. Hutchings D, Clarson S, Sakr A. Studies of the mechanical properties of free films prepared using an ethylcellulose pseudolatex coating system. *Int J Pharm* 1994; 104:203–213.

40. Gutiérrez-Rocca JC, McGinity JW. Influence of water soluble and insoluble plasticizers on the physical and mechanical properties of acrylic resin copolymers, *Int J Pharm* 1994; 103:293–301.

41. Bodemeier R, Paeratakul O. Mechanical properties of dry and wet cellulosic and acrylic films prepared from aqueous colloidal polymer dispersions used in the coating of solid dosage forms. *Pharm Res* 1994; 11(6):882–888.

42. Aulton ME, Abdul-Razzak MH. The mechanical properties of hydroxypropylmethylcellulose films derived from aqueous systems. Part 1: The influence of plasticizers. *Drug Dev Ind Pharm* 1981; 7(6):649–668.

43. Hutchings DE, Sakr A. Influence of pH and plasticizers on drug release from ethylcellulose pseudolatex coated pellets. *J Pharm Sci* 1994; 83(10):1386–1390.

44. Lippold BH, Sutter BK, Lippold BC. Parameters controlling drug release from pellets coated with aqueous ethyl cellulose dispersion. *Int J Pharm* 1989; 54:15–25.

45. Bodmeier R, Paeratakul O. Propranolol HCl release from acrylic films prepared from aqueous latexes. *Int J Pharm* 1990; 59:197–204.

46. Muhammad NA, Boisvert W, Harris MR et al. Modifying the release properties of Eudragit L30D. *Drug Dev Ind Pharm* 1991; 17(18):2497–2509.

47. Aitken-Nichol C, Zhang F, McGinity JW. Hot melt extrustion of acrylic films. *Pharm Res* 1996; 13(5):804–808.

48. Zhu Y, Shah NH, Malick AW et al. Solid-state plasticization of an acrylic polymer with chlorpheniramine maleate and triethyl citrate. *Int J Pharm* 2002; 241:301–310.

49. Okhamafe AO, York P. Studies of interaction phenomena in aqueous-based film coatings containing soluble additives using thermal analysis techniques. *J Pharm Sci* 1988; 77(5):438–443.

50. Bruce LD, Petereit H, Beckert T et al. Properties of enteric coated sodium valproate pellets. *Int J Pharm* 2003; 264:85–96.

51. Okhamafe AO, York P. Thermal characterization of drug/polymer and excipient/polymer interactions in some film coating formulation. *J Pharm Pharmacol* 1989; 41:1–6.

52. Dittgen M. Relationship between film properties and drug release from acrylic films. *Drug Dev Ind Pharm* 1985; 11(2/3):269–279.

53. Guo J. Effects of plasticizers on water permeation and mechanical properties of cellulose acetate: Antiplasticization in slightly plasticized polymer film. *Drug Dev Ind Pharm* 1993; 19(13):1541–1555.

54. Couchman PR, Karasz FE. *Macromolecules* 1978; 11:117–119.

55. Rowe RC. The cracking of film coatings on film-coated tablets—A theoretical approach with practical implications. *J Pharm Pharmacol* 1981; 33:423–426.

56. Aulton ME, Abdul-Razzak MH. The mechanical properties of hydroxypropylmethylcellulose films derived from aqueous systems. Part 2: The influence of solid inclusions. *Drug Dev Ind Pharm* 1984; 10(4):541–561.

57. Okhamafe AO, York P. Relationship between stress, interaction and the mechanical properties of some pigmented tablet coating films. *Drug Dev Ind Pharm* 1985; 11(1):131–146.

58. Bianchini R, Resciniti M, Vecchio C. Technological evaluation of aqueous enteric coating systems with and without insoluble additives. *Drug Dev Ind Pharm* 1991; 17(13):1779–1794.

59. Porter SC. The effect of additives on the properties of an aqueous film coating. *Pharm Tech* 1980; March, 67–75.

60. Maul KA, Schmidt PC. Influence of different-shaped pigments on bisacodyl release from Eudragit L30D. *Int J Pharm* 1995; 118:103–112.

61. Wan LSC, Lai WF. The influence of antitack additives on drug release from film-coated granules. *Int J Pharm* 1993; 94:39–47.

62. Schultz P, Ingunn T, Kleinebudde P. A new multiparticulate delayed release system. Part II: Coating formulation and properties of free films. *J Control Release* 1997; 47:191–199.

63. Gibson SHM, Rowe RC, White FFT. Mechanical properties of pigmented tablet coating formulations and their resistance to cracking. I. Static mechanical measurement *Int J Pharm* 1986; 48:63–77.

64. Felton LA, McGinity JW. Influence of insoluble excipients on film coating systems. *Drug Dev Ind Pharm* 2002; 28(3):225–243.

65. Bodmeier R, Paeratakul O. The effect of curing on drug release and morphological properties of ethylcellulose pseudolatex-coated beads. *Drug Dev Ind Pharm* 1994; 20(9):1517–1533.

66. Bodmeier R, Paeratakul O. Evaluation of drug-containing polymer films prepared from aqueous latexes. *Pharm Res* 1989; 6(8):725–730.

67. Dashevsky, A, Ahmed, A et al. Effect of water-soluble polymers on the physical stability of aqueous polymeric dispersions and their implications on drug release from coated pellets. *Drug Dev Ind Pharm* 2010; 36(2):152–160.

68. Nesbitt RU, Mahjour M, Mills NL et al. Effect of substrate on mass release from ethylcellulose latex coated pellets. *J Control Release* 1994; 32:71–77.
69. Dangel C, Kolter K, Reich HB et al. Aqueous enteric coatings with methacrylic acid copolymer type C on acidic and basic drugs in tablets and pellets, part I: Acetylsalicylic acid tablets and crystals. *Pharm Tech* 2000; March, 64–70.
70. Dangel C, Kolter K, Reich HB et al. Aqueous enteric coatings with methacrylic acid copolymer type C on acidic and basic drugs in tablets and pellets, part II: Dosage forms containing indomethacin and diclofenac sodium. *Pharm Tech* 2000; April, 36–42.
71. Crotts G, Sheth A, Twist J et al. I. Development of an enteric coating formulation and process for tablets primarily composed of a highly water-soluble, organic acid. *Eur J Pharm Biopharm* 2001; 51:71–76.
72. Ozturk SS, Palsson BO, Donohoe B et al. Kinetics of release from enteric-coated tablets. *Pharm Res* 1988; 5(9):550–564.
73. Dressman JB, Amidon GL. Radiotelemetric method for evaluating enteric coatings. *J Pharm Sci* 1984; 73:935–938.
74. Doherty C, York P. Microenvironmental pH control on drug dissolution. *Int J Pharm* 1989; 50:223–232.
75. Yang ST, Ghebre-Sellassie I. The effect of product bed temperature on the microstructure of Aquacoat-based controlled-release coatings. *Int J Pharm* 1990; 60:109–124.
76. Lovgren KI, Pilbrant AG, Yasumura M et al. Pharmaceutical preparation for oral use. US Patent 4,786,505, November 22, 1988.
77. Thoma K, Bechtold K. Influence of aqueous coatings on the stability of enteric coated pellets and tablets. *Eur J Pharm Biopharm* 1999; 47:39–50.
78. Dangel C, Schepky G, Reich HB et al. Comparative studies with Kollicoat MAE 30D and Kollicoat MAE 30 DP in aqueous spray dispersions and enteric coatings on highly swellable caffeine cores. *Drug Dev Ind Pharm* 2000; 26(4):415–421.
79. Yuan J, Clipse NM, Wu SH. The effects of alternating combinations of enteric coating and HPMC as inner and outer coatings on the performance of coated aspirin tablets. *Pharm Tech*, November 2003; 1–7.
80. Tiwari R, Murthy R, Agarwal, S. The influence of hot melt subcoat and polymer coat combination on highly water soluble sustained release multiparticulate formulation. *Int J Drug Del* 2013; 5:131–136.
81. Yang Z, Lu Y, Tang X. Pseudoephedrine hydrochloride sustained-release pellets prepared by a combination of hot-melt subcoating and polymer coating. *Drug Dev Ind Pharm* 2008; 34(12):1323–1330.
82. Zheng W, Cerea M, Sauer D et al. Properties of theophylline tablets powder-coated with methacrylate ester copolymers. *J Drug Del Sci Technol* 2004; 14(4):319–325.
83. Sauer D, Zheng W, Coots L et al. Influence of processing parameters and formulation factors on the drug release from tablets powder-coated with Eudragit L30D-55. *Eur J Pharm Biopharm* 2007; 67:464–475.

14 Physical Aging of Polymers and Its Effect on the Stability of Solid Oral Dosage Forms

Linda A. Felton, Shawn A. Kucera, and James W. McGinity

CONTENTS

INTRODUCTION

Film coating is an effective method to modify drug release from tablets and pellets. Aqueous-based coating technology has become more popular due to environmental and regulatory requirements that restrict the use of organic solvents in production. The formation of thin, transparent films from aqueous-based latex or pseudolatex dispersions occurs with the simultaneous evaporation of water [1,2]. Figure 14.1 is an illustration of film formation from such systems. During the coating process (Stage I), water evaporates from the film-coated substrate at a constant rate. The latex particles begin to pack together and fuse to form a continuous film. As the colloidal particles begin to fuse and coalesce, as seen in Stage II, the rate of water evaporation decreases. By Stage III, film formation is considered complete; however, it is during Stage III that changes occur in the drug-release rate due to physical aging of the polymeric film coating.

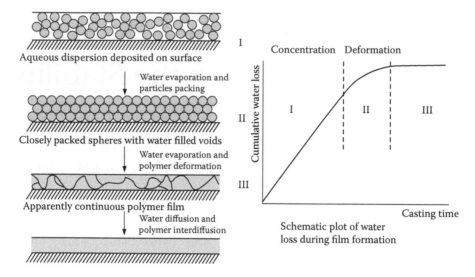

FIGURE 14.1 The formation of thin films from polymeric lattices occurs with the simultaneous evaporation of water.

Physical aging, or enthalpy relaxation, has been known to polymer scientists for many years. All amorphous polymers show physical aging, when the material becomes more rigid, brittle, and dense with time [3]. Struik [4] discussed the early work of Simon [5] who had shown that amorphous materials were not in thermodynamic equilibrium at temperatures below their glass transition temperature. The dynamic state is a result of the materials possessing a volume, enthalpy, and entropy that are greater than in the equilibrium state as shown in Figure 14.2. The free volume concept states that the transport mobility of particles in a closely packed system primarily depends on the degree of packing or the free volume.

When the polymer is cooled to a temperature below its glass transition temperature (T_g), the mobility will be small but not zero. At this stage, the free volume is greater than it would be at equilibrium and the volume will slowly decrease [4,6]. This contraction is accompanied by a decrease in the polymer chain mobility, which leads to a densification of the polymer, influencing both porosity and tortuosity [6,7].

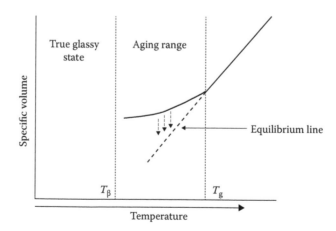

FIGURE 14.2 Graphical representation of the origin of physical aging. T_g is the glass transition temperature of the polymer, and T_β is the temperature of the highest secondary transition. (Adapted from Struik, Scope of the work. In: L. C. E. Struik, ed., *Physical Aging in Amorphous Polymers and Other Materials*. New York: Elsevier Scientific Publishing Company.)

Diffusion of a drug molecule through a thin film is governed by Fick's first law of diffusion (Equation 14.1):

$$Q = \frac{D \times S \times (C_1 - C_2) \times t}{h} \tag{14.1}$$

where Q (the amount of drug diffused over a period of time, t) is a function of h, the film thickness; S is the surface area available for diffusion; C_1, the concentration of drug in the donor compartment; C_2, the concentration of drug in the acceptor compartment, and D, the diffusion coefficient of the drug. The physical aging of a polymeric film results in a change in the diffusion coefficient [7,8], which can be shown by the Iyer equation (Equation 14.2):

$$D = \frac{D_w \times e}{\tau} \tag{14.2}$$

where D_w is the diffusion coefficient of the drug in water, and D is the diffusion coefficient of the drug and is a function of both the film's porosity, e, and tortuosity, τ. As a film ages, it becomes more dense [3], resulting in a decrease in film porosity and an increase in tortuosity, thus causing a decrease in the dissolution rate of drug from film-coated dosage forms over time [7].

This chapter examines the causes of physical aging in polymers used in the coating of pharmaceutical dosage forms as well as methods of quantifying this problem and factors that influence aging. The chapter highlights some approaches that may inhibit physical aging and prevent changes in drug-release rates from coated dosage forms over time.

METHODS OF QUANTIFYING PHYSICAL AGING

The physical changes in pharmaceutical polymers resulting from aging can be evaluated and quantified by a number of analytical methods, including measurement of the T_g, typically done by differential scanning calorimetry (DSC), analysis of mechanical properties or film permeability, dissolution of drug from a coated dosage form, and free volume measurements. The presence of drug crystals on the surface of the coating, which can also indicate polymer aging, can be studied using powder X-ray diffraction (PXRD) and scanning electron microscopy (SEM).

MECHANICAL ANALYSIS

When a polymer is cooled below its glass transition temperature, the amorphous material has a higher specific volume, enthalpy, and entropy than the equilibrium state would possess at the same temperature [9]. The structural changes in the glassy state due to relaxation of the polymer can manifest changes in the physical properties that are of critical importance to pharmaceutical scientists. These changes include decreases in elongation [7,9,10] and creep compliance [11] as well as increases in elastic modulus [9,10] and tensile stress [7,10,12]. These parameters are all quantifiable by examining the physical–mechanical properties of polymeric films as a function of time and storage conditions.

Unilateral Stress–Strain

A simple method to examine the physical–mechanical properties of polymeric films is by unilateral stress–strain experiments [6,7,10–19]. Changes in the internal structures of polymers strongly affect their physical and mechanical properties [20], and the results from stress–strain experiments allow the researcher to gather information on the tensile properties, modulus, and elongation of thin films [7,12]. The industry standard for these measurements is published by the American Society for Testing Materials (ASTM) D 882-02: Standard Test Method for Tensile Properties of Thin Plastic Sheeting [21]. The specimen to be tested should have a thickness of less than 1.0 mm and a width between 5.0 and 25.4 mm and should be at least 50 mm longer than the grip separation. The specimens used for the test should have an overall uniform thickness within 10%. The film specimen is placed in the grips of an instrument such as an Instron testing device (Norwood, MA). One grip is fixed, and the other is moved at a constant rate. As the movable grip is extended, the film is subjected to strain, which is recorded by a computer having specialized software packages. These software packages allow for the automatic calculation of such parameters as tensile strength at break or maximum load, percentage of elongation and the elastic or Young's modulus of a film specimen. More details on tensile testing can be found in Chapter 4.

Creep Compliance

Creep testing is another common method that allows scientists to measure the changes in the physical–mechanical properties of a polymer as it ages [6,11,15,18,22–28]. Creep is the progressive deformation of a material at a constant load. Creep tests measure the dimensional changes that occur over time under a constant static load that is applied to the specimen at a set temperature [29]. The industry standard for creep testing is ASTM D 2990-01: Standard Test Methods for Tensile, Compressive, and Flexural and Creep-Rupture of Plastics [29]. Specimens for tensile creep measurements should conform to the same standards as those used in unilateral stress–strain experiments.

The creep of a specimen occurs in three stages. Following an initial rapid elongation upon application of a load, the creep rate decreases rapidly with time during Stage 1. Stage 2 is denoted by the attainment of a steady state with respect to creep rate. Stage 3 is characterized by a rapid increase in creep rate followed by fracture of the specimen. Graphically, when plotted as a log–log plot, the creep compliance of a material is linear in relation to time [29]. During physical aging, creep compliance decreases as indicated by an increase in the slope of creep modulus versus time on a log–log plot.

Membrane Permeability

Measuring the vapor permeability of a film as a function of time and storage conditions has been previously used to qualitatively analyze physical aging in thin polymeric films [7,11,12,14,16,25,30–38]. As the film undergoes further gradual coalescence, its permeability to a gas will decrease due to increases in film density and tortuosity. As physical aging progresses, a decrease in water vapor transmission rate is typically observed.

The water vapor transmission rate is the steady flow of water vapor per unit time through a unit area under specific conditions of temperature and humidity [39]. A useful guideline is the ASTM E 96/E 96 M-05. The guideline describes two methods for determining the moisture vapor permeability of a thin film. One method, known as the desiccant method, involves placing a thin polymer film over the opening of a cup containing anhydrous calcium chloride as a desiccant. The film is secured and the apparatus is placed in a constant-temperature, constant-humidity environment. The cup is weighed periodically, and a graph of weight versus time is plotted. The second method is called the water method, in which the cup contains a saturated salt solution of a known relative humidity (RH) rather than a desiccant. With this method, the permeability of the film is evaluated by quantifying the transfer of water vapor from the cup through the specimen to a controlled atmosphere over time.

The rate of water vapor transmission can be calculated using Equation 14.3:

$$WVT = \frac{G}{tA} \tag{14.3}$$

where WVT is the water vapor transmission in g/hr m^2, G/t is the slope of the line from the weight change versus time plot, and A is the surface area of the film. These data can be used to calculate the permeability of a thin film. Permeability is simply the arithmetic product of permeance and thickness, and permeance is the rate of water vapor transmission through the film as a function of vapor pressure differences between the two surfaces. Permeance is a performance measure of the film whereas permeability is a property of the material. The permeance of the film can be calculated using Equation 14.4:

$$Permeance = \frac{WVT}{\Delta p} = \frac{WVT}{S(R_1 - R_2)} \tag{14.4}$$

where S is the saturation vapor pressure at test temperature, R_1 is the RH at the source (in the chamber for the dessicant method and in the cup for the water method), and R_2 is the RH at the vapor sink. The permeability of the film is calculated by multiplying the thickness of the film by its permeance.

FREE VOLUME MEASUREMENTS

Ellipsometry

Ellipsometry is an optical technique for measuring the dielectric properties (i.e., refractive index) of thin films [35–37,40–42]. Huang and Paul first reported on the use of ellipsometry in monitoring the physical aging of thin glassy films by changes in refractive index [40]. This method has the advantage that no damage is done to the film specimen, allowing the same sample to be examined throughout an aging study.

The Lorentz–Lorenz parameter (L) is derived from the Lorentz–Lorenz equation (Equation 14.5) [40]:

$$L = \frac{n^2 - 1}{n^2 + 2} = \frac{\rho N_{av} \alpha}{3 M_0 \varepsilon_0} \tag{14.5}$$

The equation shows that the refractive index (n) is directly related to r, the density of the polymer, where N_{av} is Avogadro's number, α is the average polarizability of the polymer repeat unit, M_0 is the molecular weight of the polymer repeat unit, and ε_0 is the permittivity of free space.

The Lorentz–Lorenz parameter (Equation 14.5) can also be related to the density of a polymer by Equation 14.6 [41]:

$$L = \rho C \tag{14.6}$$

where ρ is the density of the material and C is a material constant from the bulk values of refractive index and density at 25°C [41]. The fractional free volume, f, at any time is then determined by Equation 14.7:

$$f = \frac{V - V_0}{V} = 1 - \rho V_0 = 1 - \frac{L}{C} V_0 \tag{14.7}$$

where $V = 1/\rho$ is the specific volume at that aging time, V_0 is the occupied volume of the polymer computed from the van der Waals volume of the polymer, V_w, by the Bondi method, where $V_0 = 1.3\ V_w$ [41].

Positron Annihilation Spectroscopy

Another method used to quantify the changes in free volume due to the physical aging of polymeric films is by the use of positron annihilation spectroscopy (PALS) [18,43–46]. This method is able to measure the free volume as well as the free volume distribution in a polymeric film [18,43,47]. The positron is a particle that has the same properties as an electron but with an opposite charge. When a positron and an electron meet, it is likely that a positronium atom will form [47]. There are two possible positronium "states" that can exist: the para-positronium (p-Ps) and the ortho-positronium (o-Ps). Although the p-Ps state has a very short life of about 125 ps [47] in a vacuum, the o-Ps has a relatively long lifetime of about 142 ns [47] under the same conditions and a lifetime of about 1 to 10 ns in a polymer [43]. When the o-Ps atom annihilates, three gamma rays are emitted, which can be detected to determine the lifetime of the particle.

In a PALS experiment, a radioactive sample of ^{22}NaCl [23,43,47] is used to inject lone positrons into the polymer sample. The lifetime of the positron in the sample [l] is therefore due to the electron density at the location of the positron according to Equation 14.8 [43]:

$$\lambda = C \int \rho_+ \rho_- \, dV \tag{14.8}$$

where C is a constant and ρ_+ and ρ_- are the positron and electron densities, respectively. The lifetime of the o-Ps particle (in nanoseconds), τ, is described by Equation 14.9 [43,44,47]:

$$\tau = \frac{1}{2} \left[1 - \frac{R}{R + \Delta R} + \frac{1}{2\pi} \sin\left(\frac{2\pi R}{R + \Delta R} \right) \right]^{-1} \tag{14.9}$$

where R is the radius of the spherical free volume holes and ΔR represents the thickness of the electron layer, which is a constant of 1.656 Å [23,47]. Thus, there is a direct correlation between the lifetime of the o-Ps and the size of the free volume voids in the polymer matrix, as seen in Figure 14.3 [23].

THERMAL AND MICROSCOPIC ANALYSIS

DSC is a common analytical method used to determine various polymer properties, including melting temperature, degree of crystallinity, T_g, and enthalpy of transition. The technique is widely used

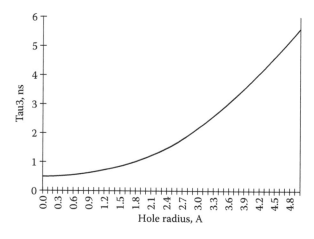

FIGURE 14.3 *Ortho*-Positron lifetime versus hole radius according to Equation 14.9. (From D. M. Bigg, *Polym Eng Sci*, 36, 737–743, 1996.)

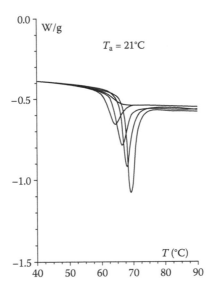

FIGURE 14.4 Heat flux (W/g) versus temperature T for an organic powder coating physically aged at 21°C. From left-hand to right-hand side: aging times, $t = 0, 7, 24, 63$, and 159 days. (From D. Y. Perera, *Prog Organic Coat*, 44, 55–62, 2002.)

to investigate excipient–polymer interactions and evaluate the effectiveness of plasticizing agents in polymeric films. A film sample and a reference are heated at a programmed rate and more energy is absorbed (or emitted) in the sample during a phase change. The energy or heat flow is plotted against temperature or time, and software programs are used to determine the desired property. During physical aging, there is a decrease in enthalpy or enthalpy of relaxation, which can be measured by DSC. This parameter is commonly used to study physical aging and can be determined by integrating the endothermal peak present in the T_g region during the initial scan [48]. As the polymer ages, both the peak size and the temperature corresponding to its maximum will increase [18,48], as seen in Figure 14.4.

Polymer films may contain various additives, such as endogenous emulsifiers, active pharmaceutical ingredients, or excipients that either improve processability or modify drug release from the coated dosage forms. In some cases, the polymer may be stored at a temperature above the T_g. At this point, the specific volume of the polymer is large as is the molecular mobility of the polymer, and it is possible for the additive components to crystallize during storage. Techniques such as DSC, PXRD, and SEM can be used to scan the polymeric films to determine if crystal growth is present.

VARIABLES THAT INFLUENCE PHYSICAL AGING

PLASTICIZERS

Plasticizers reduce the intermolecular attractions between polymer chains to increase the flexibility of the resulting film and enhance the formation of thin films from aqueous lattices. The selection of a plasticizer is of the utmost importance when formulating a coating dispersion. Plasticizers must remain in the film, exhibiting little or no tendency for migration or volatilization. Moreover, plasticizers must be compatible with the polymer. Using a plasticizer that is incompatible with aqueous latex may result in poor film formation and instabilities with respect to drug release over time during storage. Furthermore, when blending polymers, a plasticizer that has a higher affinity for one polymer system can result in non-uniform distribution within the resulting film and subsequent redistribution during curing and/or storage, which ultimately affects drug release [49].

TABLE 14.1

Effect of Aging on Eudragit® RL and RS 100 Films Plasticized with 0.5% Glyceryl Triacetate

Film	Age (Days)	Decrease in Plasticizer Content (%)	Decrease in Elongation at Break (%)	Change in Permeation Rate (%)
Eudragit RL	180	52	15	+31
Eudragit RS	180	48	39	−53

Source: W. Anderson, S. A. M. Abdel-Aziz, *J Pharm Pharmacol*, Suppl 22, 28, 1976.

Once incorporated, the plasticizing agent should remain in the polymeric matrix in order to produce a stable film. The permeability and mechanical strength of Eudragit RS and RL films were found to be a function of the plasticizer remaining in the film, as shown in Table 14.1 [31]. Both films exhibited a decrease in plasticizer content after six months of storage at 25°C/0% RH and a concomitant decrease in the elongation at break. In contrast, the permeability of RS films decreased during this time period, and the RL films demonstrated an increase in permeability. These results were attributed to the volatilization of the plasticizer. The loss of plasticizer was less critical for the more hydrophilic RL polymer with the void space being quickly filled by the permeant solution, thus resulting in an increase in permeation.

The selection of the appropriate plasticizer is of considerable importance. For example, Lecomte et al. [50] demonstrated that drug release was strongly dependent on the type of plasticizer incorporated in the coating and that the plasticizer influenced the underlying mechanism of drug release. Futhermore, Yang et al. [51] showed that the type of plasticizer had a significant effect on drug release when films were not fully coalesced. The amount of plasticizer used is also important. Incorporation of inadequate amounts of plasticizer can result in polymer films that are brittle or that require longer curing times to exhibit stable films. The degree of coalescence of latex particles at the end of the coating process is a function of the concentration of plasticizer in the formulation with higher concentrations of plasticizer producing enhanced or more complete film formation. For example, Amighi and Moes [49] showed that the time required to achieve a stable release rate of theophylline from pellets coated with Eudragit RS 30 D containing 5% Pharmacoat 606 at storage conditions of 40°C and 50% RH ranged between six months and 10 days, depending on concentration of TEC used in the coating formulation.

CURING AND STORAGE CONDITIONS

After completion of the coating process, coated dosage forms are often stored at elevated temperatures to promote further gradual coalescence of the film, a process known as curing. Curing of film-coated dosage forms is an important component in the mechanism of film formation from aqueous latices. The film-formation process from these aqueous dispersions relies on capillary forces to draw together and deform the latex particles and is influenced by the amount of water in the polymeric film. As the amount of water in the polymer film increases, the T_g of the film is lowered, resulting in an increased mobility of the polymer chains, which, in turn, enhances the further coalescence of the latex particles. As the humidity of the environment is decreased, the amount of water in the polymeric film is also reduced, and consequently, the capillary forces that facilitate film formation are not present. The larger the particle size of the latex, the more critical high humidity becomes during curing to achieve polymer coalescence [52]. Figure 14.5 shows that stable drug release profiles can be obtained when products coated with aqueous polymeric dispersions have been sufficiently cured to fully coalesce the latex particles.

Both curing temperature and curing time significantly affect the drug-release rate, but curing temperature is of greater consequence [54–56]. As illustrated by Figure 14.6, there is a decrease in the dissolution rate of diphenhydramine from pellets coated with a 10% weight gain of Eudragit NE 30 D

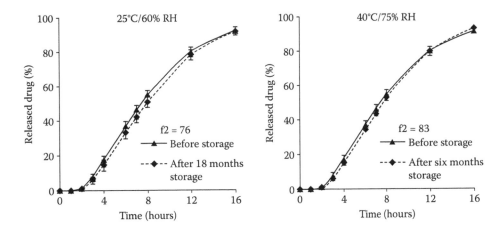

FIGURE 14.5 Storage stability of tablets coated with 10% ethylcellulose:PVA–PEG graft copolymer (90:10) and cured for 24 hours at 60°C and ambient relative humidity, stored during 18 months under ambient conditions (25°C/60% relative humidity) and six months stored under stress conditions (40°C/75% relative humidity). (From C. Gendre et al., *Int J Pharm*, 2013, 448–453, 2013.)

at all three curing temperatures investigated. The decrease in the release rate of the product stored at 30°C was small (when compared to other temperatures) and not significantly affected by length of curing time. However, as temperature and storage time were increased, the changes observed in the dissolution rate were amplified. It is suggested that in order for the polymer to achieve a stable energetic state, energy is required to overcome existing barriers that cause the stable state to be kinetically disfavored. At higher temperatures, more polymer molecules can overcome this energy barrier and reach the stable state, which is reflected by a slower release rate. On the other hand, at lower curing temperatures, fewer molecules can achieve the stable state, meaning that changes in drug release would be expected to occur slowly over time until this stable state is reached.

Curing temperature and time are also important with dry powder coating techniques. These novel coating application processes are solventless, and their advantages are discussed elsewhere [57]. For these systems, film formation occurs predominately during the curing phase after the dry polymer particles are coated on to the substrate. Kablitz and Urbanetz [58] showed that the curing temperature ideally should be close to or above the T_g of the coating formulation. The authors also suggested that longer curing times at slightly lower temperatures could also facilitate film formation.

ENDOGENOUS EXCIPIENTS

The presence of endogenous excipients in aqueous coating systems is often necessary to stabilize the dispersion during storage. In other cases, excipients are used in the emulsion polymerization process of aqueous lattices as is the case of nonoxynol 100 in Eudragit NE 30 D dispersions. However, the presence of this emulsifier can lead to serious stability issues, such as an increase in drug dissolution rate during storage [59]. Due to the relatively high melting point of the surfactant (~ 60°C), it is possible for the material to crystallize within the film during storage at room temperature. Studies have shown that crystallization of the surfactant affects the dissolution rate of the drug from coated dosage forms [59]. Further gradual coalescence and drug release from coated pellets were influenced by increasing amounts of nonoxynol 100 in the coating dispersions [60]. When a commercially available Eudragit NE 30 D dispersion (1.5% nonoxynol 100) was used to coat pellets, the drug release rate diminished by 10% over two months of storage at room temperature, and a decrease of only 5% was observed when the nonoxynol 100 concentration was 5%. However, when the surfactant concentration was increased to 10%, there was first a decrease in the dissolution rate of the drug as a result of the initial swelling of the polymer, after which the dissolution rate

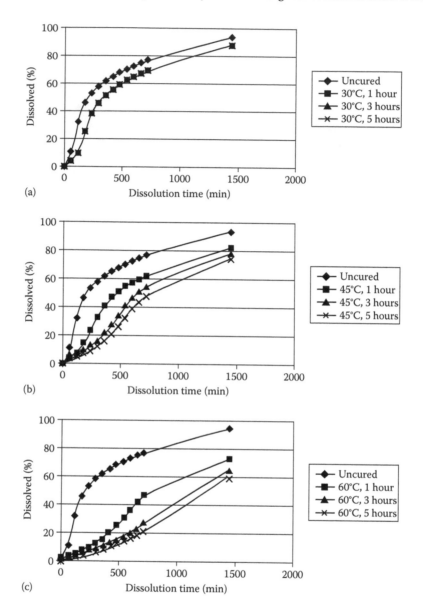

FIGURE 14.6   The dissolution profiles of diphenhydramine HCl pellets coated with 10% Eudragit® NE 30 D and cured at (a) 30°C, (b) 45°C, (c) 60°C. (From A. Y. Lin et al., *Pharm Dev Tech*, 8, 277–287, 2003.)

increased. This phenomenon was the result of further coalescence of the polymer, which decreased the drug-release rate, coupled with the dissolution of surfactant crystals, which caused large pores in the film and enhanced the release of the drug [60].

The crystallization of nonoxynol 100 in Eudragit NE 30 D free films has also been followed via calorimetric studies [61]. These studies showed the melting point of nonoxynol 100 as a single endothermic peak at around 55°C for freshly cast Eudragit NE 30 D films. The films were then stored at ambient conditions (25°C/< 35% RH), 25°C/60% RH, and 40°C/75% RH and analyzed using DSC after one, two, and four weeks of storage. As time progressed, all films showed an increase in the magnitude of the melting point endotherm of nonoxynol 100 (Figure 14.7), indicating crystal growth of the surfactant in the film, which agrees with earlier data published by Lin and Augsburger [60]. The study also concluded that lower temperatures caused a higher degree of crystallization of the emulsifying agent.

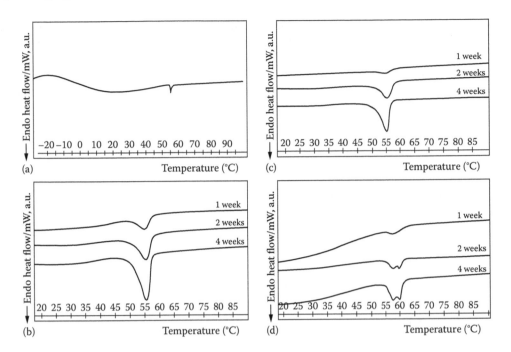

FIGURE 14.7   DSC curves of fresh films (a). DSC curves of films stored at (b) 25°C/60% RH; (c) room temperature/<35% RH; (d) 40°C/75% RH. RH, relative humidity. (From J. Bajdik et al., *J Therm Anal Calorimet*, 73, 607–613, 2003.)

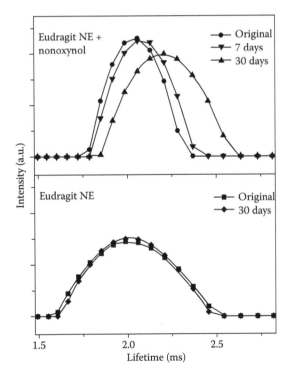

FIGURE 14.8   The effect of storage on Eudragit® NE films with and without nonoxynol 100. The relative pressure of water was 0.73 (RH = 73%) in every case. RH, relative humidity. (From R. Zelko et al., *J Pharm Biomed Anal*, 40, 249–254, 2006.)

Positron annihilation lifetime spectroscopy was used to measure the distribution of free volume holes in cast films of Eudragit NE 30 D with and without nonoxynol 100 [43]. Figure 14.8 shows that when the emulsifying agent was not present in the film, the size distribution of free volume holes remained unchanged when stored for 30 days at 25°C/75% RH. When nonoxynol 100 was present in the film, however, the size distribution of the free volume holes narrowed and was more uniform following initial sample preparation. During one month of storage at high humidity conditions, water initiated an absorption–dissolution transition in these films, and the size distribution of the free volume holes in the polymer increased. This report confirms earlier studies that indicated that nonoxynol 100 affects the long-term stability of Eudragit NE 30 D films [56,60].

METHODS USED TO STABILIZE/PREVENT AGING

As discussed above, changes in drug release occur over time due to aging and a number of variables influence the aging process [62]. Various strategies have been investigated to stabilize polymer films and minimize or prevent changes in drug release over time. These methods can be divided into two general approaches: fully coalesce the film before storage and prevent polymer chain relaxation. Examples of these approaches are presented below.

Increased Plasticizer Concentration

Plasticizers lower both the glass transition temperature and the minimum filmformation temperature of the polymer. Furthermore, the degree of coalescence of latex particles at the completion of the coating process increases as the amount of plasticizer in the formulation increases, due to the plasticizer's ability to weaken polymeric intermolecular attractions, thus allowing the polymer molecules to move more readily, increasing the flexibility of the polymer. For example, Amighi and Moës showed that theophylline pellets coated with a formulation containing Eudragit RS 30 D, 5% Pharmacoat 606, 50% talc, and 30% TEC exhibited virtually no change in dissolution rate upon storage, whereas significant changes in drug release were observed for the coating formulations containing only 10% or 20% TEC [49].

Although liquid plasticizers may be lost through evaporation during coating or subsequent storage, solid-state plasticizers have the distinct advantage of remaining in the film throughout the life of the dosage form. Studies have been conducted in which nonpareil beads were coated with Eudragit RS 30 D containing 40% ibuprofen as the active ingredient and a solid-state plasticizer [63]. The coated beads were cured at 40°C for a period of 24 hours and then stored at 23°C and 0% RH. No significant difference was found between the initial drug release rate and the release profiles of the stored samples (Figure 14.9). The authors reported that the presence of ibuprofen in the coating also served as an anti-adherent, preventing the agglomeration of pellets during the coating process and subsequent storage.

Curing and Storage

The conditions at which dosage forms are cured as well as stored can have a significant effect on the stability of the polymeric film. When dosage forms are cured at high temperatures, the time required to reach a fully coalesced film decreases in comparison to curing at lower temperatures [64]. At temperatures above the T_g of the film, the mobility of the polymer chains increases and latex coalescence is accelerated so that films are nearly completely coalesced.

The role of humidity during curing has been discussed above. It is worthwhile to note that humidity in the environment during storage can also significantly influence drug release from coated dosage forms. Water vapor in the atmosphere that is adsorbed by the polymeric film can act as a plasticizer, increasing the molecular mobility of the polymer chains and aiding in the densification and further coalescence of the polymer. Kranz and Gutsche [65], for example, showed that the dissolution rate

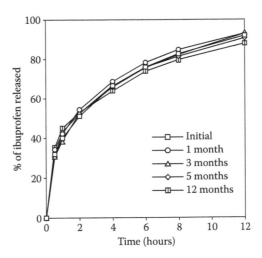

FIGURE 14.9 Effect of storage time at 23°C on the dissolution rate of ibuprofen from nonpareil beads coated with 10% Eudragit® RS 30 D polymer containing 40% ibuprofen (*n* = 6). (From C. Wu, J. W. McGinity, *AAPS Pharm Sci Tech*, 2, 4, article 24, 2001.)

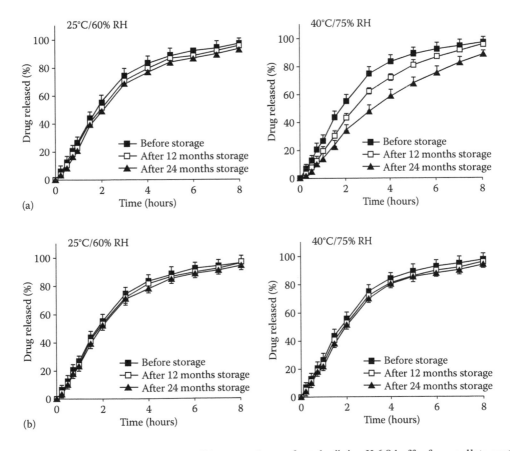

FIGURE 14.10 Influence of storage conditions on release of vatalanib in pH 6.8 buffer from pellets coated with ethylcellulose:Kollicoat MAE 30 D polymer blend (65:35) in (a) open containers and (b) closed containers. Coating was at a 10% level and pellets were cured for two days at 60°C. (From H. Kranz, S. Gutsche, *Int J Pharm*, 112–119, 2009.)

of an ethylcellulose:Kollicoat MAE 30 D polymer blend was stable over a 24-month period when stored in open containers at 25°C/60% RH as shown in Figure 14.10. In contrast, storage in open container storage at the more stressful 40°C/75% RH resulted in a decrease in drug release over the same time frame. When stored in *closed* containers, however, the same coated pellets showed no change in drug release over the two-year period, irrespective of the storage condition.

USE OF POLYMER BLENDS

The addition of a miscible, high-glass-transition polymer is another approach that has been shown to stabilize drug release from sustained release coatings. Stabilization occurs when the molecular mobility of the polymer film decreases on increasing the T_g of the polymer blend. High glass transition temperature polymers also serve as a framework to resist the densification and further coalescence of the much lower glass transition temperature continuous phase. As an example, Eudragit L 100-55 was found to be miscible with Eudragit RS 30 D [66], and although the enteric polymer increased the drug release rate from coated theophylline pellets as the pH of the dissolution media increased, the product exhibited no physical aging when stored at 40°C, that is, a static

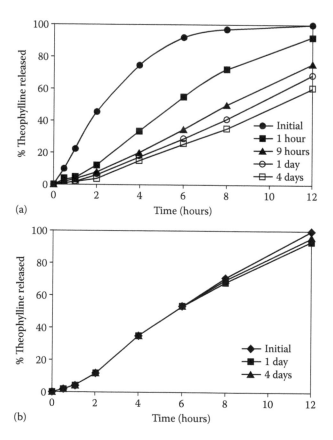

FIGURE 14.11 Influence of storage time at 40°C on the release of theophylline from pellets coated with (a) 12% Eudragit® RS 30 D plasticized with 17.5% TEC and (b) 12% Eudragit RS 30 D/L 100-55 (3:1) plasticized with 17.5% TEC, over-coated with 2% Eudragit RD 100. Dissolution performed in pH 7.4 phosphate buffer medium using the USP method 2 (*n* = 6). RH, relative humidity; TEC, triethyl citrate. (From C. Wu, J. W. McGinity, *Pharm Dev Tech*, 8, 103–110, 2003.)

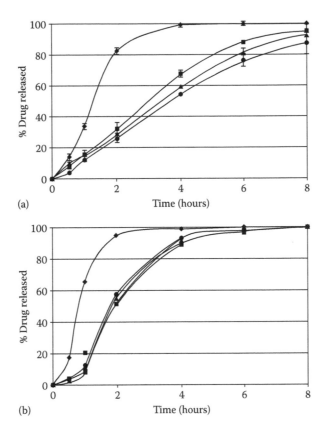

(a)

(b)

FIGURE 14.12 (a) Influence of curing time on the release of PPA-HCl from pellets coated with Eudragit NE 30 D, stored at 60°C (USP 24, apparatus 2, 500 ml of pH 1.2 HCl, 37°C, 100 rpm, $n = 3$): ◆, initial; ■, four hours; ▲, 24 hours; ●, 72 hours. (b) Influence of curing time on the release of PPA-HCL from pellets coated with Eudragit NE 30 D containing 16.7% Eudragit L 30 D-55 stored at 60°C: ◆, initial; ■, for four hours; ▲, for 10 hours; ●, for 24 hours, ≤, for five days. PPA, phenylpropanolamine. (From W. Zheng, J. W. McGinity, *Drug Dev Ind Pharm*, 29, 357–366, 2003.)

drug release profile over time (Figure 14.11). Another study [67] showed that the addition of 16.7% Eudragit L 30 D-55 to Eudragit NE 30 D decreased the tackiness of the films and, when cured at 60°C, the drug-release rate of the coated pellets stabilized after four hours of storage (Figure 14.12).

Hydrophilic, water-soluble polymers have found use in stabilizing sustained-release polymers in coating applications. It has been shown that these excipients form boundaries that inhibit the further coalescence of the functional polymer. For example, Siepmann and coworkers used small amounts of polyvinyl alcohol–polyethylene glycol graft copolymers to stabilize drug release from ethylcellulose films [68,69]. In another study, hydroxyethylcellulose (HEC) was used to stabilize the release rate from Eudragit RS 30 D films [7] as shown in Figure 14.13. Atomic force microscopy was used to characterize the surface morphology of the cast films. Films of Eudragit RS 30 D (Figure 14.14a) exhibited a smooth, regular surface where all latex particle boundaries had disappeared. In contrast, a rough surface was observed for acrylic films containing 10% HEC (Figure 14.14b). The hydrophilic polymer had surrounded the hydrophobic acrylic latex particles and prevented the further coalescence and densification of the film. The HEC allowed the film structure to be retained during storage and stabilized the permeability and mechanical properties of the film,

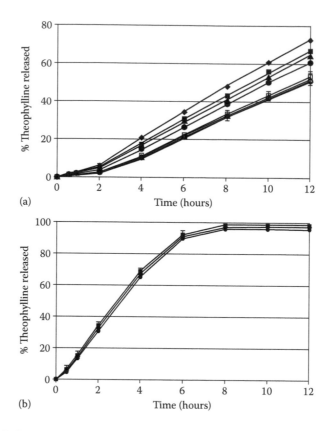

(a)

(b)

FIGURE 14.13 (a) Influence of storage time on the release rate of theophylline from pellets coated with Eudragit® RS 30 D stored at 25°C/60% RH in open high-density polyethylene containers (USP 26 apparatus 2900 ml, 50 mM phosphate buffer, pH 7.4, 37°C, 50 RPM, $n = 3$): ◆, initial; ■, 1 week; ▲, 2 weeks; ●, 4 weeks; ◇, 4 months. (b) Influence of storage time on the release rate of theophylline from pellets coated with Eudragit® RS 30 D containing 10% HEC stored at 25°C/60% RH in open HDPE containers (USP 26 apparatus 2900 ml, 50 mM phosphate buffer, pH 7.4, 37°C, 50 RPM, $n = 3$): initial, 1 week, 2 weeks, 4 weeks, 4 months. RH, relative humidity; HDPE, high-density polythylene. (From W. Zheng et al., *Eur J Pharm Biopharm*, 59, 147–154, 2005.)

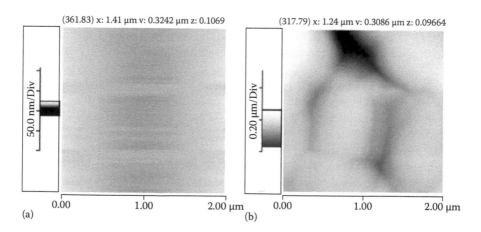

(a)   (b)

FIGURE 14.14 AFM image of cast polymeric films: (a) Eudragit® RS 30 D and (b) Eudragit® RS 30 D containing 10% HEC. AFM, atomic force microscopy; HEC, hydroxyethylcellulose. (From W. Zheng et al., *Eur J Pharm Biopharm*, 59, 147–154, 2005.)

OTHER ADDITIVES IN FILM COATING FORMULATIONS

High Solids Content

Talc is traditionally used as an anti-tacking agent in the coating formulation and is usually present at concentrations of 50% to 100%. Generally, the addition of higher amounts of talc is seldom used because high levels of this hydrophobic material could alter drug release from the dosage form. However, it has been shown that the inclusion of up to 200% talc can be used to successfully formulate coated pellets with a sustained drug-release rate [70]. When this amount of talc was added to a 95:5 blend of Eudragit RS/RL 30 D plasticized with TEC, the acrylic polymer functioned as an effective binder for the talc, resulting in a continuous film coat. Although film formation was

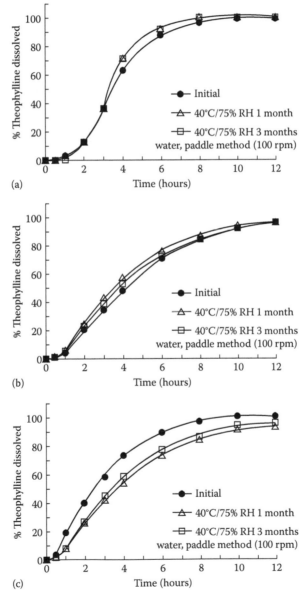

FIGURE 14.15 Stability of theophylline release rate from pellets coated with Eudragit® RS/RL containing 200% talc after storage at 40°C/75% RH with (a) 10% TEC, (b) 20% TEC, and (c) 30% TEC. RH, relative humidity; TEC, triethyl citrate. (From T. Maejima, J. W. McGinity, *Pharm Dev Tech*, 6, 211–221, 2001.)

incomplete, the coating still provided a sustained release of the drug. The high talc content of the films also resulted in no agglomeration of the coated pellets during curing at 60°C or storage at 40°C/75% RH in open containers. The authors stated that the addition of 10% or 20% TEC to the coating formulation resulted in dosage forms that were physically stable and showed no significant change in drug-release rate during storage for three months (Figure 14.15).

Use of Albumin

Albumin was investigated to stabilize drug release from Eudragit RS/RL 30 D films [12]. The addition of 10% albumin resulted in the destabilization of the film, exhibiting significant aging during a three-month period as demonstrated by substantial changes in drug release over time (Figure 14.16). This was attributed to the intermolecular interactions between the protein and polymer. The pH of the dispersion was above the isoelectric point of the albumin, resulting in attractive forces between the quaternary ammonia groups on the polymer and the negatively charged protein. This destabilization was also the result of increased water absorption by the polymeric film. However, when the Eudragit dispersion was first acidified to a pH of 2.5 and then the albumin was added, there was no change in drug release when the coated dosage forms were stored at both 40°C/75%

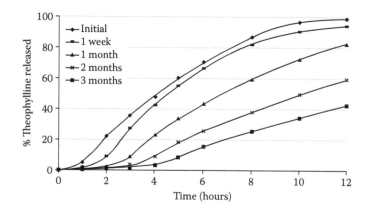

FIGURE 14.16 The influence of albumin on the release of theophylline from pellets coated with Eudragit® RS/RL 30 D (15% WG) containing 10% albumin and stored at 40°C/75% RH in open containers (*n* = 3). RH, relative humidity. (From S. A. Kucera et al., *Drug Dev Ind Pharm*, 33, 717–726, 2007.)

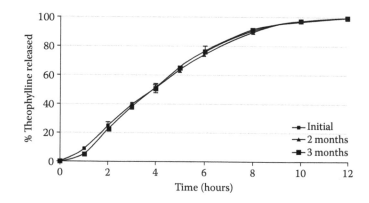

FIGURE 14.17 The influence of dispersion pH (2.5) on the release of theophylline from pellets coated with Eudragit® RS/RL 30 D (15% WG) containing 10% albumin and stored at 40°C/75% RH in hermetically sealed HDPE containers with desiccant (*n* = 3). RH, relative humidity; HDPE, high-density polyethylene. (From S. A. Kucera et al., *Drug Dev Ind Pharm*, 33, 717–726, 2007.)

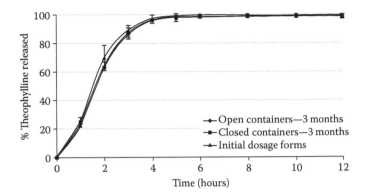

FIGURE 14.18   The effect of gelatin on the release of theophylline from pellets coated with Eudragit® RS/RL 30 D containing 10% gelatin and stored at 40°C/75% RH in open and closed containers (*n* = 3). RH, relative humidity. (From S. A. Kucera et al., *Drug Dev Ind Pharm*, 33, 717–726, 2007.)

RH (Figure 14.17) and 25°C/60% RH in aluminum induction–sealed high-density polyethylene containers. Electrostatic forces between the positively charged species, as well as the elimination of moisture, were responsible for the stabilization. In the same study, gelatin was also investigated as a possible protein to stabilize drug release. Although 10% gelatin in the coating dispersion did prevent aging effects, the drug-release rate was noticeably faster than when albumin was used in the films (Figure 14.18). The increase in drug release rate was attributed to the hydration and subsequent dissolution of the gelatin in the film, resulting in areas of high diffusion.

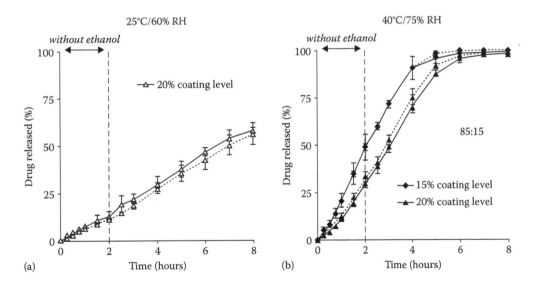

FIGURE 14.19   Storage stability of theophylline pellets coated with ethylcellulose/guar gum blends: (a) 93:7 blend ratio, 20% coating level, before and after six months storage in open containers at 25°C/60%RH, two hours exposure to 0.1 M HCl followed by phosphate buffer pH 7.4; (b) 85:15 blend ratio, 15% or 20% coating level, before and after six months storage at 40°C/75% RH, two hours exposure to 0.1 M HCl/ethanol 60:40, followed by phosphate buffer pH 7.4. The solid curves show drug release before storage, the dotted curves after storage. (From Y. Rosiaux et al., *Eur J Pharm Biopharm*, 1250–1258, 2013.)

Guar Gum

Guar gum is a polysaccharide used as a bio-adhesive material, a suspending agent, a binder and a disintegrant in tablet and capsule formulations, and as a viscosity-increasing agent [71]. Recently, guar gum was combined with ethylcellulose to provide sustained release formulations that are ethanol-resistant [72]. These researchers investigated a number variables to identify crucial formulation parameters for preparing these novel coatings and found that formulations containing the guar gum were stable for six months even when stored in open containers as shown in Figure 14.19.

SUMMARY

Physical aging is a phenomenon that affects all polymers. Simply utilizing alternative coating systems or polymers is not the solution to formulations that exhibit these stability issues. The subject has been extensively discussed in the chemical engineering literature and is an important consideration during formulation development for pharmaceutical scientists. The physical aging of polymers has been shown to cause changes in the physical–mechanical, permeability, and drug release properties of polymeric films due to a densification and decrease in free volume of the polymer as it relaxes to an equilibrated thermodynamic state. Because the coating of oral dosage forms with aqueous polymeric latices is one of the most widely used methods for controlling drug release, the stability of these systems is of the utmost importance. Aging has been shown to be influenced by a number of factors including humidity and temperature during storage as well as excipients in the coating formulation. A number of techniques have been used to stabilize polymeric films and prevent aging. Care must be taken to both plan for and identify potential aging issues during the early stages of product development. This includes determining the mechanism or mechanisms of destabilization, identifying the most appropriate stabilizer for the coating formulation, and ensuring that the coated dosage forms are cured to a point that film formation from the aqueous latex is complete.

REFERENCES

1. Lin F, Meier DJ. A study of latex film formation by atomic force microscopy. 1. A comparison of wet and dry conditions. *Langmuir* 1995; 11:2726–2733.
2. Lippold BC, Pages RM. Film formation, reproducibility of production and curing with respect to release stability of functional coatings from aqueous polymer dispersions. *Pharmazie* 2001; 56:5–17.
3. Greiner R, Schwarzl FR. Volume relaxation and physical aging of amorphous polymers I. Theory of volume relaxation after single temperature jumps. *Colloid Polym Sci* 1989; V267:39–47.
4. Struik LCE. Scope of the work (Chap 1). In: Struik LCE, ed., Physical Aging in Amorphous Polymers and Other Materials. New York: Elsevier Scientific Publishing Company, 1978, p. 1.
5. Simon F. Z. *Anorg Allgem Chem* 1931; 23:219.
6. Guo J-H. Aging processes in pharmaceutical polymers. *Pharm Sci Technol Today* 1999; 2:478–483.
7. Zheng W, Sauer D, McGinity JW. Influence of hydroxyethylcellulose on the drug release properties of theophylline pellets coated with Eudragit® RS 30 D. *Eur J Pharm Biopharm* 2005; 59:147–154.
8. Iyer U, Hong W-H, Das N, Ghebre-Sellaissie I. Comparative evaluation of three organic solvent and dispersion-based ethylcellulose coating formulations. *Pharm Tech* 1990; 14:68–86.
9. Priestley RD, Ellison CJ, Broadbelt LJ, Torkelson JM. Structural relaxation of polymer glasses at surfaces, interfaces, and in between. *Science* 2005; 309:456–459.
10. Gutierrez-Rocca JC, McGinity JW. Influence of physical aging on the physical–mechanical properties of acrylic resin films cast from aqueous dispersions and organic solutions. *Drug Dev Ind Pharm* 1993; 19:315–332.
11. Guo J-H, Robertson RE, Amidon GL. Influence of physical aging on mechanical properties of polymer free films: the prediction of long-term aging effects on the water permeability and dissolution rate of polymer film-coated tablets. *Pharm Res* 1991; 8:1500–1504.
12. Kucera SA, Shah NH, Malick AW, Infeld MA, McGinity JW. The use of proteins to minimize the physical aging of EUDRAGIT® sustained release films. *Drug Dev Ind Pharm* 2007; 33:717–726.
13. Sinko CM, Yee AF, Amidon GL. Prediction of physical aging in controlled-release coatings: The application of the relaxation coupling model to glassy cellulose acetate. *Pharm Res* 1991; V8:698–705.

14. Guo J-H, Robertson RE, Amidon GL. An investigation into the mechanical and transport properties of aqueous latex films: a new hypothesis for the film-forming mechanism of aqueous dispersion system. *Pharm Res* 1993; V10:405–410.

15. Matsumoto DS. Time-temperature superposition and physical aging in amorphous polymers. *Polym Eng Sci* 1988; 28:1313–1317.

16. Heng PWS, Chan LW, Ong KT. Influence of storage conditions and type of plasticizers on ethylcellulose and acrylate films formed from aqueous dispersions. *J Pharm Pharm Sci* 2003; 6:334–344.

17. Omari DM, Sallam A, Abd-Elbary A, El-Samaligy M. Lactic acid-induced modifications in films of Eudragit RL and RS aqueous dispersions. *Int J Pharm* 2004; 274:85–96.

18. Hutchinson JM. Physical aging of polymers. *Prog Polym Sci* 1995; 20:703–760.

19. Dai C-A, Liu M-W. The effect of crystallinity and aging enthalpy on the mechanical properties of gelatin films. *Mater Sci Eng* 2006; 423:121–127.

20. Drozdov AD. Physical aging in amorphous polymers far below the glass transition temperature. *Comput Mater Sci* 1999; 15:422–434.

21. ASTM. ASTM D 882-02: Standard Test Method for Tensile Properties of Thin Plastic Sheeting, 2002.

22. Drozdov AD. A constitutive model for physical ageing in amorphous glassy polymers. *Modell Simul Mater Sci Eng* 1999; 7:1045–1060.

23. Bigg DM. A review of positron annihilation lifetime spectroscopy as applied to the physical aging of polymers. *Polym Eng Sci* 1996; 36:737–743.

24. Montes H, Viasnoff V, Jurine S, Lequeux F. Ageing in glassy polymers under various thermal histories. *J Stat Mech Theor Exp* 2006; P03003.

25. McCaig MS, Paul DR. Effect of film thickness on the changes in gas permeability of a glassy polyarylate due to physical aging. Part I. Experimental observations. *Polymer* 2000; 41:629–637.

26. Barbero EJ, Ford KJ. Equivalent time temperature model for physical aging and temperature effects on polymer creep and relaxation. *J Eng Mater Technol* 2004; 126:413–419.

27. Pasricha A, Dillard DA, Tuttle ME. Effect of physical aging and variable sress history on the strain response of polymeric composites. *Composite Sci Technol* 1997; 57:1271–1279.

28. Sinko CM, Yee AF, Amidon GL. The effect of physical aging on the dissolution rate of anionic polyelectrolytes. *Pharm Res* 1990; V7:648–653.

29. ASTM. ASTM D 2990-01: Standard Test Methods for Tensile, Compressive, and Flexuraly Creep and Creep-Rupture of Plastics, 2001.

30. Ageeva MG. Moisture-resistant film coatings for orally administered medicinal forms. *Pharm Chem J* 1970; 4:342–346.

31. Anderson W, Abdel-Aziz SAM. Ageing effects in cast acrylate-methacrylate film. *J Pharm Pharmacol* 1976; (suppl 22):28.

32. Chowhan ZT, Amaro AA, Chi L-H. Comparative evaluations of aqueous film coated tablet formulations by high humidity Aging. *Drug Dev Ind Pharm* 1982; 8:713–737.

33. Guo J-H. A theoretical and experimental study of the additive effects of physical aging and antiplasticization on the water permeability of polymer film coatings. *J Pharm Sci* 1994; 83:447–449.

34. Heinämäki JT, Lehtola V-M, Nikupaavo P, Yliruusi JK. The mechanical and moisture permeability properties of aqueous-based hydroxypropyl methylcellulose coating systems plasticized with polyethylene glycol. *Int J Pharm* 1994; 112:191–196.

35. Huang Y, Paul DR. Physical aging of thin glassy polymer films monitored by gas permeability. *Polymer* 2004; 45:8377–8393.

36. Huang Y, Paul DR. Experimental methods for tracking physical aging of thin glassy polymer films by gas permeation. *J Membr Sci* 2004; 244:167–178.

37. Huang Y, Paul DR. Effect of temperature on physical aging of thin glassy polymer films. *Macromolecules* 2005; 38:10148–10154.

38. Tiemblo P, Guzman J, Riande E, Mijangos C, Reinecke H. Effect of physical aging on the gas transport properties of PVC and PVC modified with pyridine groups. *Polymer* 2001; 42:4817–4824.

39. ASTM. ASTM E 96/E 96 M-05: Standard Test Methods for Water Vapor Transmission of Materials, 2005.

40. Huang Y, Paul DR. Physical aging of thin glassy polymer films monitored by optical properties. *Macromolecules* 2006; 39:1554–1559.

41. Huang Y, Wang X, Paul DR. Physical aging of thin glassy polymer films: Free volume interpretation. *J Membr Sci* 2006; 277:219–229.

42. Kawana S, Jones RAL. Effect of physical ageing in thin glassy polymer films. *Eur Phys J E* 2003; V10:223–230.

43. Zelko R, Orban A, Suvegh K. Tracking of the physical ageing of amorphous pharmaceutical polymeric excipients by positron annihilation spectroscopy. *J Pharm Biomed Anal* 2006; 40:249–254.

44. Zelko R, Orban A, Suvegh K, Riedl Z, Racz I. Effect of plasticizer on the dynamic surface tension and the free volume of Eudragit systems. *Int J Pharm* 2002; 244:81–86.

45. Kobayashi Y, Zheng W, Meyer EF, McGervey JD, Jamieson AM, Simha R. Free volume and physical aging of poly(vinyl acetate) studied by positron annihilation. *Macromolecules* 1989; 22:2302–2306.

46. Chang G-W, Jamieson AM, Yu Z, McGervey JD. Physical aging in the mechanical properties of miscible polymer blends. *J Appl Polym Sci* 1997; 63:483–496.

47. Cangialosi D, Schut H, van Veen A, Picken SJ. Positron annihilation lifetime spectroscopy for measuring free volume during physical aging of polycarbonate. *Macromolecules* 2003; 36:142–147.

48. Perera DY. Effect of thermal and hygroscopic history on physical ageing of organic coatings. *Prog Organic Coat* 2002; 44:55–62.

49. Amighi K, Moës AJ. Influence of plasticizer concentration and storage conditions on the drug release rate from EUDRAGIT® RS 30 D film-coated sustained-release theophylline pellets. *Eur J Pharm Biopharm* 1996; 42:29–35.

50. Lecomte F, Siepmann J, Walther M, MacRae RJ, Bodmeier R. Polymer blends used for the aqueous coating of solid dosage forms: importance of the type of plasticizer. *J Control Rel* 2004; 99:1–13.

51. Yang QW, Flament MP, Siepmann F, Busignies V, Leclerc B, Herry C, Tchoreloff P, Siepmann J. Curing of aqueous polymeric film coatings: Importance of the coating level and type of plasticizer. *Eur J Pharm Biopharm* 2010; 362–370.

52. Siepmann F, Siepmann J, Walther M, MacRae R, Bodmeier R. Aqueous HPMCAS coatings: Effects of formulation and processing parameters on drug release and mass transport mechanisms. *Eur J Pharm Biopharm* 2006; 262–269.

53. Gendre C, Genty M, Fayard B, Tfayli A, Boiret M, Lecoq O, Baron M, Chaminade P, Pean JM. Comparative static curing versus dynamic curing on tablet coating. *Int J Pharm* 2013; 448–453.

54. Shao ZJ, Moralesi L, Diaz S, Muhammadi NA. Drug release from Kollicoat® SR 30 D-coated nonpareil beads: evaluation of coating level, plasticizer type, and curing condition. *AAPS Pharm Sci Tech* 2002; 3(2):article 15 (online only).

55. Lin AY, Muhammad NA, Pope D, Augsburger LL. A study on the effects of curing and storage conditions on controlled release diphenhydramine HCl pellets coated with Eudragit® NE 30 D. *Pharm Dev Tech* 2003; 8:277–287.

56. Billa N, Yuen K-H, Peh K-K. Diclofenac release from Eudragit-containing matrices and effects of thermal treatment. *Drug Dev Ind Pharm* 1998; 24:45–50.

57. Sauer S, Cerea M, DiNunzio J, McGinity J. Dry powder coating of pharmaceuticals: A review. *Int J Pharm* 2013; 457:488–502.

58. Kablitz CD, Urbanetz NA. Characterization of the film formation of the dry coating process. *Eur J Pharm Biopharm* 2007; 449–457.

59. Amighi KA, Moës AJ. Influence of curing conditions in the drug release rate from Eudagrit® NE 30 D film coated sustained-release theophylline pellets. *STP Pharm Sci* 1997; 7(2):141–147.

60. Lin AY, Augsburger LL. Study of crystallization of endogenous surfactant in Eudragit® NE 30 D-free films and its influence on drug-release properties of controlled-release diphenhydramine HCl pellets coated with Eudragit NE 30 D. *AAPS Pharm Sci* 2001; 3(2):article 14 (online only).

61. Bajdik J, Pintye-Hodi K, Regdon GJ, Fazekas P, Szabo-Revesz P, Eros I. The effect of storage on the behaviour of Eudragit® NE free film. *J Therm Anal Calorimet* 2003; 73:607–613.

62. Siepmann J, Siepmann F. Stability of aqueous polymeric controlled release film coatings. *Int J Pharm* 2013; 437–445.

63. Wu C, McGinity JW. Influence of ibuprofen as a solid-state plasticizer in Eudragit® RS 30 D on the physicochemical properties of coated beads. *AAPS Pharm Sci Tech* 2001; 2(4):article 24 (online only).

64. Wesseling M, Bodmeier R. Influence of plasticization time, curing conditions, storage time, and core properties on the drug release from aquacoat-coated pellets. *Pharm Dev Tech* 2001; 6:325–331.55.

65. Kranz H, Gutsche S. Evaluation of the drug release patterns and long term stability of aqueous and organic coated pellets by using blends of enteric and gastrointestinal insoluble polymers. *Int J Pharm* 2009; 112–119.

66. Wu C, McGinity JW. Influence of an enteric polymer on drug release rates of theophylline from pellets coated with Eudragit® RS 30 D. *Pharm Dev Tech* 2003; 8:103–110.

67. Zheng W, McGinity JW. Influence of Eudragit® NE 30 D blended with Eudragit® L 30 D-55 on the release of phenylpropanolamine hydrochloride from coated pellets. *Drug Dev Ind Pharm* 2003; 29:357–366.

68. Siepmann F, Muschert S. Leclercq B, Carlin B, Siepmann J. How to improve the storage stability of aqueous polymeric film coatings. *J Control Rel* 2008; 126:26–33.
69. Muschert S, Siepmann F, Leclercq B, Siepmann J. Dynamic and static curing of theylcellulose:PVA-PEG graft copolymer film coatings. *Eur J Pharm Biopharm* 2011; 455–461.
70. Maejima T, McGinity JW. Influence of film additives on stabilizing drug release rates from pellets coated with acrylic polymers. *Pharm Dev Tech* 2001; 6:211–221.
71. Rowe RC, Sheskey PJ, Cook WG, Fenton M., eds. Handbook of Pharmaceutical Excipients, 7th ed: Pharmaceutical Press, London, 2012.
72. Rosiaux Y, Velghe C, Muschert S, Chokshi R, Leclercq B, Siepmann F, Siepmman J. Ethanol-resistant ethylcellulose/guar gum coatings—Importance of formulation parameters. *Eur J Pharm Biopharm* 2013; 1250–1258.

Index

Page numbers followed by f and t indicate figures and tables, respectively.